SOLVING
LEAST SQUARES
PROBLEMS

Prentice-Hall
Series in Automatic Computation

MARTIN, *Teleprocessing Network Organization*

MARTIN AND NORMAN, *The Computerized Society*

MATHISON AND WALKER, *Computers and Telecommunications: Issues in Public Policy*

MCKEEMAN, et al., *A Compiler Generator*

MEYERS, *Time-Sharing Computation in the Social Sciences*

MINSKY, *Computation: Finite and Infinite Machines*

NIEVERGELT, et al., *Computer Approaches to Mathematical Problems*

PLANE AND MCMILLAN, *Discrete Optimization: Integer Programming and Network Analysis for Management Decisions*

PRITSKER AND KIVIAT, *Simulation with GASP II: a FORTRAN-Based Simulation Language*

PYLYSHYN, editor, *Perspectives on the Computer Revolution*

RICH, *International Sorting Methods Illustrated with PL/1 Programs*

RUSTIN, editor, *Algorithm Specification*

RUSTIN, editor, *Computer Networks*

RUSTIN, editor, *Data Base Systems*

RUSTIN, editor, *Debugging Techniques in Large Systems*

RUSTIN, editor, *Design and Optimization of Compilers*

RUSTIN, editor, *Formal Semantics of Programming Languages*

SACKMAN AND CITRENBAUM, editors, *On-Line Planning: Towards Creative Problem-Solving*

SALTON, editor, *The SMART Retrieval System: Experiments in Automatic Document Processing*

SAMMET, *Programming Languages: History and Fundamentals*

SCHAEFER, *A Mathematical Theory of Global Program Optimization*

SCHULTZ, *Spline Analysis*

SCHWARZ, et al., *Numerical Analysis of Symmetric Matrices*

SHAW, *The Logical Design of Operating Systems*

SHERMAN, *Techniques in Computer Programming*

SIMON AND SIKLOSSY, editors, *Representation and Meaning: Experiments with Information Processing Systems*

STERBENZ, *Floating-Point Computation*

STERLING AND POLLACK, *Introduction to Statistical Data Processing*

STOUTEMYER, *PL/1 Programming for Engineering and Science*

STRANG AND FIX, *An Analysis of the Element Method*

STROUD, *Approximate Calculation of Multiple Integrals*

TAVISS, editor, *The Computer Impact*

TRAUB, *Iterative Methods for the Solution of Equations*

UHR, *Pattern Recognition, Learning, and Thought: Computer-Programmed Models of Higher Mental Processes*

VAN TASSEL, *Computer Security Management*

VARGA, *Matrix Iterative Analysis*

WAITE, *Implementing Software for Non-Numeric Application*

WILKINSON, *Rounding Errors in Algebraic Processes*

WIRTH, *Systematic Programming: An Introduction*

SOLVING
LEAST SQUARES
PROBLEMS

CHARLES L. LAWSON

Jet Propulsion Laboratory
California Institute of Technology

RICHARD J. HANSON

Department of Mathematics and Computer Science
Washington State University

PRENTICE-HALL, INC.

ENGLEWOOD CLIFFS, NEW JERSEY

Library of Congress Cataloging in Publication Data

LAWSON, CHARLES L.
  Solving least squares problems.

  (Prentice-Hall series in automatic computation)
  Bibliography.
  1. Electronic data processing—Least squares.
I. Hanson, Richard J., joint author.
II. Title.
QA275.L38   519.4   73–20119
ISBN 0–13–822585–0

10 9 8 7 6 5 4 3 2 1

Printed in the United States of America

PRENTICE-HALL INTERNATIONAL, INC., *London*
PRENTICE-HALL OF AUSTRALIA, PTY. LTD., *Sydney*
PRENTICE-HALL OF CANADA, LTD., *Toronto*
PRENTICE-HALL OF INDIA PRIVATE LIMITED, *New Delhi*
PRENTICE-HALL OF JAPAN, INC., *Tokyo*

CONTENTS

PREFACE

This book brings together a body of information on solving least squares problems whose practical development has taken place mainly during the past decade. This information is valuable to the scientist, engineer, or student who must analyze and solve systems of linear algebraic equations. These systems may be overdetermined, underdetermined, or exactly determined and may or may not be consistent. They may also include both linear equality and inequality constraints.

Practitioners in specific fields have developed techniques and nomenclature for the least squares problems of their own discipline. The material presented in this book can unify this divergence of methods.

Essentially all real problems are nonlinear. Many of the methods of reaching an understanding of nonlinear problems or computing using nonlinear models involve the local replacement of the nonlinear problem by a linear one. Specifically, various methods of analyzing and solving the nonlinear least squares problem involve solving a sequence of linear least squares problems. One essential requirement of these methods is the capability of computing solutions to (possibly poorly determined) linear least squares problems which are reasonable in the context of the nonlinear problem.

For the reader whose immediate concern is with a particular application we suggest first reading Chapter 25. Following this the reader may find that some of the Fortran programs and subroutines of Appendix C have direct applicability to his problem.

There have been many contributors to theoretical and practical development of least squares techniques since the early work of Gauss. We particularly cite Professor G. H. Golub who, through broad involvement in both theoretical and practical aspects of the problem, has contributed many significant ideas and algorithms relating to the least squares problem.

The first author introduced the second author to the least squares problem in 1966. Since that time the authors have worked very closely together in the adaptation of the stable mathematical methods to practical computational problems at JPL. This book was written as part of this collaborative effort.

We wish to express our appreciation to the late Professor G. E. Forsythe for extending to us the opportunity of writing this book. We thank the management of the Jet Propulsion Laboratory, California Institute of Technology, for their encouragement and support. We thank Drs. F. T. Krogh and D. Carta for reading the manuscript and providing constructive comments. A number of improvements and simplifications of mathematical proofs were due to Dr. Krogh.

Our thanks go to Mrs. Roberta Cerami for her careful and cheerful typing of the manuscript in all of its stages.

CHARLES L. LAWSON
RICHARD J. HANSON

1 INTRODUCTION

This book is intended to be used as both a text and a reference for persons who are investigating the solutions of linear least squares problems. Such least squares problems often occur as a component part of some larger computational problem. As an example, the determination of the orbit of a spacecraft is often considered mathematically as a multipoint boundary value problem in the field of ordinary differential equations, but the computation of the orbital parameters usually amounts to a nonlinear least squares estimation problem using various linearization schemes.

More generally, almost any problem with sufficient data to overdetermine a solution calls for some type of approximation method. Most frequently *least squares* is the approximation criterion chosen.

There are a number of auxiliary requirements which often arise in least squares computations which merit identification for algorithmic development. Examples:

A problem may require certain equality or inequality relations between variables. A problem may involve very large volumes of data so that the allocation of computer storage is of major concern.

Many times the purpose of least squares computation is not merely to find some set of numbers that "solve" the problem, but rather the investigator wishes to obtain additional quantitative information describing the relationship of the solution parameters to the data. In particular, there may be a family of different solution parameters that satisfy the stated conditions almost equally well and the investigator may wish to identify this indeterminacy and make a choice in this family to meet some additional conditions.

This book presents numerical methods for obtaining least squares solu-

1

tions keeping the points above in mind. These methods have been used successfully by many engineers and scientists, together with the authors, in the NASA unmanned space program.

The least squares problem that we are considering here is known by different names by those in different scientific disciplines. For example, mathematicians may regard the (least squares) problem as finding the closest point in a given subspace to a given point in a function space. Numerical analysts have also tended to use this framework, which tends to ignore data errors. It follows that the opportunity to exploit the often present arbitrariness of the solution is disregarded.

Statisticians introduce probability distributions into their conception of the problem and use such terms as *regression analysis* to describe this area. Engineers reach this problem by studying such topics as *parameter estimation, filtering*, or *process identification*.

The salient point is this: When these problems (formulated in any of these contexts) reach the stage of numerical computation, they contain the same central problem, namely, a sequence of linear least squares systems.

This basic linear least squares problem can be stated as follows:

PROBLEM   LS

Given a real m × n *matrix* A *of rank* k ≤ *min* (m, n), *and given a real* m-*vector* b, *find a real* n-*vector* x_0 *minimizing the euclidean length of* Ax − b.

(The reader can refer to Appendix A for the definitions of any unfamiliar linear algebraic terms.) We shall use the symbolism $Ax \cong b$ to denote Problem LS.

This problem can also be posed and studied for the case of complex A and b. The complex case arises much less frequently in practice than the real case and the theory and computational methods for the real case generalize quite directly to the complex case.

In addition to the statement of Problem LS (in a given context) there is an additional condition: The numerical data that constitute A and b have only a limited number of accurate digits and after those digits the data are completely uncertain *and hence arbitrary*! In practice this is the usual state of affairs. This is due, in part, to the limited accuracy of measurements or observations. It is important that explicit note be taken of this situation and that it be used to advantage to obtain an appropriate approximate solution to the problem. We shall discuss methods for accomplishing this, particularly in Chapters 25 and 26.

Consider, as an important example, the case in which the linear least squares problem arises from a nonlinear one and the solution vector x_0 is to be used as a correction to be added to a current nominal solution of the nonlinear problem. The linearized problem will be a useful approximation

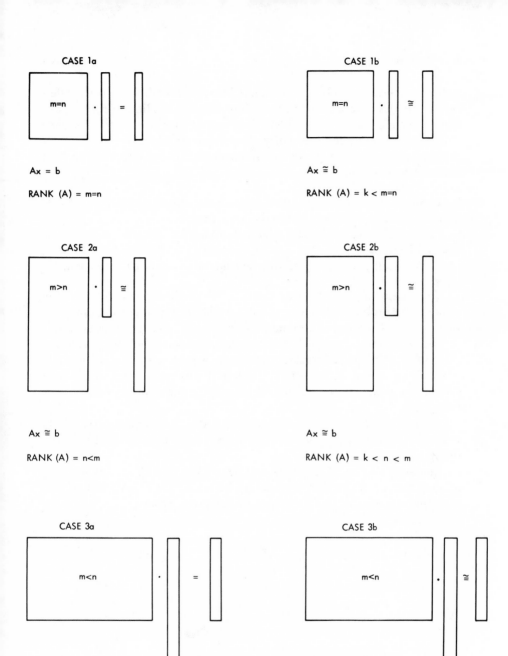

CASE 1a

$Ax = b$

RANK (A) = m=n

CASE 1b

$Ax \cong b$

RANK (A) = k < m=n

CASE 2a

$Ax \cong b$

RANK (A) = n<m

CASE 2b

$Ax \cong b$

RANK (A) = k < n < m

CASE 3a

$Ax = b$

RANK (A) = m < n

CASE 3b

$Ax \cong b$

RANK (A) = k < m < n

Fig. 1.1 The six cases of Problem **LS** depending on the relative sizes of *m*, *n*, and Rank (*A*).

to the nonlinear one only in some limited neighborhood. If there are different vectors that give satisfactorily small residuals in the linear problem, one may prefer the one of smallest length to increase the likelihood of staying within the neighborhood of validity of the linear problem.

The following three points are fundamental to our approach to least squares computation.

1. Since the data defining Problem LS are uncertain, we can change these data to suit our needs within this uncertainty.

2. We shall use orthogonal elimination matrices directly on the linear system of Problem LS.

3. Practicality, for computer implementation, is stressed throughout.

We shall briefly comment on the first two of these points. Our goal in changing the problem, as stated in point 1, will be to avoid the situation above where a "small" change in the data produces "large" changes in the solution.

Orthogonal transformation matrices, mentioned in point 2, have a natural place in least squares computations because they leave the euclidean length of a vector invariant. Furthermore, it is desirable to use them because of their stability in computation with regard to propagating data errors or uncertainty.

We have made no assumption regarding the relative sizes of m and n. In our subsequent discussion of Problem LS it will be convenient to identify the six cases illustrated in Fig. 1.1.

The main interest of this book will be in Case 2a with special attention to the situation where data uncertainty leads to Case 2b. Algorithms and discussions applicable to all six cases will be given, however.

2 ANALYSIS OF THE LEAST SQUARES PROBLEM

The central point of this chapter will be an analysis of Problem LS based on a certain decomposition of an $m \times n$ matrix A in the form HRK^T where H and K are orthogonal matrices. The definition of an orthogonal matrix and other linear algebraic concepts are summarized in Appendix A.

Our interest in this general type of decomposition $A = HRK^T$ is motivated by the practicality and usefulness of certain specific computable decompositions of this type that are introduced in Chapters 3 and 4.

An important property of orthogonal matrices is the preservation of euclidean length under multiplication. Thus for any m-vector y and any $m \times m$ orthogonal matrix Q,

$$(2.1) \qquad \|Qy\| = \|y\|$$

In the context of Problem LS, the least squares problem of minimizing the euclidean length of $Ax - b$, we have

$$(2.2) \qquad \|Q(Ax - b)\| = \|QAx - Qb\| = \|Ax - b\|$$

for any $m \times m$ orthogonal matrix Q and any n-vector x.

The use of orthogonal transformations allows us to express the solution of Problem LS in the following way:

(2.3) THEOREM

Suppose that A *is an* m $\times$ n *matrix of rank* k *and that*

$$(2.4) \qquad A = HRK^T$$

5

where

(a) H *is an* m × m *orthogonal matrix.*
(b) R *is an* m × n *matrix of the form*

$$R = \begin{bmatrix} R_{11} & 0 \\ 0 & 0 \end{bmatrix}$$

(c) R_{11} *is a* k × k *matrix of rank* k.
(d) K *is an* n × n *orthogonal matrix.*

Define the vector

(2.5)
$$H^T b = g = \begin{bmatrix} g_1 \\ g_2 \end{bmatrix} \begin{matrix} \} \, k \\ \} \, m-k \end{matrix}$$

and introduce the new variable

(2.6)
$$K^T x = y = \begin{bmatrix} y_1 \\ y_2 \end{bmatrix} \begin{matrix} \} \, k \\ \} \, n-k \end{matrix}$$

Define $\tilde{y}_1$ *to be the unique solution of*

(2.7)
$$R_{11} y_1 = g_1$$

(1) *Then all solutions to the problem of minimizing* $\| Ax - b \|$ *are of the form*

(2.8)
$$\hat{x} = K \begin{bmatrix} \tilde{y}_1 \\ y_2 \end{bmatrix}$$

where y_2 *is arbitrary.*
(2) *Any such* $\hat{x}$ *gives rise to the same residual vector* r *satisfying*

(2.9)
$$r = b - A\hat{x} = H \begin{bmatrix} 0 \\ g_2 \end{bmatrix}$$

(3) *The norm of* r *satisfies*

(2.10)
$$\| r \| = \| b - A\hat{x} \| = \| g_2 \|$$

(4) *The unique solution of minimum length is*

(2.11)
$$\tilde{x} = K \begin{bmatrix} \tilde{y}_1 \\ 0 \end{bmatrix}$$

Proof: Replacing A by the right side of Eq. (2.4) and using Eq. (2.2) yields

$$(2.12) \qquad \| Ax - b \|^2 = \| HRK^T x - b \|^2$$
$$= \| RK^T x - H^T b \|^2$$

Using Eq. (2.5) to (2.8) we see that

$$(2.13) \qquad \| Ax - b \|^2 = \| R_{11} y_1 - g_1 \|^2 + \| g_2 \|^2$$

for all x.

The right member of Eq. (2.13) has the minimum value $\| g_2 \|^2$ when

$$(2.14) \qquad R_{11} y_1 = g_1$$

Equation (2.14) has a unique solution $\tilde{y}_1$ since R_{11} is of rank k.

The most general solution for y is given by

$$(2.15) \qquad \tilde{y} = \begin{bmatrix} \tilde{y}_1 \\ y_2 \end{bmatrix}$$

where y_2 is arbitrary.

Then with $\hat{x}$ defined by Eq. (2.8) one obtains

$$b - A\hat{x} = b - HRK^T\hat{x} = H(g - R\tilde{y}) = H \begin{bmatrix} 0 \\ g_2 \end{bmatrix}$$

which establishes Eq. (2.9).

Clearly the minimum length vector $\tilde{y}$, of the form given in Eq. (2.15), is that one with $y_2 = 0$. From Eq. (2.8) it follows that the solution $\tilde{x}$ of minimum euclidean length is given by

$$(2.16) \qquad \tilde{x} = K \begin{bmatrix} \tilde{y}_1 \\ 0 \end{bmatrix}$$

This completes the proof of Theorem (2.3).

Any decomposition of an $m \times n$ matrix $A = HRK^T$, as in Eq. (2.4), will be called an *orthogonal decomposition* of A.

In case $k = n$ or $k = m$, quantities having dimensions involving $(n - k)$ or $(m - k)$ are respectively absent. In particular, the solution to Problem LS is unique in case $k = n$.

Notice that the solution of minimum length, the set of all solutions, and the minimum value, for the problem of minimizing $\| Ax - b \|$, are all unique; they do not depend on the particular orthogonal decomposition.

EXERCISE

(2.17) Suppose $A_{m \times n}$, $m \geq n$, is of rank n with $Q_1 R_1 = A = Q_2 R_2$. If the Q_i are $m \times n$ with orthogonal column vectors and the R_i are $n \times n$ upper triangular, then there exists a diagonal matrix D with diagonal elements plus or minus one such that $Q_2 D = Q_1$ and $D R_1 = R_2$.

3 ORTHOGONAL DECOMPOSITION BY CERTAIN ELEMENTARY ORTHOGONAL TRANSFORMATIONS

In the previous chapter it was shown that Problem LS could be solved assuming an orthogonal decomposition of the $m \times n$ matrix A. In this chapter we shall establish the existence of such a decomposition for any matrix A using explicit orthogonal transformations.

Two elementary orthogonal transformations that will be useful are presented in the following two lemmas.

(3.1) LEMMA

Given an m-vector v *(which is not the zero vector), there exists an orthogonal matrix* Q *such that*

(3.2) $$Qv = -\sigma \|v\| e_1$$

with

$$e_1 = \begin{bmatrix} 1 \\ 0 \\ \cdot \\ \cdot \\ \cdot \\ 0 \end{bmatrix}$$

and

$$\sigma = \begin{cases} +1 & \text{if } v_1 \geq 0 \\ -1 & \text{if } v_1 < 0 \end{cases}$$

where v_1 *is the first component of* v.

9

Proof: Define

$$u = v + \sigma \|v\| e_1$$

and

(3.3)
$$Q = I_m - \frac{2uu^T}{u^T u}$$

The proof is completed by verifying directly that the matrix Q of Eq. (3.3) is symmetric and orthogonal and that the remaining statement of Lemma (3.1) is satisfied.

The transformation defined in Eq. (3.3) was used by A. S. Householder in discussing certain eigenvalue problems [Householder (1958)]. For this reason the matrix Q is called a *Householder transformation matrix*.

This transformation can be viewed geometrically as a reflection in the $(m-1)$-dimensional subspace, S, orthogonal to the vector u. By this it is meant that $Qu = -u$ and $Qs = s$ for all $s \in S$.

In the special case that one wishes to transform only one element of the vector v to zero, the following *Givens transformation* [Givens (1954)] is often used. Since this transformation only modifies two components of a vector, it suffices to discuss its operation on a 2-vector.

(3.4)   LEMMA

Given a 2-vector $v = (v_1, v_2)^T$ *(with either* $v_1 \neq 0$ *or* $v_2 \neq 0$*), there is a 2 $\times$ 2 orthogonal matrix.*

(3.5)
$$G = \begin{pmatrix} c & s \\ -s & c \end{pmatrix}$$

with

(3.6)
$$c^2 + s^2 = 1$$

such that

(3.7)
$$Gv = \begin{bmatrix} (v_1^2 + v_2^2)^{1/2} \\ 0 \end{bmatrix}$$

Proof: Simply put

(3.8)
$$c = \frac{v_1}{(v_1^2 + v_2^2)^{1/2}}$$

and

(3.9)
$$s = \frac{v_2}{(v_1^2 + v_2^2)^{1/2}}$$

Then the matrix G of Eq. (3.5) can be easily shown to be orthogonal and satisfy Eq. (3.7), completing the proof of Lemma (3.4).

We also remark that we may choose G to be symmetric as well as orthogonal by defining

$$(3.10) \qquad G = \begin{pmatrix} c & s \\ s & -c \end{pmatrix}$$

with c and s as in Eq. (3.8) and (3.9).

(3.11) THEOREM

> Let A be an $m \times n$ matrix. There is an $m \times m$ orthogonal matrix Q such that $QA = R$ is zero below the main diagonal.

Let an $m \times m$ orthogonal matrix Q_1 be chosen as in Lemma (3.1) so that the first column of $Q_1 A$ is zeroed in components 2 through m. Next choose an $(m - 1) \times (m - 1)$ orthogonal matrix P_2, which when applied to the $(m - 1)$ vector consisting of components 2 through m of the second column of $Q_1 A$ results in zeros in components 3 through m. The transformation matrix

$$Q_2 = \begin{pmatrix} 1 & 0 \\ 0 & P_2 \end{pmatrix}$$

is orthogonal and $Q_2 Q_1 A$ has zeros below the diagonal in both the first two columns.

Continuing in this way, a product of at most n orthogonal transformations can be constructed that will transform A to upper triangular form. These remarks can be formalized to provide a finite induction proof of Theorem (3.11).

The algorithmic details of the construction of these transformations will be given in Chapter 10. This decomposition of a matrix as a product of an orthogonal matrix and an upper triangular matrix is called a *QR decomposition*. It plays an important role in a number of computational algorithms of linear algebra [Francis (1961); Golub and Businger (1965)].

For Cases 1a and 2a of Fig. 1.1, in which Rank $(A) = n$, Theorem (3.11) establishes the existence of an orthogonal decomposition of A. Thus from Theorem (3.11) we can write

$$(3.12) \qquad A = Q^T R = Q^T R I_n$$

where the matrices Q^T, R, and I_n in this representation have the properties required of the matrices H, R, and K^T of an orthogonal decomposition for a matrix A, as given in Theorem (2.3).

In the case that the rank of A is m, Case 3a of Fig. 1.1, Theorem (3.11) allows us to write

$$(3.13) \qquad\qquad A^T = Q^T R$$

so that

$$(3.14) \qquad\qquad A = R^T Q = I_m R^T Q$$

Here, I_m, R^T, and Q in this representation have the properties required of the matrices H, R, and K^T for an orthogonal decomposition of a matrix of rank m, as given in Theorem (2.3).

In Cases 1b, 2b, and 3b of Fig. 1.1, the matrix R obtained in Theorem (3.11) is not necessarily in the form required for an orthogonal decomposition.

We proceed to discuss additional transformations that will obtain the orthogonal decomposition for these cases.

(3.15) THEOREM

> Let A be an m $\times$ n *matrix whose rank* k *satisfies* k $<$ n $\leq$ m. *There is an* m $\times$ m *orthogonal matrix* Q *and an* n $\times$ n *permutation matrix* P *such that*

$$(3.16) \qquad\qquad QAP = \begin{bmatrix} R_{k \times k} & T_{k \times (n-k)} \\ 0_{(m-k) \times k} & 0_{(m-k) \times (n-k)} \end{bmatrix}$$

> *Here* R *is a* k $\times$ k *upper triangular matrix of rank* k.

Proof: Let the permutation matrix P be selected so that the first k columns of the matrix AP are linearly independent. By application of Theorem (3.11) there is an $m \times m$ orthogonal matrix such that QAP is upper triangular. Since the first k columns of AP are linearly independent, the same will be true of the first k columns of QAP.

All the elements in rows $k + 1$ through m of columns $k + 1$ through n of QAP will be zero. Otherwise the rank of QAP would be greater than k, which would contradict the fact that A is of rank k. Therefore QAP has the form indicated in the right member of Eq. (3.16). This completes the proof of Theorem (3.15).

The submatrix $[R : T]$ in the right member of Eq. (3.16) can be further transformed to the compact form required of the matrix R in Theorem (2.3). This transformation is described by the following lemma:

(3.17) Lemma

> Let [R: T] *be a* k × n *matrix where* R *is of rank* k. *There is an* n × n *orthogonal matrix* W *such that*

(3.18) $$[R:T]W = [\hat{R}_{k \times k} : 0_{k \times (n-k)}]$$

> *Here* $\hat{R}$ *is a lower triangular matrix of rank* k.

Lemma (3.17) follows from Theorem (3.15) by identification of the present n, k, $[R:T]$, and W, respectively, with m, n, A^T, and Q^T of Theorem (3.15).

Lemma (3.17) can be used together with Theorem (3.15) to prove the following theorem:

(3.19) Theorem

> Let A *be an* m × n *matrix of rank* k. *Then there is an* m × m *orthogonal matrix* H *and an* n × n *orthogonal matrix* K *such that*

(3.20) $$H^T A K = R, \qquad A = H R K^T$$

> *where*

(3.21) $$R = \begin{bmatrix} R_{11} & 0 \\ 0 & 0 \end{bmatrix}$$

> *Here* R_{11} *is a* k × k *nonsingular triangular matrix.*

Note that by an appropriate choice of H and K of Eq. (3.20) we can arrange that R_{11} of Eq. (3.21) be either upper triangular or lower triangular.

In summary, it has been shown that there are constructive procedures for producing orthogonal decompositions of the form $A = H R K^T$ in each of the six cases displayed in Fig. 1.1. In all the cases the rank k submatrix R_{11} of Theorem (2.3) is obtained in triangular form. Thus it is particularly easy to compute the solution of Eq. (2.7).

As already noted, in Cases 1a and 2a, the matrix K^T of the decomposition can be taken to be the identity matrix I_n. Likewise, in Case 3a, the matrix H can be taken to be the identity matrix I_m.

To illustrate these orthogonal decompositions we give a numerical example for each of the six cases of Fig. 1.1.

(3.22) Case 1a *Square, nonsingular*

$$m = n = 3, \qquad \text{Rank } (A) = 3$$

$$QA = R$$

$$Q = \begin{bmatrix} -0.4800 & -0.4129 & -0.7740 \\ 0.6616 & -0.7498 & -0.0103 \\ -0.5761 & -0.5170 & 0.6331 \end{bmatrix}$$

$$A = \begin{bmatrix} 0.4087 & 0.1593 & 0.6593 \\ 0.3515 & 0.9665 & 0.6245 \\ 0.6590 & 0.9343 & 0.9039 \end{bmatrix}$$

$$R = \begin{bmatrix} -0.8514 & -1.1987 & -1.2740 \\ 0.0 & -0.6289 & -0.0414 \\ 0.0 & 0.0 & -0.1304 \end{bmatrix}$$

(3.23)   CASE 2a *Overdetermined, full rank*

$$m = 3, \qquad n = 2, \qquad \text{Rank}(A) = 2$$

$$QA = \begin{bmatrix} R \\ 0 \end{bmatrix}$$

$$Q = \begin{bmatrix} -0.4744 & -0.4993 & -0.7250 \\ 0.5840 & -0.7947 & 0.1652 \\ -0.6587 & -0.3450 & 0.6687 \end{bmatrix}$$

$$A = \begin{bmatrix} 0.4087 & 0.1594 \\ 0.4302 & 0.3516 \\ 0.6246 & 0.3384 \end{bmatrix}$$

$$R = \begin{bmatrix} -0.8615 & -0.4965 \\ 0.0 & -0.1304 \end{bmatrix}$$

(3.24)   CASE 3a *Underdetermined, full rank*

$$m = 2, \qquad n = 3, \qquad \text{Rank}(A) = 2$$

$$AQ = [R : 0]$$

$$A = \begin{bmatrix} 0.4087 & 0.4301 & 0.6246 \\ 0.1594 & 0.3515 & 0.3384 \end{bmatrix}$$

$$Q = \begin{bmatrix} -0.4744 & 0.5840 & -0.6587 \\ -0.4993 & -0.7947 & -0.3450 \\ -0.7250 & 0.1652 & 0.6687 \end{bmatrix}$$

$$R = \begin{bmatrix} -0.8615 & 0.0 \\ -0.4965 & -0.1304 \end{bmatrix}$$

(3.25)   CASE 1b *Square, singular*

$$m = n = 5, \qquad \text{Rank } (A) = 3$$

$$QAK = \begin{bmatrix} R & 0 \\ 0 & 0 \end{bmatrix}$$

$$Q = \begin{bmatrix} -0.0986 & -0.6926 & -0.5937 & -0.0522 & -0.3941 \\ 0.3748 & 0.0072 & -0.5424 & 0.3529 & 0.6639 \\ -0.1244 & 0.2438 & -0.3484 & -0.8633 & 0.2419 \\ 0.2401 & -0.6653 & 0.4813 & -0.2980 & 0.4234 \\ 0.8813 & 0.1351 & -0.0160 & -0.1966 & -0.4076 \end{bmatrix}$$

$$A = \begin{bmatrix} 0.1376 & 0.4086 & 0.1594 & 0.4390 & 0.4113 \\ 0.9665 & 0.6246 & 0.3383 & 0.7221 & 0.8746 \\ 0.8285 & 0.0661 & 0.9112 & 0.6266 & 0.2327 \\ 0.0728 & 0.3485 & 0.8560 & 0.8348 & 0.2474 \\ 0.5500 & 0.9198 & 0.0080 & 0.7610 & 1.0506 \end{bmatrix}$$

$$K = \begin{bmatrix} -0.9660 & 0.0 & 0.0 & 0.0569 & -0.2520 \\ 0.1279 & -0.6389 & 0.0 & -0.4688 & -0.5963 \\ -0.0709 & 0.1861 & -0.8383 & -0.4806 & 0.1632 \\ 0.0726 & -0.4109 & -0.5354 & 0.7253 & -0.1144 \\ -0.2002 & -0.6231 & 0.1027 & -0.1411 & 0.7357 \end{bmatrix}$$

$$R = \begin{bmatrix} 1.4446 & 1.6867 & 1.2530 \\ 0.0 & -1.3389 & -0.1486 \\ 0.0 & 0.0 & 1.1831 \end{bmatrix}$$

(3.26)   CASE 2b *Overdetermined, rank deficient*

$$m = 6, \qquad n = 5, \qquad \text{Rank } (A) = 3$$

$$QAK = \begin{bmatrix} R & 0 \\ 0 & 0 \end{bmatrix}$$

$$Q = \begin{bmatrix} -0.0894 & -0.6279 & -0.5382 & -0.0473 & -0.3573 & -0.4221 \\ 0.3768 & 0.0510 & -0.4986 & 0.3522 & 0.6813 & -0.1360 \\ -0.1235 & 0.2301 & -0.3633 & -0.8609 & 0.2386 & 0.0417 \\ 0.3354 & -0.6906 & 0.4092 & -0.2991 & 0.3539 & 0.1685 \\ 0.7849 & 0.1389 & -0.1992 & -0.1088 & -0.4751 & 0.2956 \\ -0.3259 & -0.2324 & -0.3500 & 0.1768 & 0.0153 & 0.8281 \end{bmatrix}$$

$$A = \begin{bmatrix} 0.1376 & 0.4087 & 0.1593 & 0.4308 & 0.4163 \\ 0.9667 & 0.6246 & 0.3384 & 0.8397 & 0.8029 \\ 0.8286 & 0.0661 & 0.9111 & 0.7495 & 0.1577 \\ 0.0728 & 0.3485 & 0.8560 & 0.8068 & 0.2644 \\ 0.5501 & 0.9198 & 0.0080 & 0.7910 & 1.0323 \\ 0.6498 & 0.2725 & 0.3599 & 0.5350 & 0.3801 \end{bmatrix}$$

$$K = \begin{bmatrix} -0.9846 & 0.0 & 0.0 & -0.0868 & -0.1520 \\ 0.1343 & -0.6392 & 0.0 & -0.4230 & -0.6281 \\ 0.0177 & 0.1650 & -0.8460 & -0.4808 & 0.1598 \\ -0.0467 & -0.3820 & -0.5248 & 0.7488 & -0.1256 \\ -0.1006 & -0.6468 & 0.0942 & -0.1469 & 0.7356 \end{bmatrix}$$

$$R = \begin{bmatrix} 1.5636 & 1.7748 & 1.4574 \\ 0.0 & -1.3537 & -0.1233 \\ 0.0 & 0.0 & 1.1736 \end{bmatrix}$$

(3.27)   CASE 3b *Underdetermined, rank deficient*

$$m = 4, \qquad n = 5, \qquad \text{Rank}(A) = 3$$

$$QAK = \begin{bmatrix} R & 0 \\ 0 & 0 \end{bmatrix}$$

$$Q = \begin{bmatrix} -0.9757 & 0.0103 & 0.1390 & -0.1693 \\ 0.0 & -0.9989 & 0.0298 & -0.0363 \\ 0.0 & 0.0 & -0.7729 & -0.6345 \\ -0.2193 & -0.0458 & -0.6184 & 0.7533 \end{bmatrix}$$

$$A = \begin{bmatrix} 0.4087 & 0.1593 & 0.6594 & 0.4302 & 0.3516 \\ 0.6246 & 0.3383 & 0.6591 & 0.9342 & 0.9038 \\ 0.0661 & 0.9112 & 0.6898 & 0.1931 & 0.1498 \\ 0.2112 & 0.8150 & 0.7983 & 0.3406 & 0.2803 \end{bmatrix}$$

$$K = \begin{bmatrix} -0.4225 & 0.0249 & 0.3156 & -0.0047 & -0.8493 \\ -0.1647 & -0.1529 & -0.9349 & -0.0530 & -0.2696 \\ -0.6816 & 0.6377 & -0.0851 & 0.1253 & 0.3254 \\ -0.4447 & -0.4483 & 0.1187 & -0.7222 & 0.2561 \\ -0.3634 & -0.6070 & 0.0713 & 0.6782 & 0.1858 \end{bmatrix}$$

$$R = \begin{bmatrix} 0.9915 & 0.0 & 0.0 \\ 1.5246 & 0.5840 & 0.0 \\ 1.2573 & -0.1388 & 1.1077 \end{bmatrix}$$

EXERCISES

(3.28) Find an eigenvalue–eigenvector decomposition of the Householder matrix $H = I - 2ww^T$, $\|w\| = 1$.

(3.29) Find an eigenvalue–eigenvector decomposition of the Givens reflection matrix of Eq. (3.10).

(3.30) Show that G of Eq. (3.10) is a Householder transformation matrix.

4 ORTHOGONAL DECOMPOSITION BY SINGULAR VALUE DECOMPOSITION

In this chapter we shall present another practically useful orthogonal decomposition of the $m \times n$ matrix A. In the previous chapter the matrix A was expressed as a product HRK^T where R was a certain rectangular matrix whose nonzero elements belonged to a nonsingular triangular submatrix. We shall show here that this nonsingular submatrix of R can be further simplified so that it constitutes a nonsingular diagonal matrix. This decomposition is particularly useful in analyzing the effect of data errors as they influence solutions to Problem LS.

This decomposition is closely related to the eigenvalue–eigenvector decomposition of the symmetric nonnegative definite matrices $A^T A$ and AA^T. The standard facts concerning the eigenvalue–eigenvector decomposition of symmetric nonnegative definite matrices are summarized in Appendix A.

(4.1) THEOREM   (*Singular Value Decomposition*)

Let A *be an* m $\times$ n *matrix of rank* k. *Then there is an* m $\times$ m *orthogonal matrix* U, *an* n $\times$ n *orthogonal matrix* V, *and an* m $\times$ n *diagonal matrix* S *such that*

(4.2) $$U^T A V = S, \qquad A = USV^T$$

Here the diagonal entries of S *can be arranged to be nonincreasing; all these entries are nonnegative and exactly* k *of them are strictly positive.*

The diagonal entries of S are called the *singular values* of A. It will be convenient to give the proof of Theorem (4.1) first for the special case in which $m = n = \text{Rank } (A)$. The more general statement follows easily from this special case.

(4.3)   LEMMA

Let A be an n × n matrix of rank n. Then there exists an n × n orthogonal matrix U, an n × n orthogonal matrix V, and an n × n diagonal matrix S such that

(4.4) $U^T A V = S, \quad A = U S V^T$

The successive diagonal entries of S are positive and nonincreasing.

Proof of Lemma (4.3): The positive definite symmetric matrix $A^T A$ has an eigenvalue–eigenvector decomposition

(4.5) $A^T A = V D V^T$

where the $n \times n$ matrix V is orthogonal and the matrix D is diagonal with positive nonincreasing diagonal entries.

Define the diagonal matrix S to be the $n \times n$ matrix whose diagonal terms are the positive square roots of the respective diagonal entries of D. Thus

(4.6) $D = S^T S = S^2$

and

(4.7) $S^{-1} D S^{-1} = I_n$

Define the $n \times n$ matrix

(4.8) $U = A V S^{-1}$

From Eq. (4.5), (4.7), and (4.8) and the orthogonality of V,

(4.9) $\begin{aligned} U^T U &= S^{-1} V^T A^T A V S^{-1} \\ &= S^{-1} D S^{-1} \\ &= I_n \end{aligned}$

so that U is orthogonal.

From Eq. (4.8) and the fact that V is orthogonal,

(4.10) $\begin{aligned} U S V^T &= A V S^{-1} S V^T \\ &= A V V^T \\ &= A \end{aligned}$

This completes the proof of Lemma (4.3).

Proof of Theorem (4.1): Let

(4.11) $$A = HRK^T$$

where H, R, and K^T have the properties noted in Theorem (3.19).

Since the $k \times k$ matrix R_{11} of Eq. (3.21) is nonsingular, Lemma (4.3) shows that we may write

(4.12) $$R_{11} = \tilde{U}\tilde{S}\tilde{V}^T$$

Here $\tilde{U}$ and $\tilde{V}$ are $k \times k$ orthogonal matrices, while $\tilde{S}$ is a nonsingular diagonal matrix with positive and nonincreasing diagonal elements.

It follows from Eq. (4.12) that the matrix R of Eq. (3.21) may be written as

(4.13) $$R = \hat{U}S\hat{V}^T$$

where $\hat{U}$ is the $m \times m$ orthogonal matrix

(4.14) $$\hat{U} = \begin{bmatrix} \tilde{U} & 0 \\ 0 & I_{m-k} \end{bmatrix}$$

$\hat{V}$ is the $n \times n$ orthogonal matrix

(4.15) $$\hat{V} = \begin{bmatrix} \tilde{V} & 0 \\ 0 & I_{n-k} \end{bmatrix}$$

and S is the $m \times n$ diagonal matrix

(4.16) $$S = \begin{bmatrix} \tilde{S} & 0 \\ 0 & 0 \end{bmatrix}$$

Then defining U and V as

(4.17) $$U = H\hat{U}$$
(4.18) $$V = K\hat{V}$$

we see from Eqs. (4.11)–(4.18) that

(4.19) $$A = USV^T$$

where U, S, and V have the properties stated in Theorem (4.1). This completes the proof of Theorem (4.1).

Notice that the singular values of the matrix A are uniquely determined even though the orthogonal matrices U and V of Eq. (4.19) are not unique.

Let σ be a singular value of A with multiplicity l. Thus among the ordered singular values there is an index i such that

$$\sigma = s_j \quad \text{for } j = i, i+1, \ldots, i+l-1$$

and

$$\sigma \neq s_j \quad \text{for } j < i \text{ or } j \geq i+l$$

The l-dimensional subspace T of n-space spanned by the column vectors v_j, $j = i, i+1, \ldots, i+l-1$, of V is uniquely determined by A but the specific set of orthogonal basis vectors of T constituting columns $i, i+1, \ldots, i+l-1$ of V is not uniquely determined.

More specifically, let $k = \min(m, n)$ and let Q be a $k \times k$ orthogonal matrix of the form

$$Q = \begin{bmatrix} I_{i-1} & 0 & 0 \\ 0 & P_{l \times l} & 0 \\ 0 & 0 & I_{k-l-i+1} \end{bmatrix}$$

where P is $l \times l$ and orthogonal. If $A = USV^T$ is a singular value decomposition of A and $s_i = s_{i+1} = \cdots = s_{i+l-1}$, then $\tilde{U}S\tilde{V}^T$ with

$$\tilde{U} = U \begin{bmatrix} Q & 0 \\ 0 & I_{m-k} \end{bmatrix} \text{ and } \tilde{V} = V \begin{bmatrix} Q & 0 \\ 0 & I_{n-k} \end{bmatrix}$$

is also a singular value decomposition of A.

The following numerical example gives the singular value decomposition of the matrix A introduced previously as Example (3.23) for Case 2a.

(4.20)
$$A = U \begin{bmatrix} S \\ 0 \end{bmatrix} V^T$$

$$U = \begin{bmatrix} -0.4347 & 0.6141 & -0.6587 \\ -0.5509 & -0.7599 & -0.3450 \\ -0.7125 & 0.2129 & 0.6687 \end{bmatrix}$$

$$S = \begin{bmatrix} 0.9965 & 0.0 \\ 0.0 & 0.1128 \end{bmatrix}$$

$$V = \begin{bmatrix} -0.8626 & 0.5058 \\ -0.5058 & -0.8626 \end{bmatrix}$$

EXERCISES

(4.21) If A is $n \times n$ and symmetric with distinct singular values, then
   (a) A has distinct (real) eigenvalues.
   (b) A singular value decomposition of A can be used to obtain an eigenvalue–eigenvector decomposition of A and conversely. What if the singular values are not distinct?

(4.22) If s_1 is the largest singular value of A, then $\|A\| = s_1$.

(4.23) If R is an $n \times n$ nonsingular matrix and s_n is its smallest singular value, then $\|R^{-1}\| = s_n^{-1}$.

(4.24) Let s_1 and s_n, respectively, denote the largest and smallest singular values of a matrix $A_{m \times n}$ of rank n. Show that $s_n\|x\| \le \|Ax\| \le s_1\|x\|$ for all n-vectors x.

(4.25) Let $s_1, \ldots, s_k$ be the nonzero singular values of A. Then

$$\|A\|_F = \left(\sum_{i=1}^{k} s_i^2 \right)^{1/2}$$

(4.26) Let $A = USV^T$ be a singular value decomposition of A. Show that the column vectors of U are eigenvectors of the symmetric matrix AA^T.

5 PERTURBATION THEOREMS FOR SINGULAR VALUES

The singular values of a matrix are very stable with respect to changes in the elements of the matrix. Perturbations of the elements of a matrix produce perturbations of the same, or smaller, magnitude in the singular values. The purpose of this chapter is to present Theorems (5.7) and (5.10), which provide precise statements about this stability, and Theorem (5.12), which gives bounds on the perturbation of singular values due to removing a column, or row, of a matrix.

These theorems are direct consequences of corresponding theorems about the stability of eigenvalues of a symmetric matrix. We first quote the three relevant eigenvalue theorems.

(5.1) THEOREM

Let B, A, *and* E *be* $n \times n$ *symmetric matrices with* $B - A = E$. *Denote their respective eigenvalues by* β_i, α_i, *and* ϵ_i, $i = 1, \ldots, n$, *each set labeled in nonincreasing order. Then*

$$\epsilon_n \leq \beta_i - \alpha_i \leq \epsilon_1 \qquad i = 1, \ldots, n$$

A weaker conclusion that is often useful because it requires less detailed information about E *is*

$$|\beta_i - \alpha_i| \leq \max_j |\epsilon_j| \equiv \|E\| \qquad i = 1, \ldots, n$$

(5.2) THEOREM   (*Wielandt–Hoffman*)

With the same hypotheses as in Theorem (5.1),

$$\sum_{i=1}^{n} (\beta_i - \alpha_i)^2 \leq \sum_{i=1}^{n} \epsilon_i^2 \equiv \sum_{i=1}^{n} \sum_{j=1}^{n} e_{ij}^2 \equiv \|E\|_F^2$$

(5.3) THEOREM

> Let A be an n × n *symmetric matrix with eigenvalues* $\alpha_1 \geq \alpha_2 \geq \cdots$
> $\geq \alpha_n$. Let k *be an integer,* $1 \leq k \leq n$. Let B *be the* (n − 1) × (n − 1)
> *symmetric matrix obtained by deleting the kth row and column from* A.
> *Then the ordered eigenvalues* β_i *of* B *interlace with those of* A *as follows:*

$$\alpha_1 \geq \beta_1 \geq \alpha_2 \geq \beta_2 \geq \cdots \geq \beta_{n-1} \geq \alpha_n$$

The reader is referred to Wilkinson (1965a), pp. 99–109, for discussion
and proofs of these three theorems as well as the Courant–Fischer Minmax
Theorem on which Theorems (5.1) and (5.3) depend. Theorem (5.2) is due to
Hoffman and Wielandt (1953) and is further treated in Wilkinson (1970).

To derive singular value perturbation theorems from these eigenvalue
perturbation theorems, we shall make use of the relationship between the
singular value decomposition of a matrix A and the eigenvalue–eigenvector
decomposition of the symmetric matrix

(5.4)
$$C = \begin{bmatrix} 0 & A \\ A^T & 0 \end{bmatrix}$$

If A is square with a singular value decomposition $A = USV^T$, it is easily
verified that C has an eigenvalue–eigenvector decomposition

(5.5)
$$C = \begin{bmatrix} \tilde{U} & -\tilde{U} \\ \tilde{V} & \tilde{V} \end{bmatrix} \cdot \begin{bmatrix} S & 0 \\ 0 & -S \end{bmatrix} \cdot \begin{bmatrix} \tilde{U}^T & \tilde{V}^T \\ -\tilde{U}^T & \tilde{V}^T \end{bmatrix}$$

where

$$\tilde{U} = 2^{-1/2} U$$

and

$$\tilde{V} = 2^{-1/2} V$$

If $A_{m \times n}$, $m \geq n$, has a singular value decomposition

$$A = [U_{m \times n}^{(1)} : U_{m \times (m-n)}^{(2)}] \begin{bmatrix} S_{n \times n} \\ 0_{(m-n) \times n} \end{bmatrix} V_{n \times n}^T$$

then C, as defined in Eq. (5.4), has the eigenvalue–eigenvector decomposition

$$C = P \begin{bmatrix} S & 0 & 0 \\ 0 & -S & 0 \\ 0 & 0 & 0_{(m-n) \times (m-n)} \end{bmatrix} P^T$$

where

$$P = \begin{bmatrix} \tilde{U}^{(1)} & -\tilde{U}^{(1)} & U^{(2)} \\ \tilde{V} & \tilde{V} & 0_{n \times (m-n)} \end{bmatrix}$$

$$\tilde{U}^{(1)} = 2^{-1/2} U^{(1)}$$

and

$$\tilde{V} = 2^{-1/2} V$$

Clearly, analogous results hold for the case of $m < n$. For later reference we state the following theorem based on the discussion above:

(5.6)   THEOREM

Let A be an $m \times n$ matrix and $k = min$ (m, n). Let C be the $(m + n) \times (m + n)$ symmetric matrix defined by Eq. (5.4). If the singular values of A are $s_1, \ldots, s_k$, then the eigenvalues of C are $s_1, \ldots, s_k, -s_1, \ldots, -s_k$, and zero repeated $|m - n|$ times.

We may now state the following three theorems regarding the perturbation of singular values.

(5.7)   THEOREM

Let B, A, and E be $m \times n$ matrices with $B - A = E$. Denote their respective singular values by β_i, α_i, and ϵ_i, $i = 1, \ldots, k$; $k = min$ (m, n), each set labeled in nonincreasing order. Then

(5.8) $|\beta_i - \alpha_i| \leq \epsilon_1 \equiv \|E\|$ $i = 1, \ldots, k$

Proof: Introduce the three symmetric matrices

(5.9) $\tilde{B} = \begin{bmatrix} 0 & B \\ B^T & 0 \end{bmatrix}$, $\tilde{A} = \begin{bmatrix} 0 & A \\ A^T & 0 \end{bmatrix}$,   and $\tilde{E} = \begin{bmatrix} 0 & E \\ E^T & 0 \end{bmatrix}$

Then

$$\tilde{B} - \tilde{A} = \tilde{E}$$

The eigenvalues of these matrices are related to the singular values of B, A, and E as stated in Theorem (5.6). Applying Theorem (5.1) to the matrices $\tilde{B}$, $\tilde{A}$, and $\tilde{E}$, we obtain the result in Eq. (5.8).

(5.10)   THEOREM

With hypotheses as in Theorem (5.7), there follows the inequality

(5.11) $\sum_{i=1}^{k} (\beta_i - \alpha_i)^2 \leq \sum_{i=1}^{k} \epsilon_i^2 \equiv \sum_{i=1}^{m} \sum_{j=1}^{n} e_{ij}^2 \equiv \|E\|_F^2$

Proof: Introducing Eq. (5.9) and using Theorems (5.6) and (5.2), we obtain

$$2 \sum_{i=1}^{k} (\beta_i - \alpha_i)^2 \leq \| \tilde{E} \|_F^2 = 2\| E \|_F^2$$

which is equivalent to Eq. (5.11).

(5.12)   THEOREM

Let A *be an* m × n *matrix. Let* k *be an integer,* $1 \leq k \leq n$. *Let* B *be the* m × (n − 1) *matrix resulting from the deletion of column* k *from* A. *Then the ordered singular values* β_i *of* B *interlace with those* α_i *of* A *as follows:*

CASE 1 $m \geq n$

(5.13) $\alpha_1 \geq \beta_1 \geq \alpha_2 \geq \beta_2 \geq \cdots \geq \beta_{n-1} \geq \alpha_n \geq 0$

CASE 2 $m < n$

(5.14) $\alpha_1 \geq \beta_1 \geq \alpha_2 \geq \beta_2 \geq \cdots \geq \alpha_m \geq \beta_m \geq 0$

Proof: The results in Eq. (5.13) and (5.14) follow directly from application of Theorem (5.3) to the symmetric matrices $\hat{A} = A^T A$ and $\hat{B} = B^T B$. In Case 1 the eigenvalues of $\hat{A}$ and $\hat{B}$ are α_i^2, $i = 1, \ldots, n$, and β_i^2, $i = 1, \ldots, n - 1$, respectively. In Case 2 the eigenvalues of $\hat{A}$ are α_i^2, $i = 1, \ldots, m$, and zero repeated $n - m$ times, while the eigenvalues of $\hat{B}$ are β_i^2, $i = 1, \ldots, m$, and zero repeated $n - 1 - m$ times.

EXERCISES

(5.15)   *Problem* [Eckart and Young (1936)]:   Given an $m \times n$ matrix A, of rank k, and a nonnegative integer $r < k$, find an $m \times n$ matrix B of rank r that minimizes $\| B - A \|_F$.

Solution: Let $A = USV^T$ be a singular value decomposition of A with ordered singular values $s_1 \geq s_2 \geq \cdots \geq s_k > 0$. Let $\tilde{S}$ be constructed from S by replacing s_{r+1} through s_k by zeros. Show that $\tilde{B} = U\tilde{S}V^T$ solves the stated problem, and give an expression for $\| \tilde{B} - A \|_F$ in terms of the singular values of A.

Remark: The proof is straightforward using Theorem (5.10). The proof given by Eckart and Young contains the essential ideas needed to prove Theorem (5.10).

(5.16)   Solve the problem of Exercise (5.15) with $\| \cdot \|_F$ replaced by $\| \cdot \|$.

(5.17) Define $\kappa(A)$ to be the ratio of the largest singular value of A to the smallest nonzero singular value of A. (This "condition number" of A will be used in subsequent chapters.) Show that if Rank $(A_{m \times n}) = n$ and if $B_{m \times r}$ is a matrix obtained by deleting $(n - r)$ columns from A, then $\kappa(B) \le \kappa(A)$.

6 BOUNDS FOR THE CONDITION NUMBER OF A TRIANGULAR MATRIX

In the discussion of the practical solution of least squares problems in Chapter 25, the need will arise to estimate the largest and smallest nonzero singular values of a matrix A since the ratio of these two quantities (the condition number of A) has an interpretation of being an error magnification factor. This interpretation of the condition number will be presented in Chapter 9.

The most direct approach is to compute the singular values of A (see Chapters 4 and 18 and Appendix C). One may, however, wish to obtain bounds on the condition number without computing the singular values. In this chapter we present some theorems and examples relating to this problem.

Most of the algorithms which are described in this book produce as an intermediate step a nonsingular triangular matrix, say, R, which has the same nonzero singular values as the original matrix A. In general, a nonsingular triangular matrix is a better starting point for the bounding of singular values than is a full matrix. Therefore we shall deal only with bounds for the singular values of a nonsingular triangular matrix R.

Denote the ordered singular values of the $n \times n$ nonsingular triangular matrix R by $s_1 \geq s_2 \geq \cdots \geq s_n > 0$. As noted in Exercise (4.22), $s_1 = \|R\|$. Therefore, one easily computable lower bound for s_1 is

$$(6.1) \qquad s_1 \geq \max_{ij} |r_{ij}|$$

Furthermore since R is triangular, the reciprocals of the diagonal elements of R are elements of R^{-1}. Therefore

$$(6.2) \qquad \|R^{-1}\| \geq \max_i |r_{ii}^{-1}|$$

From Exercise (4.23), $s_n^{-1} = \|R^{-1}\|$, and thus

(6.3) $$s_n \leq \min_i |r_{ii}|$$

Therefore, a lower bound p for the condition number $\kappa = s_1/s_n$ is available as

(6.4) $$\kappa \geq p \equiv \frac{\max_{ij} |r_{ij}|}{\min_i |r_{ii}|}$$

This lower bound p, while of some practical use, cannot in general be regarded as a reliable estimate of κ. In fact, κ can be substantially larger than p.

As our first example [Kahan (1966b), p. 790], let the upper triangular $n \times n$ matrix R be defined by

(6.5) $$r_{ij} = \begin{cases} 0 & \text{if } j < i \\ 1 & \text{if } j = i \\ -1 & \text{if } j > i \end{cases}$$

For example, for $n = 4$ we have

(6.6) $$R = \begin{bmatrix} 1 & -1 & -1 & -1 \\ 0 & 1 & -1 & -1 \\ 0 & 0 & 1 & -1 \\ 0 & 0 & 0 & 1 \end{bmatrix}$$

Using Eq. (6.4) we find $p = 1$ as a lower bound for κ, which tells us nothing at all. For this matrix a more realistic upper bound for s_n can be obtained by noting the effect of R on the vector y with components

(6.7) $$y_i = 2^{1-i} \quad \text{for } i = 1, \ldots, n-1$$
$$y_n = y_{n-1}$$

It is easily verified that $Ry = z$ where

(6.8) $$z_i = 0 \quad \text{for } i = 1, \ldots, n-1$$
$$z_n = 2^{2-n}$$

Then, using the inequalities

(6.9) $$s_n^{-1} = \|R^{-1}\| \geq \frac{\|y\|}{\|z\|} \geq \frac{1}{2^{2-n}}$$

we obtain

(6.10) $\kappa = \dfrac{s_1}{s_n} \geq 2^{n-2}$

The inequality of Eq. (6.9) may also be interpreted as showing that R is close to a singular matrix in the sense that there exists a matrix E such that $\|E\| \leq 2^{2-n}$ and $R - E$ is singular. It is easily verified that such a matrix is given by $E = zy^T/y^Ty$.

To illustrate this example for the case of $n = 4$, we note that $y = (1, \frac{1}{2}, \frac{1}{4}, \frac{1}{4})^T$, $z = (0, 0, 0, \frac{1}{4})^T$, $s_4 \leq \frac{1}{4}$, and $\kappa \geq 4$. Furthermore, subtracting the matrix

$$E = \begin{bmatrix} 0 & 0 & 0 & 0 \\ 0 & 0 & 0 & 0 \\ 0 & 0 & 0 & 0 \\ \frac{2}{11} & \frac{1}{11} & \frac{1}{22} & \frac{1}{22} \end{bmatrix}$$

from the matrix R of Eq. (6.6) gives a singular matrix.

This example illustrates the possible danger of assuming a matrix is well conditioned even when the ρ of Eq. (6.4) is small and the matrix is "innocent looking."

Recall that we are primarily interested in triangular matrices that arise from algorithms for solving systems of linear equations. The matrix defined by Eq. (6.5) has been discussed in the literature because it is invariant under Gaussian elimination with either full or partial pivoting. It is not invariant, however, under elimination by Householder transformations using column interchanges (such as the Algorithm HFTI, which will be described in Chapter 14).

As a numerical experiment in this connection we applied Algorithm HFTI to the matrix R of Eq. (6.5). Let $\tilde{R}$ denote the triangular matrix resulting from this operation. We also computed the singular values of R. The computing was done on a UNIVAC 1108 with mixed precision (see Chapter 17) characterized by single and double precision arithmetic, respectively, of

$$\eta = 2^{-27} \doteq 0.745 \times 10^{-8} \quad \text{and} \quad \omega = 2^{-58} \doteq 0.347 \times 10^{-17}$$

Table 6.1 lists the computed values of the last diagonal element $\tilde{r}_{nn}$ and the smallest singular value s_n as well as the ratio $s_n/\tilde{r}_{nn}$ for $n = 20$ through 26. Observe that in these cases $\tilde{r}_{nn}$ provides a good estimate of the size of s_n.

It is nevertheless true that there exist $n \times n$ matrices with unit column norms which could be produced by Algorithm HFTI and whose smallest singular value is smaller than the smallest diagonal element by a factor of approximately 2^{1-n}.

Table 6.1   LAST DIAGONAL ELEMENT OF $\tilde{R}$ AND LAST SINGULAR VALUE OF R

n	$\tilde{r}_{nn} \times 10^8$	$s_n \times 10^8$	$s_n/\tilde{r}_{nn}$
20	330.	286.	0.867
21	165.	143.	0.867
22	82.5	71.6	0.868
23	40.5	35.8	0.884
24	20.5	17.9	0.873
25	8.96	8.95	0.999
26	4.59	4.52	0.985

The following example of such a matrix is due to Kahan (1966b), pp. 791–792. Define the $n \times n$ matrix R by

$$r_{ij} = \begin{cases} 0 & \text{if } i > j \\ s^{i-1} & \text{if } i = j \\ -cs^{i-1} & \text{if } i < j \end{cases}$$

where s and c are positive numbers satisfying $s^2 + c^2 = 1$. For example, if $n = 4$, we have

(6.11)
$$R = \begin{bmatrix} 1 & -c & -c & -c \\ 0 & s & -cs & -cs \\ 0 & 0 & s^2 & -cs^2 \\ 0 & 0 & 0 & s^3 \end{bmatrix}$$

For general n, the matrix R is upper triangular, the column vectors are all of unit euclidean length, and the inequalities

(6.12)
$$r_{kk}^2 \geq \sum_{i=k}^{j} r_{ij}^2 \qquad \text{for } k = 1, \ldots, n-1$$
$$\text{and } j = k+1, \ldots, n$$

are satisfied. As established in Chapter 14, these inequalities imply that R is a matrix that could result from applying the Algorithm HFTI to some matrix A.

Let $T = R^{-1}$. The elements of T are given by

$$t_{ij} = \begin{cases} 0 & \text{if } j < i \\ \dfrac{1}{s^{j-1}} & \text{if } j = i \\ \dfrac{c}{s^{j-1}} & \text{if } j = i+1 \\ \dfrac{c(1+c)^{j-i-1}}{s^{j-1}} & \text{if } j - i \geq 2 \end{cases}$$

For example, if $n = 4$, we have

$$R^{-1} = T = \begin{bmatrix} 1 & \dfrac{c}{s} & \dfrac{c(1+c)}{s^2} & \dfrac{c(1+c)^2}{s^3} \\[2ex] 0 & \dfrac{1}{s} & \dfrac{c}{s^2} & \dfrac{c(1+c)}{s^3} \\[2ex] 0 & 0 & \dfrac{1}{s^2} & \dfrac{c}{s^3} \\[2ex] 0 & 0 & 0 & \dfrac{1}{s^3} \end{bmatrix}$$

Let τ_n denote the last column vector of R^{-1}. Then

$$\|\tau_n\|^2 = \frac{1 + c^2[(1+c)^{2n-2} - 1]}{(1+c)^2 - 1} r_{nn}^{-2}$$

As $s \to 0^+$ and $c = (1 - s^2)^{1/2} \to 1^-$, the product $\|\tau_n\|^2 r_{nn}^2$ approaches $(4^{n-1} + 2)/3$ from below. Thus there is some value of s, depending on n, for which

$$\|\tau_n\|^2 r_{nn}^2 \geq \frac{4^{n-1}}{3}$$

Then, using Exercise (4.24),

$$s_n \leq \frac{\|R\tau_n\|}{\|\tau_n\|} = \frac{1}{\|\tau_n\|}$$

we have

$$s_n \leq 3^{1/2} 2^{1-n} |r_{nn}| \doteq (1.73) 2^{1-n} |r_{nn}|$$

The following theorem shows that this case represents nearly the minimum value attainable by $s_n/|r_{nn}|$.

(6.13)   THEOREM   [Stated without proof in Faddeev, et al. (1968)]

Let A be an m × n matrix of rank n. Assume all column vectors of A have unit euclidean length. Let R be an upper triangular matrix produced by Householder triangularization of A with interchange of columns as in Algorithm HFTI (14.9). Then s_n, the smallest singular value of A, is related to r_{nn}, the last diagonal element of R, by

(6.14) $s_n \leq |r_{nn}|$

and

(6.15) $s_n \geq 3(4^n + 6n - 1)^{-1/2} |r_{nn}| \geq 2^{1-n} |r_{nn}|$

Proof: From the way in which R is constructed, its column vectors have unit euclidean length and it has the same singular values as A. Furthermore, as a result of the use of column interchanges, the following inequalities will hold:

$$(6.16) \qquad r_{kk}^2 \geq \sum_{i=k}^{j} r_{ij}^2 \qquad \text{for } k = 1, \ldots, n-1$$
$$\text{and } j = k+1, \ldots, n$$

From Eq. (6.16) it follows that

$$(6.17) \qquad |r_{kk}| \geq |r_{ij}| \qquad \text{for } i \geq k, j > k, k = 1, \ldots, n-1$$

Since

$$(6.18) \qquad s_n = \|R^{-1}\|^{-1}$$

the assertions in Eq. (6.14) and (6.15) are equivalent to the assertions

$$(6.19) \qquad \|R^{-1}\| \geq |r_{nn}^{-1}|$$

and

$$(6.20) \qquad \|R^{-1}\| \leq 3^{-1}(4^n + 6n - 1)^{1/2} |r_{nn}^{-1}|$$

The assertion in Eq. (6.19) follows from the fact that r_{nn}^{-1} is an element of R^{-1}.

To establish Eq. (6.20), we shall develop upper bounds on the magnitudes of all elements of R^{-1} and then compute the Frobenious norm of the bounding matrix $\tilde{M}$.

Define $T = R^{-1}$. It will be convenient to group the elements of T into diagonals parallel to the main diagonal. Introduce bounding parameters for the diagonals as follows:

$$(6.21) \quad g_k = \max \{|t_{i,i+k}|, i = 1, \ldots, n-k\} \qquad \text{for } k = 0, 1, \ldots, n-1$$

Then

$$(6.22) \qquad g_0 = \max_i |t_{ii}| = \max_i |r_{ii}^{-1}| = |r_{nn}^{-1}|$$

The elements in the kth superdiagonal of T are expressible in terms of elements of the main diagonal and preceding superdiagonals of T and elements of R as follows:

$$(6.23) \qquad t_{i,i+k} = -\frac{\sum_{l=1}^{k} r_{i,i+l} t_{i+l,i+k}}{r_{ii}} \qquad \begin{array}{l} 1 \leq i \leq n-1 \\ 1 \leq k \leq n-i \end{array}$$

Since $|r_{i,i+1}| \leq |r_{ii}|$, we may use Eq. (6.21) and (6.23) to obtain

(6.24)
$$g_k = \max_i |t_{i,i+k}|$$
$$\leq \max_i \sum_{i=1}^{k} |t_{i+1,i+k}| \leq \sum_{i=1}^{k} g_{k-1}$$
$$= \sum_{j=0}^{k-1} g_j$$

It is then easily verified by induction that

(6.25) $g_k \leq 2^{k-1} g_0 = 2^{k-1} |r_{nn}^{-1}|$ for $k = 1, \ldots, n$

Define an $n \times n$ upper triangular matrix M by

(6.26)
$$m_{i,j} = \begin{cases} 0 & \text{if } j < i \\ 1 & \text{if } j = i \\ 2^{j-i-1} & \text{if } j > i \end{cases}$$

and define

(6.27)
$$\tilde{M} = |r_{nn}^{-1}| M$$

For example, if $n = 4$,

$$M = \begin{bmatrix} 1 & 1 & 2 & 4 \\ 0 & 1 & 1 & 2 \\ 0 & 0 & 1 & 1 \\ 0 & 0 & 0 & 1 \end{bmatrix}$$

From Eq. (6.21) and (6.25) to (6.27), it follows that the elements of R^{-1} ($= T$) are individually bounded in magnitude by the corresponding elements of $\tilde{M}$. Thus

(6.28) $s_n^{-1} = \|R^{-1}\| \leq \|R^{-1}\|_F \leq \|\tilde{M}\|_F = |r_{nn}^{-1}| \cdot \|M\|_F$

To compute the Frobenious norm of M, we note that for $j \geq 2$ the sum of squares of the elements of column j is

(6.29)
$$w_j^2 = \left(1 + \sum_{i=0}^{j-2} 4^i\right) = \frac{4^{j-1} + 2}{3}$$

This expression is also correct for $j = 1$, assuming the sum in the middle member is defined as zero.

Then

(6.30) $\|R^{-1}\|_F^2 = \|T\|_F^2 \leq r_{nn}^{-2} \|M\|_F^2 = r_{nn}^{-2} \sum_{j=1}^{n} w_j^2 = r_{nn}^{-2} \dfrac{4^n + 6n - 1}{9}$

The inequalities in Eq. (6.28) and (6.30) together establish the inequality in Eq. (6.20) and complete the proof of Theorem (6.13).

The following more general theorem has also been stated by Faddeev, Kublanovskaya, and Faddeeva (1968).

(6.31) THEOREM

Let A and R be matrices as given in Theorem (6.13). The singular values $s_1 \geq s_2 \geq \cdots \geq s_n$ of A are related to the diagonal elements r_{ii} of R by the inequalities

(6.32) $$2^{1-i}|r_{ii}| \leq 3(4^i + 6i - 1)^{-1/2}|r_{ii}| \leq s_i \leq (n - i + 1)^{1/2}|r_{ii}|$$
$$\text{for } i = 1, \ldots, n$$

Proof: Define R_j to be the leading principal submatrix of R of order j. Denote the ith singular value of R_j by $s_i^{(j)}$. Using Theorem (5.12) we have the inequalities

(6.33) $$s_i = s_i^{(n)} \geq s_i^{(n-1)} \geq \cdots \geq s_i^{(i)}$$

Theorem (6.13) may be applied to R_i to obtain

(6.34) $$s_i^{(i)} \geq 3(4^i + 6i - 1)^{-1/2}|r_{ii}|$$

which with Eq. (6.33) establishes the lower bound for s_i stated in Eq. (6.32).

Define W_j to be the principal submatrix of R of order $n + 1 - j$ consisting of the intersections of rows and columns j through n. Note that the leading diagonal element of W_i is r_{ii}. Using the properties in Eq. (6.16) and Exercise (6.37), one obtains

(6.35) $$\| W_i \| \leq (n + 1 - i)^{1/2}|r_{ii}|$$

Denote the ith singular value of W_j by $\omega_i^{(j)}$. Then from Theorem (5.12) we may write

(6.36) $$s_i = \omega_i^{(1)} \leq \omega_{i-1}^{(2)} \leq \cdots \leq \omega_1^{(i)}$$

Since $\omega_1^{(i)} = \| W_i \|$, Eq. (6.35) and (6.36) may be combined to obtain the upper bound for s_i in Eq. (6.32), completing the proof of Theorem (6.31).

EXERCISE

(6.37) Let A be an $m \times n$ matrix with column vectors a_j. Then $\| A \| \leq n^{1/2} \max_j \| a_j \|$.

7

THE PSEUDOINVERSE

If A is an $n \times n$ nonsingular matrix, the solution of the problem $Ax = b$ can be written as $x = A^{-1}b$ where A^{-1} is the (unique) inverse matrix for A. The inverse matrix is a very useful mathematical concept even though efficient and reliable contemporary methods [Forsythe and Moler (1967); Wilkinson (1965a)] for computing the solution of $Ax = b$ do not involve explicit computation of A^{-1}.

For Problem LS there arises the question as to whether there exists some $n \times m$ matrix Z, uniquely determined by A, such that the (unique) minimum length solution of Problem LS is given by $x = Zb$. This is indeed the case and this matrix Z is called the pseudoinverse of A. As noted above for the inverse, the pseudoinverse is a useful mathematical concept but one would not usually compute the pseudoinverse explicitly in the process of computing a solution of Problem LS.

The following two theorems lead to a constructive definition of the pseudoinverse of an $m \times n$ matrix A.

(7.1) THEOREM

Let A *be an* m $\times$ n *matrix of rank* k *with an orthogonal decomposition* A = HRK^T *as in the hypotheses of Theorem (2.3). Then the unique minimum length solution of Problem LS is given by*

(7.2)
$$x = K \begin{pmatrix} R_{11}^{-1} & 0 \\ 0 & 0 \end{pmatrix} H^T b$$

Proof: The conclusion, Eq. (7.2), is an alternative way of writing Eq. (2.5) to (2.7) and Eq. (2.11).

(7.3) THEOREM

Let $A_{m \times n} = HRK^T$ as in Theorem (7.1). Define

$$Z = K \begin{pmatrix} R_{11}^{-1} & 0 \\ 0 & 0 \end{pmatrix} H^T$$

Then Z is uniquely defined by A; it does not depend on the particular orthogonal decomposition of A.

Proof: For each j, $1 \leq j \leq m$, the jth column z_j of Z can be written as $z_j = Ze_j$, where e_j is the jth column of the identity matrix I_m. From Theorem (7.1), z_j is the unique minimum length solution of the least squares problem $Ax_j \cong e_j$. This completes the proof of Theorem (7.3).

In view of Theorems (7.1) and (7.3) we make the following definition.

(7.4) DEFINITION

For a general $m \times n$ matrix A, the *pseudoinverse* of A, denoted by A^+, is the $n \times m$ matrix whose jth column z_j is the unique minimum length solution of the least squares problem

$$Ax_j \cong e_j$$

where e_j is the jth column of the identity matrix I_m.

This definition along with Theorems (7.1) and (7.3) immediately allow us to write the minimum length solution to Problem LS as

(7.5) $$\bar{x} = A^+b$$

The following two cases are worthy of special note. For a square nonsingular matrix B, the pseudoinverse of B is the inverse of B:

(7.6) $$B^+ = B^{-1}$$

For an $m \times n$ matrix,

$$R = \begin{pmatrix} R_{11} & 0 \\ 0 & 0 \end{pmatrix}$$

with R_{11} a $k \times k$ nonsingular matrix, the pseudoinverse of R is the $n \times m$ matrix

(7.7) $$R^+ = \begin{pmatrix} R_{11}^{-1} & 0 \\ 0 & 0 \end{pmatrix}$$

The pseudoinverse of an $m \times n$ matrix A can be characterized in other ways. Each of the following two theorems provides an alternative characterization of the pseudoinverse.

(7.8) THEOREM

> If $A = HRK^T$ is any orthogonal decomposition of A, as given in Theorem (2.3), then $A^+ = KR^+H^T$ where R^+ is given in Eq. (7.7).

(7.9) THEOREM   [The Penrose Conditions; Penrose (1955)]

> The $n \times m$ matrix A^+ is the unique matrix X that satisfies the following four conditions:
> (a) $AXA = A$
> (b) $XAX = X$
> (c) $(AX)^T = AX$
> (d) $(XA)^T = XA$

Proofs of these two theorems are left as exercises.

The explicit representation of A^+ given in Theorem (7.8) is particularly useful for computation. For example, if one has the orthogonal decomposition A of Eq. (3.20), with R_{11} of Eq. (3.21) nonsingular and triangular, then

(7.10)
$$A^+ = K\begin{pmatrix} R_{11}^{-1} & 0 \\ 0 & 0 \end{pmatrix}H^T$$

If one has the singular value decomposition of Eq. (4.2), as given in Theorem (4.1), then

(7.11)
$$A^+ = VS^+U^T$$

As mentioned at the beginning of this chapter, there is usually no need to construct the pseudoinverse of a matrix A explicitly. For example, if the purpose of the computation is the solution of Problem LS, then it is more economical in computer time and storage to compute the solution with the following three steps [which are essentially repeated from the proof of Theorem (2.3)].

(7.12)
$$g = H^Tb$$

(7.13) Solve $R_{11}y_1 = g_1$ for y_1.

(7.14)
$$x = K\begin{pmatrix} y_1 \\ 0 \end{pmatrix} = A^+b$$

EXERCISES

(7.15) Verify Eq. (7.6) and (7.7).

(7.16) Prove Theorem 7.8.

(7.17) [Penrose (1955)] Prove Theorem 7.9.

(7.18) Prove $(A^+)^T = (A^T)^+$ and $(A^+)^+ = A$.

(7.19) Prove: If $Q_{m \times n}$ has orthonormal columns or orthonormal rows, then $Q^+ = Q^T$.

   If $Q_{m \times n}$ is of rank n and satisfies $Q^+ = Q^T$, then Q has orthonormal columns.

(7.20) [Penrose (1955)] If U and V have orthonormal columns, then $(UAV^T)^+ = VA^+U^T$.

(7.21) (a) Let $A = USV^T$ be a singular value decomposition of A. Write singular value decompositions for the four matrices $P_1 = A^+A$, $P_2 = I - A^+A$, $P_3 = AA^+$, and $P_4 = I - AA^+$ in terms of the matrices U, S, and V.

   (b) Deduce from part (a) that the matrices P_i are symmetric and idempotent, hence projection matrices.

   (c) Let T_i denote the subspace associated with the projection matrix P_i; that is, $T_i = \{x: P_ix = x\}$. Identify the relationship of each subspace T_i to the row or column spaces of A.

(7.22) Prove that the most general solution of $Ax \cong b$ is $x = A^+b + (I - A^+A)y$, where y is arbitrary.

(7.23) [Greville (1966)] For nonsingular matrices one has the identity $(AB)^{-1} = B^{-1}A^{-1}$. The analogous equation $(AB)^+ = B^+A^+$ is not satisfied by all matrices $A_{m \times k}$ and $B_{k \times n}$.

   (a) Exhibit two matrices A and B with $m = n = 1$ and $k = 2$ such that $(AB)^+ \neq B^+A^+$.

   (b) Prove that matrices $A_{m \times k}$ and $B_{k \times n}$ satisfy $(AB)^+ = B^+A^+$ if and only if the range space of B is an invariant space of A^TA and the range space of A^T (row space of A) is an invariant space of BB^T.

(7.24) If Rank $(A_{m \times n}) = n$, then $A^+ = (A^TA)^{-1}A^T$.

   If Rank $(A_{m \times n}) = m$, then $A^+ = A^T(AA^T)^{-1}$.

(7.25) [Penrose (1955); Graybill, et al. (1966)] Given A, the equations $XAA^T = A^T$ and $A^TAY = A^T$ are each consistent. If matrices X and Y satisfy these two equations, respectively, then XA and AY are projection matrices (symmetric and idempotent) and $A^+ = XAY$. Note the simplification of this proposition in case A is symmetric.

(7.26) [Graybill, et al. (1966)] The pseudoinverse of a rectangular matrix A can be defined in terms of the pseudoinverse of the symmetric matrix $A^T A$ or $A A^T$, whichever is more convenient, by the formulas $A^+ = (A^T A)^+ A^T$ or $A^+ = A^T (A A^T)^+$, respectively.

(7.27) [Penrose (1955)] If A is normal (i.e., satisfies $A^T A = A A^T$), then $A^+ A = A A^+$ and $(A^n)^+ = (A^+)^n$.

(7.28) [Penrose (1955)] If $A = \Sigma A_i$ with $A_i A_j^T = 0$ and $A_i^T A_j = 0$ whenever $i \neq j$, then $A^+ = \Sigma A_i^+$.

8 PERTURBATION BOUNDS FOR THE PSEUDOINVERSE

Our objective here and in Chapter 9 is to study the relationship of perturbation of the data of Problem LS to perturbation of the solution of the problem. In this chapter we develop the perturbation theorems for the pseudoinverse. These theorems are then used in Chapter 9 to study perturbation of the Problem LS.

In practice the consideration of such perturbations can arise due to the limited precision with which observable phenomena can be quantified. It is also possible to analyze the effects of round-off errors in the solution procedure as though their effects were due to perturbed data. This analysis for algorithms using Householder transformations will be described in Chapters 15, 16, and 17.

Results relating to perturbation of the pseudoinverse or the solution of Problem LS have been given by a number of authors. The treatment of this problem by Wedin (1969) seems most appropriate for our present purposes in terms of generality and the convenient form of the final results. For earlier treatments of this perturbation problem or special cases of the problem, see Golub and Wilkinson (1966), Björck (1967a and 1967b), Pereyra (1968), Stewart (1969), and Hanson and Lawson (1969).

Let A and E be $m \times n$ matrices and define the perturbed matrix

$$(8.1) \qquad \tilde{A} = A + E$$

and the residual matrix

$$(8.2) \qquad G = \tilde{A}^+ - A^+$$

We wish to determine the dependence of G on E and in particular to obtain bounds for $\|G\|$ in terms of $\|A\|$ and $\|E\|$.

It will be convenient to introduce the four projection matrices:

(8.3) $$P = A^+A = A^T A^{T+}, \qquad Q = AA^+ = A^{T+}A^T$$

(8.4) $$\tilde{P} = \tilde{A}^+\tilde{A} = \tilde{A}^T \tilde{A}^{T+}, \qquad \tilde{Q} = \tilde{A}\tilde{A}^+ = \tilde{A}^{T+}\tilde{A}^T$$

These matrices have many useful properties derivable directly from the Penrose conditions [Theorem (7.9)]. Also see Exercise (7.21) and the summary of standard properties of projection matrices given in Appendix A.

The matrices defined in Eq. (8.1) to (8.4) will be used throughout this chapter without further reference.

(8.5) THEOREM

Using definitions of Eq. (8.1) and (8.2), the matrix G satisfies

(8.6) $$G = G_1 + G_2 + G_3$$

where

(8.7) $$G_1 = -\tilde{A}^+ E A^+$$

(8.8) $$G_2 = \tilde{A}^+(I - Q) = \tilde{A}^+\tilde{A}^{T+}E^T(I - Q)$$

(8.9) $$G_3 = -(I - \tilde{P})A^+ = (I - \tilde{P})E^T A^{T+}A^+$$

These matrices are bounded as follows:

(8.10) $$\|G_1\| \leq \|E\|\cdot\|A^+\|\cdot\|\tilde{A}^+\|$$

(8.11) $$\|G_2\| \leq \|E\|\cdot\|\tilde{A}^+\|^2$$

(8.12) $$\|G_3\| \leq \|E\|\cdot\|A^+\|^2$$

Proof: Write G as a sum of eight matrices as follows:

(8.13) $$\begin{aligned} G &= [\tilde{P} + (I - \tilde{P})]\cdot(\tilde{A}^+ - A^+)\cdot[Q + (I - Q)] \\ &= \tilde{P}\tilde{A}^+Q + \tilde{P}\tilde{A}^+(I - Q) - \tilde{P}A^+Q - \tilde{P}A^+(I - Q) \\ &\quad + (I - \tilde{P})\tilde{A}^+Q + (I - \tilde{P})\tilde{A}^+(I - Q) \\ &\quad - (I - \tilde{P})A^+Q - (I - \tilde{P})A^+(I - Q) \end{aligned}$$

Using the properties

$$\tilde{P}\tilde{A}^+ = \tilde{A}^+, \qquad (I - \tilde{P})\tilde{A}^+ = 0$$
$$A^+Q = A^+, \qquad A^+(I - Q) = 0$$

Eq. (8.13) reduces to

$$\begin{aligned} G &= (\tilde{A}^+Q - \tilde{P}A^+) + \tilde{A}^+(I - Q) - (I - \tilde{P})A^+ \\ &\equiv G_1 + G_2 + G_3 \end{aligned}$$

To expose the first-order dependence of G on E, write

(8.14)
$$G_1 = \tilde{A}^+ A A^+ - \tilde{A}^+ \tilde{A} A^+ = -\tilde{A}^+ E A^+$$
$$G_2 = \tilde{A}^+ \tilde{Q}(I - Q) = \tilde{A}^+ \tilde{A}^{T+} \tilde{A}^T (I - Q)$$
$$= \tilde{A}^+ \tilde{A}^{T+} (\tilde{A}^T - A^T)(I - Q) = \tilde{A}^+ \tilde{A}^{T+} E^T (I - Q)$$
$$G_3 = -(I - \tilde{P}) P A^+ = -(I - \tilde{P}) A^T A^{T+} A^+$$
$$= -(I - \tilde{P})(A^T - \tilde{A}^T) A^{T+} A^+ = (I - \tilde{P}) E^T A^{T+} A^+$$

The bounds in Eq. (8.10) to (8.12) follow from the facts that $\|I - Q\| \leq 1$ and $\|I - \tilde{P}\| \leq 1$, completing the proof of Theorem (8.5).

We remark that for the case of real numbers a and $\tilde{a}$ (and in fact for square nonsingular matrices) one has the algebraic identity

$$\tilde{a}^{-1} - a^{-1} = a^{-1}(a - \tilde{a})\tilde{a}^{-1}$$

which suggests that Eq. (8.10) is a reasonable form to expect in a bound for $\|G\|$. The additional terms G_2 and G_3 are specifically associated with the cases of nonsquare matrices or square singular matrices in the following sense: The matrix G_2 can be nonzero only if Rank $(A) < m$ and G_3 can be nonzero only if Rank $(\tilde{A}) < n$ since $Q = I_m$ if Rank $(A) = m$ and $\tilde{P} = I_n$ if Rank $(\tilde{A}) = n$.

We next wish to replace $\|\tilde{A}^+\|$ in the right sides of Eq. (8.10), (8.11) and (8.12) by its bound in terms of $\|A^+\|$ and $\|E\|$. Such a bound is available only under the assumptions in Eq. (8.16) and (8.17) of the following theorem.

(8.15) THEOREM

Assume

(8.16)
$$\text{Rank } (A + E) \leq \text{Rank } (A) = k \geq 1$$

and

(8.17)
$$\|A^+\| \cdot \|E\| < 1$$

Let s_k denote the smallest nonzero singular value of A and $\epsilon = \|E\|$. Then

(8.18)
$$\text{Rank } (A + E) = k$$

and

(8.19)
$$\|(A + E)^+\| \leq \frac{\|A^+\|}{1 - \|A^+\| \cdot \|E\|} = \frac{1}{s_k - \epsilon}$$

Proof: The inequality of Eq. (8.17) can be written as $\epsilon/s_k < 1$ or equivalently $s_k - \epsilon > 0$. Let $\tilde{s}_k$ denote the kth singular value of $\tilde{A} = A + E$.

From Theorem (5.7)

(8.20) $$\tilde{s}_k \geq s_k - \epsilon$$

which implies that Rank $(A + E) \geq k$. With the inequality of Eq. (8.16) this establishes Eq. (8.18). The inequality of Eq. (8.20) can be written as

$$\frac{1}{\tilde{s}_k} \leq \frac{1}{s_k - \epsilon}$$

which is equivalent to the inequality of Eq. (8.19). This completes the proof of Theorem (8.15).

The conditions of Eq. (8.16) and (8.17) are necessary. It is easy to verify that $\|A^+\|$ may be unbounded if either of these conditions are not satisfied. As long as we shall be imposing Eq. (8.16) and (8.17) to obtain Eq. (8.19), we can take advantage of the condition in Eq. (8.18) to prove that $\|E\| \cdot \|\tilde{A}^+\|^2$ can be replaced by $\|E\| \cdot \|A^+\| \cdot \|\tilde{A}^+\|$ in the bound for $\|G_2\|$ given in Eq. (8.11). This is accomplished by Theorems (8.21) and (8.22).

(8.21) THEOREM

If Rank $(\tilde{A}) = $ Rank (A), then

$$\|\tilde{Q}(I - Q)\| = \|Q(I - \tilde{Q})\|$$

Proof: This proof is due to F. T. Krogh. Write singular value decompositions

$$A = USV^T$$

and

$$\tilde{A} = \tilde{U}\tilde{S}\tilde{V}^T$$

Then from Eq. (8.3) and (8.4) and the assumption that A and $\tilde{A}$ have the same rank, say, k,

$$Q = U \begin{bmatrix} I_k & 0 \\ 0 & 0 \end{bmatrix} U^T$$

and

$$\tilde{Q} = \tilde{U} \begin{bmatrix} I_k & 0 \\ 0 & 0 \end{bmatrix} \tilde{U}^T$$

Define the $m \times m$ orthogonal matrix W with submatrices W_{ij} by

$$\tilde{U}^T U = W \equiv \begin{bmatrix} W_{11} & W_{12} \\ W_{21} & W_{22} \end{bmatrix} \begin{matrix} \}k \\ \}m-k \end{matrix}$$
$$\underbrace{\phantom{W_{11}}}_{k} \underbrace{\phantom{W_{12}}}_{m-k}$$

Then

$$\| \tilde{Q}(I - Q) \| = \left\| \tilde{U} \begin{bmatrix} I_k & 0 \\ 0 & 0 \end{bmatrix} \tilde{U}^T U \begin{bmatrix} 0 & 0 \\ 0 & I_{m-k} \end{bmatrix} U^T \right\|$$

$$= \left\| \begin{bmatrix} I_k & 0 \\ 0 & 0 \end{bmatrix} W \begin{bmatrix} 0 & 0 \\ 0 & I_{m-k} \end{bmatrix} \right\| = \left\| \begin{bmatrix} 0 & W_{12} \\ 0 & 0 \end{bmatrix} \right\| = \| W_{12} \|$$

Similarly, one can verify that

$$\| Q(I - \tilde{Q}) \| = \| W_{21} \|$$

It remains to be shown that $\| W_{12} \| = \| W_{21} \|$. Let x be any $(m - k)$-vector and define

$$y = \begin{bmatrix} 0 \\ x \end{bmatrix} \begin{matrix} \}k \\ \}m - k \end{matrix}$$

Then, using the orthogonality of W,

$$\| x \|^2 = \| y \|^2 = \| Wy \|^2$$
$$= \| W_{12}x \|^2 + \| W_{22}x \|^2$$

and thus

$$\| W_{12}x \|^2 = \| x^2 \| - \| W_{22}x \|^2$$

Therefore

$$\| W_{12} \|^2 = \max_{\| x \| = 1} \| W_{12}x \|^2 = 1 - \min_{\| x \| = 1} \| W_{22}x \|^2$$
$$= 1 - s_{m-k}^2$$

where s_{m-k} is the smallest singular value of W_{22}.

Similarly, from

$$\| x \|^2 = \| y \|^2 = \| W^T y \|^2 = \| W_{21}^T x \|^2 + \| W_{22}^T x \|^2$$

one obtains

$$\| W_{21} \|^2 = 1 - \min_{\| x \| = 1} \| W_{22}^T x \|^2 = 1 - s_{m-k}^2$$

Thus $\| W_{12} \| = \| W_{21} \|$, which completes the proof of Theorem (8.21).

(8.22) THEOREM

  If Rank $(\tilde{A}) = Rank$ (A), *then the matrix* G_2 *defined in Eq. (8.8) satisfies*

(8.23) $$\| G_2 \| \leq \| E \| \cdot \| A^+ \| \cdot \| \tilde{A}^+ \|$$

Proof: Using Rank $(\tilde{A})$ = Rank (A), Theorem (8.21) permits us to write

$$\|G_2\| \leq \|\tilde{A}^+\| \cdot \|\tilde{Q}(I - Q)\|$$
$$= \|\tilde{A}^+\| \cdot \|Q(I - \tilde{Q})\|$$

Then since

$$Q(I - \tilde{Q}) = A^{T+}A^T(I - \tilde{Q})$$
$$= A^{T+}(A^T - \tilde{A}^T)(I - \tilde{Q}) = -A^{T+}E^T(I - \tilde{Q})$$

it follows that $\|G_2\|$ satisfies Eq. (8.23), which completes the proof of Theorem (8.22).

We are now in a position to establish the following theorem, which may be regarded as a more useful specialization of Theorem (8.5). By making stronger assumptions, bounds are obtained that do not include $\|\tilde{A}^+\|$.

(8.24)   **THEOREM**

Use definitions of G_i, $i = 1, 2, 3$, from Eq. (8.7) to (8.9) and, in addition, assume $\|E\| \cdot \|A^+\| < 1$ and Rank $(\tilde{A}) \leq$ Rank (A). Then Rank $(\tilde{A})$ = Rank (A) and

(8.25)
$$\|G_1\| \leq \frac{\|E\| \cdot \|A^+\|^2}{1 - \|E\| \cdot \|A^+\|}$$

(8.26)
$$\|G_2\| \leq \frac{\|E\| \cdot \|A^+\|^2}{1 - \|E\| \cdot \|A^+\|}$$

(8.27)
$$\|G_3\| \lesssim \|E\| \cdot \|A^+\|^2$$

(8.28)
$$\|G\| \leq \frac{c\|E\| \cdot \|A^+\|^2}{1 - \|E\| \cdot \|A^+\|}$$

where

(8.29) $c = \dfrac{1 + 5^{1/2}}{2} \doteq 1.618$ *if* Rank $(A) < \min(m, n)$

(8.30) $c = 2^{1/2} \doteq 1.414$ *if* Rank $(A) = \min(m, n) < \max(m, n)$

(8.31) $c = 1$ *if* Rank $(A) = m = n$

Proof: The conclusion that Rank $(\tilde{A})$ = Rank (A) is obtained from Theorem (8.15). Using the bound for $\|\tilde{A}^+\|$ obtained in that theorem, one obtains Eq. (8.25), (8.26) and (8.27) from Eq. (8.10), (8.23) and (8.12), respectively.

Thus the bound of Eq. (8.28) would follow immediately with $c = 3$. For practical purposes this would be a satisfactory result. The special result in Eq. (8.31) is also immediate since in this case $\|G_2\| = \|G_3\| = 0$. Of course this result for a square nonsingular matrix is well known and can be proved more directly [e.g., see Wilkinson (1965a), pp. 189–190].

The special results of Eq. (8.29) and (8.30) are established as follows. Let x be a unit m-vector. Define

$$x_1 = Qx$$
$$x_2 = (I - Q)x$$

Then

$$x = x_1 + x_2$$
$$1 = ||x||^2 = ||x_1||^2 + ||x_2||^2$$

and there exists a number φ such that

$$\cos \varphi = ||x_1||$$

and

$$\sin \varphi = ||x_2||$$

Let

$$\alpha \geq \frac{||E|| \cdot ||A^+||^2}{1 - ||E|| \cdot ||A^+||}$$

so that from the inequalities in Eq. (8.25) to (8.27)

$$\alpha \geq ||G_i|| \qquad i = 1, 2, 3$$

Because of the rightmost factors A^+ and $(I - Q)$ in Eq. (8.7) to (8.9), we have

$$Gx = G_1 x_1 + G_2 x_2 + G_3 x_1$$
$$\equiv y_1 + y_2 + y_3$$

Because of the leftmost factors $\tilde{A}^+$ and $(I - \tilde{P})$ in Eq. (8.7) to (8.9), the vector y_3 is orthogonal to y_1 and y_2. Thus

(8.32)
$$\begin{aligned}
||Gx||^2 &= ||y_1 + y_2||^2 + ||y_3||^2 \\
&\leq \alpha^2 [(||x_1|| + ||x_2||)^2 + ||x_1||^2] \\
&= \alpha^2 [(\cos \varphi + \sin \varphi)^2 + \cos^2 \varphi] \\
&= \alpha^2 \left(1 + \sin 2\varphi + \frac{1 + \cos 2\varphi}{2} \right) \\
&= \frac{\alpha^2 (3 + 2 \sin 2\varphi + \cos 2\varphi)}{2} \\
&\leq \frac{\alpha^2 (3 + 5^{1/2})}{2}
\end{aligned}$$

Therefore

$$\|Gx\| \leq \frac{\alpha(1 + 5^{1/2})}{2} \doteq 1.618\alpha$$

Then

$$\|G\| = \max\{\|Gx\| : \|x\| = 1\} \leq \frac{\alpha(1 + 5^{1/2})}{2}$$

which proves Eq. (8.29).

For the conclusion in Eq. (8.30) we have either Rank $(\tilde{A}) = n < m$, in which case $\tilde{P} = I_n$ so that $G_3 = 0$, or else Rank $(A) = m < n$, in which case $Q = I_m$ so that $G_2 = 0$. Thus either y_2 or y_3 is zero in Eq. (8.32) leading to

$$\|Gx\|^2 \leq 2\alpha^2$$

which establishes Eq. (8.30), completing the proof of Theorem (8.24).

Equations (8.1) to (8.9) and the theorems of this chapter may be used to prove that with appropriate hypotheses on the rank of A the elements of A^+ are differentiable functions of the elements of A.

Examples of some specific differentiation statements and formulas are given in Exercises (8.33) and (9.22)–(9.24). Note that the formulas given in these exercises generalize immediately to the case in which t is a k-dimensional variable with components $t_1, \cdots, t_k$. One simply replaces d/dt in these formulas by $\partial/\partial t_i$ for $i = 1, \ldots, k$.

Differentiation of the pseudoinverse has been used by Fletcher and Lill (1970) and by Perez and Scolnik (1972) in algorithms for constrained minimization problems. See Golub and Pereyra (1973) and Krogh (1974) for an application of differentiation of the pseudoinverse to nonlinear least squares problems in which some of the parameters occur linearly.

EXERCISE

(8.33)   [Hanson and Lawson (1969); Pavel–Parvu and Korganoff (1969)]
Let A be an $m \times n$ matrix with $m \geq n$ whose elements are differentiable functions of a real variable t. Suppose that, for $t = 0$, A is of rank n. Show that there is a real neighborhood of zero in which A^+ is a differentiable function of t and the derivative of A^+ is given by

$$\frac{dA^+}{dt} = -A^+ \frac{dA}{dt} A^+ + A^+ A^{+T} \left(\frac{dA}{dt}\right)^T (I - AA^+)$$

9 PERTURBATION BOUNDS FOR THE SOLUTION OF PROBLEM LS

In this chapter the theorems of the preceding chapter will be applied to study the effect of perturbations of A and b upon the minimum length solution x of $Ax \cong b$. We shall continue to use the definitions given in Eq. (8.1) to (8.4). Theorem (9.7) will be established without restriction on the relative sizes of m, n and $k = \text{Rank }(A)$. Following the proof of Theorem (9.7) the bounds obtained will be restated for the three special cases of $m = n = k$, $m > n = k$ and $n > m = k$.

For convenience in stating results in terms of relative perturbations we define the relative perturbations

$$(9.1) \qquad \alpha = \frac{\|E\|}{\|A\|}$$

$$(9.2) \qquad \beta = \frac{\|db\|}{\|b\|}$$

and the quantities

$$(9.3) \qquad \gamma = \frac{\|b\|}{\|A\| \cdot \|x\|} \leq \frac{\|b\|}{\|Ax\|}$$

$$(9.4) \qquad \rho = \frac{\|r\|}{\|A\| \cdot \|x\|} \leq \frac{\|r\|}{\|Ax\|} \qquad (r = b - Ax)$$

$$(9.5) \qquad \kappa = \|A\| \cdot \|A^+\|$$

and

$$(9.6) \qquad \hat{\kappa} = \frac{\kappa}{1 - \kappa \alpha} \equiv \frac{\|A\| \cdot \|A^+\|}{1 - \|E\| \cdot \|A^+\|}$$

49

The definitions in Eq. (9.1) to (9.6) of course apply only when the respective denominators are nonzero. The quantity κ of Eq. (9.5) is called the *condition number* of A.

(9.7)   THEOREM

> *Let* x *be the minimum length solution to the least squares problem* Ax $\cong$ b *with residual vector* r $=$ b $-$ Ax. *Assume* $\|E\|\cdot\|A^+\| < 1$ *and Rank* $(\tilde{A}) \leq$ *Rank* (A), *and let* x $+$ dx *be the minimum length solution to the least squares problem*

$$\tilde{A}(x + dx) \equiv (A + E)(x + dx) \cong b + db$$

> *Then*

(9.8)
$$\text{Rank } (\tilde{A}) = \text{Rank } (A)$$

(9.9)
$$\|dx\| \leq \|A^+\| \left(\frac{\|E\|\cdot\|x\|}{1 - \|E\|\cdot\|A^+\|} + \frac{\|db\|}{1 - \|E\|\cdot\|A^+\|} \right.$$
$$\left. + \frac{\|E\|\cdot\|A^+\|\cdot\|r\|}{1 - \|E\|\cdot\|A^+\|} + \|E\|\cdot\|x\| \right)$$

> *and*

(9.10)
$$\frac{\|dx\|}{\|x\|} \leq \hat{\kappa}\alpha + \hat{\kappa}\gamma\beta + \kappa\hat{\kappa}\rho\alpha + \kappa\alpha$$
$$\leq \hat{\kappa}[(2 + \kappa\rho)\alpha + \gamma\beta]$$

Proof: The conclusion in Eq. (9.8) follows from Theorem (8.15). The vectors x and $x + dx$ satisfy

$$x = A^+b$$

and

$$x + dx = \tilde{A}^+(b + db)$$

Thus

$$dx = \tilde{A}^+(b + db) - A^+b = (\tilde{A}^+ - A^+)b + \tilde{A}^+ db$$
$$= Gb + \tilde{A}^+ db$$

Note that $r = (I - Q)r = (I - Q)b$, which with Eq. (8.8) implies $G_2b = G_2r$. Then using Eq. (8.6) to (8.9) gives

(9.11)
$$dx = -\tilde{A}^+Ex + G_2r + (I - \tilde{P})E^TA^{T+}x + \tilde{A}^+ db$$

Using the bound for $\|\tilde{A}^+\|$ from Theorem (8.15) and the bound for $\|G_2\|$ from Eq. (8.26), the result in Eq. (9.9) is obtained.

Dividing the inequality in Eq. (9.9) by $\|x\|$, using the definitions in Eq. (9.1) to (9.6) gives the inequality in Eq. (9.10). This completes the proof of Theorem (9.7).

Observe that when $n = k \equiv \text{Rank}(A)$, the matrix G_3 of Eq. (8.9) is zero, which leads to the fourth term in the right member of the inequalities in Eq. (9.9) and (9.10) being zero. Similarly when $m = k \equiv \text{Rank}(A)$, the matrix G_2 of Eq. (8.8) and consequently the third term in the right member of the inequalities in Eq. (9.9) and (9.10) are zero.

Furthermore if either $n = k$ or $m = k$, then the rank of $\tilde{A}$ clearly cannot exceed the rank of A. Thus the hypothesis $\text{Rank}(\tilde{A}) \leq \text{Rank}(A)$ that was used in Theorem (9.7) is automatically satisfied when either $n = k$ or $m = k$.

These observations provide proofs for the following three theorems.

(9.12) THEOREM

Assume $m > n = k \equiv Rank(A)$ and $\|E\| \cdot \|A^+\| < 1$. Then

$$\text{Rank}(\tilde{A}) = \text{Rank}(A)$$

(9.13)
$$\|dx\| \leq \frac{\|A^+\| \left[(\|E\| (\|x\| + \|A^+\| \cdot \|r\|) + \|db\| \right]}{1 - \|E\| \cdot \|A^+\|}$$

and

(9.14)
$$\frac{\|dx\|}{\|x\|} \leq \hat{\kappa}[(1 + \kappa\rho)\alpha + \gamma\beta]$$

(9.15) THEOREM

Assume $m = n = k \equiv Rank(A)$ and $\|E\| \cdot \|A^+\| < 1$. Then

$$\text{Rank}(\tilde{A}) = \text{Rank}(A)$$

(9.16)
$$\|dx\| \leq \frac{\|A^+\| (\|E\| \cdot \|x\| + \|db\|)}{1 - \|E\| \cdot \|A^+\|}$$

and

(9.17)
$$\frac{\|dx\|}{\|x\|} \leq \hat{\kappa}(\alpha + \gamma\beta) \leq \hat{\kappa}(\alpha + \beta)$$

(9.18) THEOREM

Assume $n > m = k \equiv Rank(A)$ and $\|E\| \cdot \|A^+\| < 1$. Then

$$\text{Rank}(\tilde{A}) = \text{Rank}(A)$$

(9.19)
$$\|dx\| \leq \|A^+\| \left(\frac{\|E\| \cdot \|x\| + \|db\|}{1 - \|E\| \cdot \|A^+\|} + \|E\| \cdot \|x\| \right)$$

and

(9.20)
$$\frac{\|dx\|}{\|x\|} \leq \hat{\kappa}(\alpha + \gamma\beta) + \kappa\alpha \leq \hat{\kappa}(2\alpha + \gamma\beta) \leq \hat{\kappa}(2\alpha + \beta)$$

Alternatively, by an argument similar to that used to establish Eq. (8.30), *it can be shown that*

(9.21) $$\frac{\|dx\|}{\|x\|} \leq \hat{\kappa}(2^{1/2}\alpha + \beta)$$

The following exercises illustrate some differentiation formulas that can be derived from the results of this chapter. See the remarks preceding Exercise (8.33) for an indication of generalizations and applications of these formulas.

EXERCISES

(9.22) Let A be an $m \times n$ matrix with $m \geq n$ whose elements are differentiable functions of a real variable t. Let x be a vector function of t defined by the condition that $Ax \cong b$ for all t in a neighborhood U in which A is differentiable and Rank $(A) = n$. Show that for $t \in U$, dx/dt exists and is the solution of the least squares problem

$$A\left(\frac{dx}{dt}\right) \cong -\left(\frac{dA}{dt}\right)x + A^{+T}\left(\frac{dA}{dt}\right)^T r$$

where $r = b - Ax$.

(9.23) Further show that dx/dt is the solution of the square nonsingular system

$$A^T A\left(\frac{dx}{dt}\right) = -A^T\left(\frac{dA}{dt}\right)x + \left(\frac{dA}{dt}\right)^T r$$

(9.24) If $A = Q^T R$ where Q is $n \times m$ with orthonormal rows and R is $n \times n$ and nonsingular (a decomposition of A obtained by Householder transformations, for example), show that dx/dt satisfies

$$R\left(\frac{dx}{dt}\right) = -Q\left(\frac{dA}{dt}\right)x + (R^{-1})^T\left(\frac{dA}{dt}\right)^T r$$

10 NUMERICAL COMPUTATIONS USING ELEMENTARY ORTHOGONAL TRANSFORMATIONS

We now turn to describing computational algorithms for effecting an orthogonal decomposition of an $m \times n$ matrix A, as given in Theorem (3.19), and calculating a solution to Problem LS.

Because of the variety of applications using Householder transformations or Givens transformations, we have found it convenient and unifying to regard each of these individual transformations as computational entities. We shall describe each of these in detail. These modules will then be used in describing other, more complicated, computational procedures.

The computation of the Householder transformation can be broken into two parts: (1) the construction of the transformation and (2) its application to other vectors.

The $m \times m$ Householder orthogonal transformation can be represented in the form

$$(10.1) \qquad Q = I_m + b^{-1}uu^T$$

where u is an m-vector satisfying $\|u\| \neq 0$ and $b = -\|u\|^2/2$.

In Chapter 3 it was shown how for a given vector v a vector u could be determined so that Eq. (3.2) was satisfied. In practice, occasions arise in which one wishes to determine Q to satisfy

$$(10.2) \qquad Qv = \begin{bmatrix} v_1 \\ \cdot \\ \cdot \\ \cdot \\ v_{p-1} \\ -\sigma\left(v_p^2 + \sum_{i=l}^{m} v_i^2\right)^{1/2} \\ v_{p+1} \\ \cdot \\ \cdot \\ \cdot \\ v_{l-1} \\ 0 \\ \cdot \\ \cdot \\ \cdot \\ 0 \end{bmatrix} \equiv y$$

The computation of such a Q could be described as a row permutation followed by an ordinary Householder transformation as described in Lemma (3.1) followed by a restoring row permutation. We find it more convenient, however, to view Eq. (10.2) as defining the purpose of our basic computational module. The effect of the matrix Q in transforming v to y can be described by means of three nonnegative integer parameters, p, l, and m, as follows:

1. If $p > 1$, components 1 through $p - 1$ are to be unchanged.
2. Component p is permitted to change. This is called the *pivot element*.
3. If $p < l - 1$, components $p + 1$ through $l - 1$ are to be left unchanged.
4. If $l \leq m$, components l through m are to be zeroed.

Note that we must have

$$(10.3) \qquad\qquad 1 \leq p \leq m$$

and

$$(10.4) \qquad\qquad p < l$$

The relation $l > m$ is permitted in the Fortran implementation of this computational procedure (subroutine **H12** in Appendix C) and is interpreted to mean that Q is to be an identity matrix.

The computational steps necessary to produce an $m \times m$ orthogonal matrix Q that satisfies the conditions imposed by the integer parameters p,

l, and m can be stated as follows:

(10.5) $$s = -\sigma\left(v_p^2 + \sum_{i=l}^{m} v_i^2\right)^{1/2} \quad \text{where } \sigma = \begin{cases} +1 & \text{if } v_p \geq 0 \\ -1 & \text{if } v_p < 0 \end{cases}$$

(10.6) $u_i = 0 \qquad i = 1, \ldots, p - 1$

(10.7) $u_p = v_p - s$

(10.8) $u_i = 0 \qquad i = p + 1, \ldots, l - 1$

(10.9) $u_i = v_i \qquad i = l, \ldots, m$

(10.10) $b = su_p$

(10.11) $Q = I_m + b^{-1}uu^T \quad \text{if } b \neq 0$

$\qquad\qquad = I_m \quad \text{if } b = 0$

The fact that the matrix Q defined in Eq. (10.11) has the desired properties is established by the following three lemmas.

(10.12) LEMMA

The m-vector u *and the scalar* b *defined in Eq. (10.5) to (10.10) satisfy*

(10.13) $$b = -\frac{\|u\|^2}{2}$$

Proof: This follows from the algebraic identities

$$\begin{aligned} \|u\|^2 &= (v_p - s)^2 + \sum_{j=l}^{m} v_j^2 \\ &= v_p^2 - 2v_p s + s^2 + \sum_{j=l}^{m} v_j^2 \\ &= -2s(v_p - s) \\ &= -2su_p \\ &= -2b \end{aligned}$$

(10.14) LEMMA

The matrix Q *of Eq. (10.11) is orthogonal.*

Proof: The verification that $Q^T Q = I_m$ follows directly using Eq. (10.11) and (10.13).

(10.15) LEMMA

Let $y = Qv$. *Then*

(10.16) $y_i = v_i \qquad i = 1, \ldots, p - 1$

(10.17) $y_p = s$

$$(10.18) \qquad y_i = v_i \qquad i = p+1, \ldots, l-1$$

$$(10.19) \qquad y_i = 0 \qquad i = l, \ldots, m$$

Proof: If $v = 0$, the lemma is obviously true. For $v \neq 0$, the easily verified fact that $u^T v = -b$ while using Eq. (10.6) to (10.9) shows that

$$y = Qv = v - u$$

satisfies Eq. (10.16) to (10.19).

In actual computation it is common to produce the nonzero components of the vectors u and y in the storage previously occupied by v. The exception to this is the pth entry in the storage array; a choice must be made as to whether u_p or y_p will occupy it. We shall let y_p occupy it and store u_p in an additional location; the quantity b of Eq. (10.10) can be computed from the identity indicated there whenever it is needed.

After constructing the transformation one will generally wish to apply the transformation to a set of m-vectors, say, $c_j, j = 1, \ldots, v$. Thus it is desired to compute

$$\tilde{c}_j = Qc_j \qquad j = 1, \ldots, v$$

Using the definition of Q given by Eq. (10.11), this computation is accomplished as follows:

$$(10.20) \qquad \left. \begin{aligned} t_j &= b^{-1}(u^T c_j) \\ \tilde{c}_j &= c_j + t_j u \end{aligned} \right\} \qquad j = 1, \ldots, v$$

$$(10.21)$$

These computations will now be restated in an algorithmic form suitable for implementation as computer code. We shall define Algorithm H1(p, l, m, v, h, C, v) for constructing and optionally applying a Householder transformation and Algorithm H2(p, l, m, v, h, C, v) for optionally applying a previously constructed Householder transformation.

The input to Algorithm H1 consists of the integers p, l, m, and v; the m-vector v, and, if $v > 0$, an array C containing the m-vectors $c_j, j = 1, \ldots, v$.

The storage array C may either be an $m \times v$ array containing the vectors c_j as column vectors or a $v \times m$ array containing the vectors c_j as row vectors. These two possible storage modes will not be distinguished in describing Algorithms H1 and H2. However, in references to Algorithm H1 or H2 that occur elsewhere in this book we shall regard the column storage as normal and make special note of those cases in which the set of vectors c_j upon which Algorithm H1 or H2 is to operate are stored as row vectors in the storage array C.

Algorithm H1 computes the vector u, the number b, the vector $y = Qv$, and, if $v > 0$, the vectors $\tilde{c}_j = Qc_j, j = 1, \ldots, v$. The output of Algorithm H1 consists of the pth component of u stored in the location named h, the components l through m of u stored in these positions of the storage array named v, components 1 through $l - 1$ of y stored in these locations of the storage array named v, and, if $v > 0$, the vectors $\tilde{c}_j$, $j = 1, \ldots, v$, stored in the storage array named C.

In Algorithm H2 the quantities p, l, and m have the same meaning as for Algorithm H1. The vector v and the quantity h must contain values computed by a previous execution of Algorithm H1. These quantities define a transformation matrix Q. If $v > 0$, the array C must contain a set of m-vectors c_j, $j = 1, \ldots, v$, on input and Algorithm H2 will replace these with vectors $\tilde{c}_j = Qc_j, j = 1, \ldots, v$.

(10.22)   ALGORITHMS H1(p, l, m, v, h, C, v) [use Steps 1–11] and H2(p, l, m, v, h, C, v) [use Steps 5–11]

Step	Description
1	Set $s := \left(v_p^2 + \sum\limits_{i=l}^{m} v_i^2\right)^{1/2}$.
2	If $v_p > 0$, set $s := -s$.
3	Set $h := v_p - s$, $v_p := s$.
4	*Comment:* The construction of the transformation is complete. At Step 5 the application of the transformation to the vectors c_j begins.
5	Set $b := v_p h$.
6	If $b = 0$ or $v = 0$, go to Step 11.
7	For $j := 1, \ldots, v$, do Steps 8–10.
8	Set $s := \left(c_{pj}h + \sum\limits_{i=l}^{m} c_{ij}v_i\right)\Big/b$.
9	Set $c_{pj} := c_{pj} + sh$.
10	For $i := l, \ldots, m$, set $c_{ij} := c_{ij} + sv_i$.
11	*Comment:* (a) Algorithm H1 or H2 is completed. (b) In Step 1 the computation of the square root of sum of squares can be made resistant to underflow by computing it based on the identity $(w_l^2 + \cdots + w_m^2)^{1/2} = t[(w_l/t)^2 + \cdots + (w_m/t)^2]^{1/2}$, with $t = \max\{|w_i|, i = l, \ldots, m\}$.

The Fortran subroutine **H12** given in Appendix C implements Algorithms H1 and H2.

The other general elementary orthogonal transformation we shall discuss is the Givens rotation. The formulas for construction of the Givens rotation were given in Eq. (3.5) to (3.9). The formulas for c and s given in Eq. (3.8) and (3.9) can be reformulated to avoid unnecessary underflow or overflow. This is accomplished by computing $r = (x^2 + y^2)^{1/2}$ as

$$t = \max (|x|, |y|)$$
$$u = \min (|x|, |y|)$$

(10.23)
$$r = \begin{cases} t\left[1 + \left(\dfrac{u}{t}\right)^2\right]^{1/2} & \text{if } t \neq 0 \\ 0 & \text{if } t = 0 \end{cases}$$

On a machine that does not use normalized base 2 arithmetic, additional care should be taken to avoid loss of accuracy in computing c and s. For example, Cody (1971) has given the following reformulation of the expression in Eq. (10.23) for use with normalized base 16 arithmetic.

(10.24)
$$r = 2t\left[\frac{1}{4} + \left(\frac{u}{2t}\right)^2\right]^{1/2}$$

Algorithms $G1(v_1, v_2, c, s, r)$ and $G2(c, s, z_1, z_2)$ will be described for, respectively, constructing and applying a Givens rotation. The expression in Eq. (10.23), appropriate for base 2 arithmetic, will be used. The input to Algorithm G1 consists of the components v_1 and v_2 of the 2-vector v. The output consists of the scalars c and s, which define the matrix G of Eq. (3.5), together with the square root of sum of squares of v_1 and v_2 which is stored in the storage location named r. The storage location r can be identical with the storage location v_1 or v_2, although it is most often convenient to identify it with the storage location for v_1. The input to Algorithm G2 consists of the scalars c and s defining the matrix G of Eq. (3.5) and the components z_1 and z_2 of the 2-vector z. The output consists of the components d_1 and d_2 of the vector $d = Gz$ stored in the locations called z_1 and z_2.

(10.25) ALGORITHM $G1(v_1, v_2, c, s, r)$

Step	Description
	Step *Description*
1	If $|v_1| \leq |v_2|$, go to Step 8.
2	Set $w := v_2/v_1$.
3	Set $q := (1 + w^2)^{1/2}$.
4	Set $c := 1/q$.
5	If $v_1 < 0$, set $c := -c$.
6	Set $s := wc$.
7	Set $r := |v_1|q$ and go to Step 16.

Step	Description
8	If $v_2 \neq 0$, go to Step 10.
9	Set $c := 1$, $s := 0$, $r := 0$, and go to Step 16.
10	Set $w := v_1/v_2$.
11	Set $q := (1 + w^2)^{1/2}$.
12	Set $s := 1/q$.
13	If $v_2 < 0$, set $s := -s$.
14	Set $c := ws$.
15	Set $r := |v_2|q$.
16	*Comment:* The transformation has been constructed.

(10.26)   ALGORITHM $G2(c, s, z_1, z_2)$

Step	Description
1	Set $w := z_1 c + z_2 s$.
2	Set $z_2 := -z_1 s + z_2 c$.
3	Set $z_1 := w$.
4	*Comment:* The transformation has been applied.

The Fortran subroutines **G1** and **G2** given in Appendix C implement Algorithms G1 and G2, respectively.

Many variations of algorithmic and programming details are possible in implementing Householder or Givens transformations. Tradeoffs are possible involving execution time, accuracy, resistance to underflow or overflow, storage requirements, complexity of code, modularity of code, taking advantage of sparsity of nonzero elements, programming language, portability, etc. Two examples of such variations will be described.

Our discussion of the Householder transformation has been based on the representation given in Eq. (10.1). The Householder matrix can also be expressed as

$$(10.27) \qquad Q = I + gh^T = I + \begin{bmatrix} g_1 \\ \cdot \\ \cdot \\ \cdot \\ g_m \end{bmatrix} \cdot [1, h_2, \ldots, h_m]$$

The representations in Eq. (10.1) and (10.27) are related by the substitutions $g = b^{-1}u_1 u = s^{-1}u$ and $h = u_1^{-1}u$.

The form of Eq. (10.27) is mainly of interest for small m, in which case the need to store two vectors g and h instead of one vector u is of no consequence and the saving of two multiplications each time a product $\tilde{c} = Qc$ is computed is of some relative significance. This form is used with $m = 3$ in Martin, Peters, and Wilkinson [pp. 359–371 of Wilkinson and Reinsch (1971)] and with $m = 2$ and $m = 3$ in Moler and Stewart (1973).

Specifically with $m = 2$ the computation of $\tilde{c} = Qc$ using the expression in Eq. (10.1) [see Eq. (10.20) and (10.21)] requires five multiplications (or four multiplications and one division) and three additions, while the use of Eq. (10.27) requires only three multiplications and three additions. Thus Eq. (10.27) requires fewer operations than Eq. (10.1) and is, in fact, competitive with the Givens transformation, which, as implemented by Algorithm G2 (10.26), requires four multiplications and two additions to compute $\tilde{c} = Gc$.

The actual comparative performance of computer code based on Eq. (10.27) versus code based on the conventional Givens transformation [Algorithms G1 (10.25) and G2 (10.26)] will very likely depend on details of the code. For example, we tested two such codes that used single precision arithmetic (27-bit fraction part) on the UNIVAC 1108 computer. These were used to zero the (2, 1)-element of 100 different single precision 2×11 matrices. For each of the 1100 2-vectors transformed, the relative error was computed as $\rho = ||v' - v''||/||v'||$ where v' is the transformed vector computed by one of the two single precision codes being tested and v'' is the transformed vector computed by a double precision code. For the Householder code based on Eq. (10.27), the root-mean-square value of ρ was 1.72×2^{-27} and the maximum value of ρ was 8.70×2^{-27}. For the Givens code, the corresponding figures were 0.88×2^{-27} and 3.17×2^{-27}.

Methods of reducing the operation count for the Givens transformation have been reported by Gentleman (1972a and 1972b). These methods require that the matrix A to be operated upon and the transformed matrix $\tilde{A} = GA$ each be maintained in storage in a factored form.

For ease of description of one of the methods given by Gentleman, consider the case of a $2 \times n$ matrix A in which the (2, 1)-element is to be zeroed by left multiplication by a Givens rotation. Thus

$$(10.28) \qquad GA = \tilde{A}$$

where $\tilde{a}_{21} = 0$. Here the Givens matrix G would be constructed as

$$(10.29) \qquad r = (a_{11}^2 + a_{21}^2)^{1/2}$$

$$(10.30) \qquad G = \begin{cases} \begin{bmatrix} \dfrac{a_{11}}{r} & \dfrac{a_{21}}{r} \\[2mm] -\dfrac{a_{21}}{r} & \dfrac{a_{11}}{r} \end{bmatrix} & \text{if } r > 0 \\[6mm] \begin{bmatrix} 1 & 0 \\ 0 & 1 \end{bmatrix} & \text{if } r = 0 \end{cases}$$

In the technique to be described, instead of having the matrix A available in storage, one has matrices $D_{2 \times 2}$ and $B_{2 \times n}$ such that

(10.31) $$D = \text{Diag}\{d_1, d_2\} \qquad d_i > 0$$

and

(10.32) $$A = D^{1/2}B$$

We wish to replace D and B in storage with new matrices $\tilde{D}_{2 \times 2}$ and $\tilde{B}_{2 \times n}$ with

(10.33) $$\tilde{D} = \text{Diag}\{\tilde{d}_1, \tilde{d}_2\} \qquad \tilde{d}_i > 0$$

such that the matrix $\tilde{A}$ of Eq. (10.28) is representable as

(10.34) $$\tilde{A} = \tilde{D}^{1/2}\tilde{B} \quad \text{or} \quad \tilde{A} = -\tilde{D}^{1/2}\tilde{B}$$

We distinguish three cases:

Case I. $b_{21} = 0$
Case II. $0 < (d_2 b_{21}^2) \leq (d_1 b_{11}^2)$
Case III. $0 \leq (d_1 b_{11}^2) < (d_2 b_{21}^2)$

CASE I

We may set $\tilde{D} = D$ and $\tilde{B} = B$.

CASE II

Define

(10.35) $$t = \frac{d_2 b_{21}^2}{d_1 b_{11}^2}$$

(10.36) $$\tilde{d}_1 = \frac{d_1}{1 + t}$$

(10.37) $$\tilde{d}_2 = \frac{d_2}{1 + t}$$

(10.38) $$H = \begin{bmatrix} 1 & \dfrac{d_2 b_{21}}{d_1 b_{11}} \\ -\dfrac{b_{21}}{b_{11}} & 1 \end{bmatrix}$$

(10.39) $$\tilde{B} = HB$$

Note that $\tilde{b}_{11} = b_{11}(1 + t)$ and $\tilde{b}_{21} = 0$.

CASE III

Define

$$(10.40) \qquad t = \frac{d_1 b_{11}^2}{d_2 b_{21}^2}$$

$$(10.41) \qquad \tilde{d}_1 = \frac{d_2}{1+t}$$

$$(10.42) \qquad \tilde{d}_2 = \frac{d_1}{1+t}$$

$$(10.43) \qquad H = \begin{bmatrix} \dfrac{d_1 b_{11}}{d_2 b_{21}} & 1 \\ -1 & \dfrac{b_{11}}{b_{21}} \end{bmatrix}$$

$$(10.44) \qquad \tilde{B} = HB$$

Note that $\tilde{b}_{11} = b_{21}(1+t)$ and $\tilde{b}_{21} = 0$.

It is easily verified that the matrices $\tilde{D}$ and $\tilde{B}$ defined by this process satisfy Eq. (10.34). The saving of arithmetic in applying the transformation comes from the presence of the two unit elements in the matrix H. This permits the matrix multiplication $\tilde{B} = HB$ to be done with $2n$ additions and $2n$ multiplications in contrast to the requirement of $2n$ additions and $4n$ multiplications required for the matrix multiplication $\tilde{A} = GA$. Furthermore the square root operation usually required to construct the Givens transformation has been eliminated.

In the case of triangularizing a general $m \times n$ matrix A by a sequence of Givens rotations, one could begin with $D_1 = I_{m \times m}$ and $B_1 = A$ and produce a sequence of matrices $\{D_k\}$ and $\{B_k\}$, $k = 1, 2, \ldots$, by this procedure. In general the size of elements in D_k will decrease as k increases but the rate is limited by the fact that $\frac{1}{2} \leq (1+t)^{-1} \leq 1$. There is a corresponding but slower general increase in the size of the elements in B_k since the euclidean norm of each column vector of the product matrix $D_k^{1/2} B_k$ is invariant with respect to k.

To reduce the likelihood of overflow of elements of B or underflow of elements of D, any code implementing this version of Givens rotations should contain some procedure to monitor the size of the numbers d_i, rescaling d_i and the ith row of B whenever d_i becomes smaller than some tolerance. For example, one might set $\tau = 2^{-24}$, $\rho = \tau^{-1}$, and $\beta = \tau^{1/2}$. When a number d_i is to be operated upon, it can first be compared with τ. If $d_i < \tau$, replace d_i by ρd_i and replace $b_{ij}, j = 1, \ldots, n$, by $\beta b_{ij}, j = 1, \ldots, n$. With this choice of τ the multiplications by ρ and β will be exact operations on many computers having base 2, base 8, or base 16 floating point arithmetic.

11

COMPUTING THE SOLUTION FOR THE OVERDETERMINED OR EXACTLY DETERMINED FULL RANK PROBLEM

In Theorem (3.11) we saw that for an $m \times n$ matrix A there existed an $m \times m$ orthogonal matrix Q such that $QA = R$ is zero below the main diagonal. In this chapter we shall describe the numerical computation of the matrices Q and R using Algorithms H1 and H2 of Chapter 10.

The matrix Q is a product of Householder transformations

(11.1) $$Q = Q_n \cdots Q_1$$

where each Q_j has the form

(11.2) $$Q_j = I_m + b_j^{-1} u^{(j)} u^{(j)T} \qquad j = 1, \ldots, n$$

Rather than saving the quantity b_j that appears in Eq. (11.2), we shall recompute it when needed based on Eq. (10.10).

Introducing subscripts we have

(11.3) $$b_j = s_j u_j^{(j)} \qquad j = 1, \ldots, n$$

Each s_j is the jth diagonal entry of the R matrix and will be stored as such. The quantities $u^{(j)}$ are stored in an auxiliary array of locations named $h_j, j = 1, \ldots, n$.

The computational algorithm will construct the decomposition of Theorem (3.11). This algorithm will be known as HFT(m, n, A, h). The input to HFT consists of the integers m and n and the $m \times n$ matrix A. The output consists of the nonzero portion of the upper triangular matrix R stored in the upper triangular portion of the array named A, the scalars $u^{(j)}$ stored in the jth entry h_j of the array named h, and the remaining nonzero portions of the vectors $u^{(j)}$ stored as columns in the subdiagonal part of the jth column of the array named A.

(11.4)   ALGORITHM HFT(m, n, A, h)

Step	Description
1	For $j := 1, \ldots, n$, execute Algorithm H1($j, j+1, m, a_{1j},$ $h_j, a_{1,j+1}, n-j$) (see Algorithm 10.22).
2	*Comment:* The (forward) triangularization is computed.

In Step 1 of the algorithm above we are adopting the convention (consistent with Fortran usage) that the jth column of A can be referred to by the name of its first component, and the submatrix of A composed of columns $j+1$ through n can be referred to by the name of the first entry of column $j+1$.

In Cases 1a and 2a of Fig. 1.1 the $n \times n$ upper triangular matrix R_{11} (see Eq. 3.21) is nonsingular. Thus, in these two cases, the solution $\hat{x}$ of Problem LS can be obtained by computing

$$(11.5) \qquad\qquad g = Qb$$

partitioning g as

$$(11.6) \qquad\qquad g = \begin{bmatrix} g_1 \\ g_2 \end{bmatrix} \begin{matrix} \}n \\ \}m-n \end{matrix}$$

and solving

$$(11.7) \qquad\qquad R_{11}x = g_1$$

for the solution vector $\hat{x}$.

Using Eq. (2.9) and (2.10), the residual vector $r = b - A\hat{x}$ and its norm can be computed as

$$(11.8) \qquad\qquad r = Q_1 \cdots Q_n \begin{bmatrix} 0 \\ g_2 \end{bmatrix}$$

and

$$(11.9) \qquad\qquad p \equiv \|r\| = \|g_2\|$$

Use of Eq. (11.8) and (11.9) obviates the need to save or regenerate the data matrix $[A:b]$ for the purpose of computing residuals.

The following algorithm, HS1, will accomplish the computation of Eq. (11.5) to (11.7). The input to this algorithm will consist of the integers m and n, the arrays named A and h as they are output by the Algorithm HFT (11.4) and the array named b that holds the right-side m-vector b of Problem LS.

The output of the algorithm consists of the n-vector x replacing the first n entries of the array named b and, if $m > n$, the $(m - n)$-vector g_2 will replace entries $n + 1$ through m of the array named b.

(11.10)   ALGORITHM HS1(m, n, A, h, b)

Step	Description
1	For $j := 1, \ldots, n$, execute Algorithm H2$(j, j + 1, m, a_{1j}, h_j, b, 1)$.
2	*Comment:* In Steps 3 and 4 we shall compute the solution to the triangular system $R_{11}x = g_1$ of Eq. (11.7).
3	Set $b_n := b_n/a_{nn}$.
4	If $n \leq 1$, go to Step 6.
5	For $i := n - 1, n - 2, \ldots, 1$, set $b_i := (b_i - \sum_{j=i+1}^{n} a_{ij}b_j)/a_{ii}$.
6	*Comment:* The solution of Problem LS has been computed for Cases 1a and 2a.

The case where b represents an $m \times l$ matrix can be easily handled by changing Step 1 to execute Algorithm H2$(j, j + 1, m, a_{1j}, h_j, b, l), j = 1, \ldots, n$, and by changing Steps 3 and 5 to deal with b as an $(m \times l)$-array.

To compute the residual norm ρ [see Eq. (11.9)], one could add to Algorithm HS1

Step 7 $$\rho := \left(\sum_{i=n+1}^{m} b_i^2 \right)^{1/2}$$

To compute the residual vector [see Eq. (11.8)], one could add to Algorithm HS1

Step 8   For $i := 1, \ldots, n$, set $b_i := 0$

Step 9   For $j := n, n - 1, \ldots, 1$, execute Algorithm H2$(j, j + 1, m, a_{1j}, h_j, b, 1)$

Note that if Steps 8 and 9 are used, the solution vector x must first be moved to storage locations distinct from the array b if it is not to be overwritten. Following Step 9 the residual vector $r = b - Ax$ occupies the storage array called b.

Algorithm HS1 is based on the assumption that the matrix A of Problem LS is of rank n. *There is no test made in the algorithm to check for this.*

In practice it is important to know whether a change in the matrix A of the order of the data uncertainty could produce a matrix of rank less than n. One computational approach to this problem, involving the use of column interchanges as a first step, will be discussed in Chapter 14. Another approach, which provides more detailed information about the matrix, involves the computation of the singular value decomposition. This will be treated in Chapter 18.

Frequently, when a least squares problem is solved, there is also interest in computing the covariance matrix for the solution parameters. This topic will be treated in Chapter 12.

A Fortran implementation of Algorithms HFT and HS1 of this chapter and the covariance matrix computation COV of Chapter 12 is given in **PROG1** in Appendix C. This program constructs and solves a sequence of sample problems.

EXERCISES

(11.11) Derive a forward triangularization algorithm using Givens transformations. Count the number of adds and multiplies separately. How is this count changed if the two-multiply, two-add transformation is used?

(11.12) Determine the number of multiplications required to solve Problem LS using Algorithms HFT and HS1. Compare this count with that obtained for the Givens transformations of Exercise (11.11).

12

COMPUTATION OF THE COVARIANCE MATRIX OF THE SOLUTION PARAMETERS

The symmetric positive definite matrix

$$(12.1) \qquad C = (A^T A)^{-1} \qquad \text{Rank } (A) = n$$

or a scalar multiple of it, say, $\sigma^2 C$, has a statistical interpretation, under appropriate hypotheses, of being an estimate of the covariance matrix for the solution vector of Problem LS [e.g., see Plackett (1960)]. We shall refer to C as the *unscaled covariance matrix*.

In this chapter some algorithms for computing C will be presented. We shall also identify some commonly occurring situations in which the explicit computation of C can be avoided by reconsidering the role of C in the later analysis.

We shall not discuss the derivation or interpretation of the scalar factor σ^2 but simply note that one expression commonly used to compute σ^2 is

$$(12.2) \qquad \sigma^2 = \frac{\| A\hat{x} - b \|^2}{m - n}$$

where $\hat{x}$ is the least squares solution of $Ax \cong b$ and m and n are, respectively, the number of rows and columns of the matrix A.

We shall discuss algorithms for computing C that make use of decompositions of A. These decompositions occur in the course of three different methods for solving Problem LS that are treated in detail in this book. These three decompositions may be summarized as follows:

$$(12.3) \qquad QA = \begin{bmatrix} R \\ 0 \end{bmatrix} \qquad R_{n \times n} \text{ upper triangular} \qquad \text{(see Chapter 11)}$$

$$(12.4) \qquad \tilde{Q}AP = \begin{bmatrix} \tilde{R} \\ 0 \end{bmatrix} \qquad \tilde{R}_{n \times n} \text{ upper triangular} \qquad \text{(see Chapter 14)}$$

$$(12.5) \qquad U^T A V = \begin{bmatrix} S \\ 0 \end{bmatrix} \qquad S_{n \times n} \text{ diagonal} \qquad \text{(see Chapter 18)}$$

The decompositions of Eq. (12.3) to (12.5) are, respectively, those obtained by Algorithms HFT (11.4) and HFTI (14.9) and the singular value decomposition of Chapter 18. The matrix Q of Eq. (12.3) is $n \times n$ and orthogonal. The matrices $\tilde{Q}$ and P of Eq. (12.4) are both orthogonal, while P is a permutation matrix. The matrices U and V of Eq. (12.5) are, respectively, $m \times m$ and $n \times n$, and they are orthogonal.

It is easy to see that in the cases of Eq. (12.3) to (12.5) we have

$$(12.6) \qquad\qquad\qquad A^T A = R^T R$$

$$(12.7) \qquad\qquad\qquad A^T A = P^T \tilde{R}^T \tilde{R} P$$

$$(12.8) \qquad\qquad\qquad A^T A = V S^2 V^T$$

so that if A is of rank n, inversion of these equations, respectively, yields

$$(12.9) \qquad\qquad C = (A^T A)^{-1} = R^{-1}(R^{-1})^T$$

$$(12.10) \qquad\qquad C = (A^T A)^{-1} = P \tilde{R}^{-1}(\tilde{R}^{-1})^T P^T$$

$$(12.11) \qquad\qquad C = (A^T A)^{-1} = V S^{-2} V^T$$

We first consider the details of the computation required by Eq. (12.9) and (12.10). There are three basic steps.

1. Invert the upper triangular matrix R or $\tilde{R}$ onto itself in storage.
2. Form the upper triangular part of the symmetric matrix $R^{-1}(R^{-1})^T$ or $\tilde{R}^{-1}(\tilde{R}^{-1})^T$. This can replace R^{-1} or $\tilde{R}^{-1}$ in storage.
3. Repermute the rows and columns of the matrix $\tilde{R}^{-1}(\tilde{R}^{-1})^T$. At each step the matrix is symmetric so only the upper triangular part of it need be saved.

To derive formulas for the inversion of a triangular matrix R, denote the elements of R^{-1} by t_{ij}. Then from the identity $R^{-1}R = I$ and the fact that both R and R^{-1} are upper triangular, one obtains the equations

$$\sum_{l=i}^{j} t_{il} r_{lj} = \delta_{ij} \equiv \begin{cases} 1 & \text{if } i = j \\ 0 & \text{if } i \neq j \end{cases}$$

for $1 \leq i \leq n$, $i \leq j \leq n$.

Solving for t_{ij} gives

$$t_{ij} = \begin{cases} r_{jj}^{-1} & \text{if } j = i \\ -r_{jj}^{-1} \sum_{l=i}^{j-1} t_{il} r_{lj} & \text{if } j > i \end{cases}$$

In order to be able to replace r_{ij} by t_{ij} in computer storage when t_{ij} is computed and to reduce the number of division operations, these formulas will be used in the following form:

$$t_{ii} = r_{ii}^{-1} \qquad i = 1, \ldots, n$$

$$\left\{ t_{ij} = -t_{jj} \sum_{l=i}^{j-1} t_{il} r_{lj}, j = i + 1, \ldots, n \right\} \qquad i = 1, \ldots, n - 1$$

This set of formulas requires $(n^3/6) + O(n^2)$ operations where an operation is a multiply plus an add or a divide plus an add. The postmultiplication of R^{-1} by its transpose indicated in Eq. (12.9) also requires $(n^3/6) + O(n^2)$ operations. Thus Algorithm COV requires $(n^3/3) + O(n^2)$ operations to compute the elements of the upper triangle of $(A^T A)^{-1}$ from the elements of R.

For Algorithm COV the matrix R or $\tilde{R}$ of the right member of Eq. (12.6) or (12.7), respectively, must occupy the upper triangular part of an array named A on input. On output the upper triangular part of the matrix $(A^T A)^{-1}$ occupies the upper triangular part of the array named A. The array named p will hold the information [generated by Algorithm HFTI (14.9), for example] describing the effect of the matrix P of Eq. (12.10).

(12.12) ALGORITHM COV(A, n, p)

Step	Description
1	For $i := 1, \ldots, n$, set $a_{ii} := 1/a_{ii}$.
2	If $n = 1$, go to Step 8.
3	For $i := 1, \ldots, n - 1$, do through Step 7.
4	For $j := i + 1, \ldots, n$, do through Step 7.
5	Set $s := 0$.
6	For $l := i, \ldots, j - 1$, set $s := s + a_{il} a_{lj}$.
7	Set $a_{ij} := -a_{jj} s$.
8	For $i := 1, \ldots, n$, do through Step 12.
9	For $j := i, \ldots, n$, do through Step 12.
10	Set $s := 0$.
11	For $l := j, \ldots, n$, set $s := s + a_{il} a_{jl}$.
12	Set $a_{ij} := s$.

Step	Description
13	*Remark:* This completes the computation of the elements of the upper triangle of $(A^T A)^{-1}$ for the case of Eq. (12.3) where no permutations were used in computing R. Alternatively, in the case of Eq. (12.4) Steps 14–23 must be executed to accomplish the final premultiplication and postmultiplication by P and P^T, respectively, as indicated in Eq. (12.10). In this case we assume an array of integers p_i, $i = 1, \ldots, n$, is available recording the fact that the ith permutation performed during the triangularization of A was an interchange of columns i and p_i.
14	For $i := n, n - 1, \ldots, 1$, do through Step 22.
15	If $p_i = i$, go to Step 22.
16	Set $k := p_i$. Interchange the contents of a_{ii} and a_{kk}. If $i = 1$, go to Step 18.
17	For $l := 1, \ldots, i - 1$, interchange the contents of a_{li} and a_{lk}.
18	If $k - i = 1$, go to Step 20.
19	For $l := i + 1, \ldots, k - 1$, interchange the contents of a_{il} and a_{lk}.
20	If $k = n$, go to Step 22.
21	For $l := k + 1, \ldots, n$, interchange the contents of a_{il} and a_{kl}.
22	Continue.
23	*Remark:* The computation of the unscaled covariance matrix is completed.

Examples of Fortran implementations of this algorithm with and without column permutations are, respectively, given in the Programs **PROG1** and **PROG2** of Appendix C.

If the singular value decomposition is used, as in Eq. (12.5), the unscaled covariance matrix C satisfies Eq. (12.11). Thus the individual elements of C are given by

$$(12.13) \qquad c_{ij} = \sum_{k=1}^{n} \frac{v_{ik} v_{jk}}{s_{kk}^2}$$

If the matrix V has been obtained explicitly in an $(n \times n)$-array of storage, then VS^{-1} can replace V in storage. Next the upper triangular part of C can replace the upper triangle of VS^{-1} in storage. This requires the use of an auxiliary array of length n.

Remarks on Some Alternatives to Computing C

We wish to identify three situations in which the unscaled covariance matrix C is commonly computed.

(12.14) • The matrix C or C^{-1} may occur as an intermediate quantity in a statistical formula.

(12.15) • The matrix C may be printed (or otherwise displayed) to be inspected and interpreted by the problem originator.

(12.16) • Some subset of the elements of the matrix C may be used internally as control parameters for some automated process.

In each of these cases we shall indicate an alternative approach that may be either more informative or more economical.

First consider Case (12.14). Commonly, formulas involving C or C^{-1} contain subexpressions of the form

$$(12.17) \qquad E = BCB^T$$

or

$$(12.18) \qquad F = D^T C^{-1} D$$

The most efficient, and numerically stable, approach to computing such expressions is generally found by regarding C or C^{-1} in its factored form,

$$(12.19) \qquad C = R^{-1}(R^{-1})^T$$

or

$$(12.20) \qquad C^{-1} = R^T R$$

respectively.

Thus one would evaluate Eq. (12.17) by first solving for Y in the equation

$$YR = B$$

and then computing

$$E = YY^T$$

Similarly in the case of Eq. (12.18) one would compute

$$X = RD$$

followed by

$$F = X^T X$$

The important general principle illustrated by these observations is the fact that a factorization of C or of C^{-1} is available as shown in Eq. (12.19) and (12.20), respectively, and can be used to develop computational procedures that are more economical and more stable than those involving the explicit computation of C or C^{-1}.

These remarks above can be easily modified to include the case where the matrices C or C^{-1} involve the permutation matrix P as in Eq. (12.7) and (12.10), respectively.

In Case (12.15) the problem originator is frequently investigating correlations (or near dependencies) between various pairs of components in the solution vector x. In this context it is common to produce an auxiliary matrix,

$$(12.21) \qquad E = D^{-1}CD^{-1}$$

where D is a diagonal matrix whose ith diagonal element is given by $d_{ii} = c_{ii}^{1/2}$. Then $e_{ii} = 1$, $i = 1, \ldots, n$, and $|e_{ij}| < 1$, $i = 1, \ldots, n; j = 1, \ldots, n$.

The occurrence of an element e_{ij}, $i \neq j$, close to 1 or -1, say, $e_{ij} = 0.95$, for example, would be taken as an indication that the ith and jth components of the solution vector x are highly correlated. Algebraically this corresponds to the fact that the 2×2 principal submatrix

$$\begin{bmatrix} e_{ii} & e_{ij} \\ e_{ji} & e_{jj} \end{bmatrix}$$

is near-singular.

A weakness of this type of analysis is the fact that only dependencies between pairs of variables are easily detected. For example, there may be a set of three variables that have a mutual near dependence, whereas no two of the variables are nearly dependent. Such a set of three variables could be associated with a 3×3 principal submatrix of E such as

$$(12.22) \qquad \begin{bmatrix} 1 & -0.49 & -0.49 \\ -0.49 & 1 & -0.49 \\ -0.49 & -0.49 & 1 \end{bmatrix}$$

Here no off-diagonal element is close to 1 or -1 and thus no principal 2×2 submatrix is near-singular but the 3×3 submatrix is near-singular. This matrix becomes singular if the off-diagonal elements are changed from -0.49 to -0.50.

Dependencies involving three or more variables are very difficult to detect by visual inspection of the matrix C or E. Such dependencies are revealed, however, by the matrix V of the singular value decomposition $A = USV^T$ (or equivalently of the eigenvalue decomposition $A^TA = VS^2V^T$).

Thus if s_j is a singular value of A that is small relative to the largest singular value s_1, then the corresponding columns v_j and u_j of V and U, respectively, satisfy

(12.23) $$Av_j = s_j u_j$$

or

(12.24) $$\|Av_j\| = s_j$$

The fact that (s_j/s_1) is small may be taken as a criterion of near-singularity of A, while the vector v_j identifies a particular linear combination of columns of A that are nearly dependent. The visual inspection of columns of V associated with small singular values has been found to be a very useful technique in the analysis of ill-conditioned least squares problems.

In Case (12.16) the point to be observed is that one does not need to compute all the elements of C if only some subset of its elements are required. For example, once R^{-1} is computed, then individual elements of C may be computed independently by formulas derivable directly from Eq. (12.9). Typical examples might involve a need for only certain diagonal elements of C or only a certain principal submatrix of C.

13

COMPUTING THE SOLUTION FOR THE UNDERDETERMINED FULL RANK PROBLEM

Consider now the Problem LS, $Ax = b$, for the case $m < n$, Rank (A) $= m$ (Case 3a, Fig. 1.1).

A solution algorithm is described briefly in the following steps:

$$(13.1) \qquad AQ = [R:0]$$

$$(13.2) \qquad Ry_1 = b$$

$$(13.3) \qquad y_2 \text{ arbitrary}$$

$$(13.4) \qquad x = Qy = Q\begin{bmatrix} y_1 \\ y_2 \end{bmatrix}$$

The algorithm will compute the orthogonal matrix Q of Eq. (13.1) so that the matrix R is lower triangular and nonsingular. The existence of such matrices was established by the transposed form of Theorem (3.11).

The m-vector y_1 of Eq. (13.2) is uniquely determined. With any $(n - m)$-vector y_2 of Eq. (13.3), the vector x of Eq. (13.4) will satisfy $Ax = b$. The minimal length solution is attained by setting $y_2 = 0$ in accordance with Theorem (2.3).

There are circumstances when this underdetermined problem $Ax = b$ arises as a part of a larger optimization problem that leads to y_2 being non-zero. In particular, the problem of minimizing $\| Ex - f \|$ subject to the underdetermined equality constraints $Ax = b$ is of this type. This problem, which we call LSE, will be treated in Chapters 20 to 22.

The matrix Q of Eq. (13.1) is constructed as a product of the form

$$(13.5) \qquad Q = Q_1 \cdots Q_m$$

where each Q_j has the form

(13.6) $$Q_j = I_n + b_j^{-1} u^{(j)} u^{(j)T} \qquad j = 1, \ldots, m$$

The quantity b_j that appears in Eq. (13.6) is recomputed when needed based on Eq. (10.10). Introducing subscripts, we have

(13.7) $$b_j = s_j u_j^{(j)} \qquad j = 1, \ldots, m$$

Each s_j is the jth diagonal entry of the R matrix and will be stored as such. This algorithm will be known as HBT(m, n, A, g). The input to Algorithm HBT consists of the integers m and n and the $m \times n$ matrix A of rank m. The output consists of the nonzero portion of the lower triangular matrix R stored in the lower triangular portion of the array named A. The $u_j^{(j)}$ are stored in an auxiliary array named $g_j, j = 1, \ldots, m$. The remaining nonzero portions of the vectors $u^{(j)}$ are stored in the upper diagonal part of the jth row of the array named A.

(13.8) ALGORITHM HBT(m, n, A, g)

Step	Description
1	For $j := 1, \ldots, m$, execute Algorithm H1($j, j + 1, n, a_{j1},$ $g_j, a_{j+1,1}, m - j$) (see Algorithm 10.22).
2	*Comment:* The (backward) triangularization is computed.

In Step 1, it should be noted that $a_{j,1}$ denotes the start of a row vector in storage, while $a_{j+1,1}$ denotes the start of $m - j$ rows to be operated on by the transformations.

The following algorithm will accomplish the computation of Eq. (13.2) to (13.4). The input of this algorithm will consist of the integers m and n, the arrays named A and g as they are output by Algorithm HBT (13.8). The solution will be returned in an array named x. The array b is replaced in storage by the vector y_1 of Eq. (13.2).

(13.9) ALGORITHM HS2(m, n, A, g, b, x)

Step	Description
1	Set $b_1 := b_1 / a_{11}$.
2	For $i := 2, \ldots, m$, set $b_i := (b_i - \sum_{j=1}^{i-1} a_{ij} b_j)/a_{ii}$.
3	*Comment:* At this point the vector y_2 must be determined. If the minimal length solution is desired, put $y_2 := 0$.
4	Set $x := \begin{bmatrix} b \\ y_2 \end{bmatrix}$

Step	*Description*
5	For $j := m, m - 1, \ldots, 1$, execute Algorithm H2$(j, j + 1, n, a_{j1}, g_j, x, 1)$.
6	*Comment:* The array named x now contains a particular solution of the system $Ax = b$. If the vector y_2 of Step 4 is zero, the array x contains the solution of minimal length.

14

COMPUTING THE SOLUTION FOR PROBLEM LS WITH POSSIBLY DEFICIENT PSEUDORANK

The algorithms of Chapters 11 to 13 for full rank problems are intended for use in cases in which the matrix A is assumed to be sufficiently well conditioned so that there is no possibility that either data uncertainties or virtual perturbations of A that arise due to round-off error during the computation could replace A by a rank-deficient matrix. Bounds for these latter perturbations will be treated in Chapters 15 to 17.

There are other cases, however, when one would like to have the algorithm make a determination of how close the given matrix is to a rank-deficient matrix. If it is determined that the given matrix is so close to a rank-deficient matrix that changes in the data of the order of magnitude of the data uncertainty could convert the matrix to one of deficient rank, then the algorithm should take some appropriate action.

The algorithm should at least inform the user when this near-rank-deficient situation is detected. In addition, one may wish to have a solution vector computed by some procedure that avoids the arbitrary instability that can occur when a very ill-conditioned problem is treated as being of full rank.

One technique for stabilizing such a problem is to replace A by a nearby rank-deficient matrix, say, $\tilde{A}$, and then compute the minimal length solution to the problem $\tilde{A}x \cong b$. This replacement most commonly occurs implicitly as part of a computational algorithm. Algorithm HFTI, to be described in this chapter, is of this type. An example illustrating the use of Algorithm HFTI as well as some other stabilization methods is given in Chapter 26. In Chapter 25 other methods of stabilization are developed.

We define the *pseudorank* k of a matrix A to be the rank of the matrix $\tilde{A}$ that replaces A as a result of a specific computational algorithm. Note that pseudorank is not a unique property of the matrix A but also depends on

77

other factors, such as the details of the computational algorithm, the value of tolerance parameters used in the computation, and the effects of machine round-off errors.

Algorithm HFTI applies in particular to the rank-deficient problems identified as Cases 1b, 2b, and 3b in Fig. 1.1. Strictly speaking, however, it is the pseudorank rather than the rank of A that will determine the course of the computation.

Using Theorems (2.3) and (3.19), the mathematical relations on which Algorithm HFTI is based are the following:

$$(14.1) \qquad QAP = R \equiv \begin{bmatrix} R_{11} & R_{12} \\ 0 & R_{22} \end{bmatrix} \begin{matrix} \}k \\ \}m-k \end{matrix}$$
$$\underbrace{\phantom{R_{11}}}_{k} \underbrace{\phantom{R_{12}}}_{n-k}$$

$$(14.2) \qquad Qb = c \equiv \begin{bmatrix} c_1 \\ c_2 \end{bmatrix} \begin{matrix} \}k \\ \}m-k \end{matrix}$$

$$(14.3) \qquad [R_{11} : R_{12}]K = [W : 0]$$

$$(14.4) \qquad Wy_1 = c_1$$

$$(14.5) \qquad y_2 \text{ arbitrary}$$

$$(14.6) \qquad x = PK \begin{bmatrix} y_1 \\ y_2 \end{bmatrix} \equiv PKy$$

$$(14.7) \qquad \|b - Ax\| = \|c_2 - R_{22}y_2\| \qquad (= \|c_2\| \text{ if } y_2 = 0)$$

The algorithm will determine the orthogonal matrix Q and the permutation matrix P so that R is upper triangular and R_{11} is nonsingular.

The permutation matrix P arises implicitly as a result of the column interchange strategy used in the algorithm. The column interchange strategy is intimately related to the problem of determining the pseudorank k. An essential requirement is that the submatrix R_{11} be nonsingular. Generally, unless other criteria are suggested by special requirements of the application, one would prefer to have R_{11} reasonably well-conditioned and $\|R_{22}\|$ reasonably small.

One example of an exceptional case would be the case in which $A_{n \times n}$ is known to have zero as a distinct eigenvalue. Then one would probably wish to set $k = n - 1$ rather than have the algorithm determine k.

Another exceptional case is the weighted approach to Problem LSE (Chapter 22) where it is reasonable to permit R_{11} to be very ill-conditioned.

The column interchange strategy used in Algorithm HFTI is as follows. For the construction of the jth Householder transformation, we consider columns j through n and select that one, say, column λ, whose sum of squares of components in rows j through m is greatest. The contents of columns j

and λ are then interchanged and the jth Householder transformation is constructed to zero the elements stored in a_{ij}, $i = j + 1, \ldots, m$.

Some saving of execution time can be achieved by updating the needed sums of squares. This is possible due to the orthogonality of the Householder transformation. The details of this appear in Steps 3–10 of Algorithm HFTI.

As a result of this interchange strategy the diagonal elements of R will be nonincreasing in magnitude. In fact, they will satisfy the inequality in Eq. (6.16). In this connection, see Theorems (6.13) and (6.31).

In Algorithm HFTI and in its implementation as the Fortran subroutine **HFTI** in Appendix C, the pseudorank k is determined as the largest index j such that $|r_{jj}| > \tau$, where τ is a user-supplied nonnegative absolute tolerance parameter.

The column-interchange strategy and the appropriate selection of τ clearly depend on the initial scaling of the matrix. This topic is discussed in Chapter 25.

Choosing k to be less than $\min (m, n)$ amounts to replacing the given matrix

$$A = Q^T \begin{bmatrix} R_{11} & R_{12} \\ 0 & R_{22} \end{bmatrix} P^T$$

by the matrix

(14.8) $$\tilde{A} = Q^T \begin{bmatrix} R_{11} & R_{12} \\ 0 & 0 \end{bmatrix} P^T$$

Note that

$$\| \tilde{A} - A \| = \| R_{22} \|$$

and

$$\| \tilde{A} - A \|_F = \| R_{22} \|_F$$

Further, with the stated procedures for column interchange and pseudorank determination, one has

$$\| R_{22} \| \leq (\mu - k)^{1/2} |r_{k+1,k+1}|$$
$$\leq (\mu - k)^{1/2} \tau$$

where

$$\mu = \min (m, n)$$

The orthogonal matrix K of Eq. (14.3) is chosen so that W is $k \times k$ nonsingular and upper triangular. The k-vector y_1 is computed as the unique solution of Eq. (14.4). The $(n - k)$-vector y_2 can be given an arbitrary value but the value zero gives the minimal length solution.

The final solution x is given by Eq. (14.6). The norm of the residual can be computed using the right side of Eq. (14.7).

The input to Algorithm HFTI will consist of the data A and b stored in the arrays called A and b, respectively, the integers m and n, and the non-negative absolute tolerance parameter τ. The orthogonal matrix Q of Eq. (14.1) is a product of μ Householder transformations Q_i. The data that define these matrices occupy the lower triangular part of the array A on output plus μ additional stores in an array named h.

The array h is also used to store the squares of the lengths of columns of certain submatrices generated during the computation. These numbers are used in the selection of columns to be interchanged. This information can be (and is) overwritten by the pivot scalars for the matrices Q_i.

The permutation matrix P is constructed as a product of transposition matrices, $P = (1, p_1) \cdots (\mu, p_\mu)$. Here (i, j) denotes the permutation matrix obtained by interchanging columns i and j of I_n. The integers p_i will be recorded in an array p. The orthogonal matrix K of Eq. (14.3) is a product of k Householder transformations K_i. The data that define these matrices occupy the rectangular portion of the array A consisting of the first k rows of the last $n - k$ columns plus k additional stores in an array g.

Figure 14.1 illustrates the output storage configuration for this decomposition when $m = 6$, $n = 5$, and $k = \text{Rank}(\tilde{A}) = 3$. If one wishes to retain complete information about the original matrix A, an extra array consisting of the diagonal terms of the matrix R_{11} of Eq. (14.1) must be recorded. These would be needed to compute the matrix Q of that same equation. Since Q is applied immediately to the vector b in Algorithm HFTI this additional storage is not needed.

A:

w_{11}	w_{12}	w_{13}	k_{14}	k_{15}
u_{12}	w_{22}	w_{23}	k_{24}	k_{25}
u_{13}	u_{23}	w_{33}	k_{34}	k_{35}
u_{14}	u_{24}	u_{34}	†	†
u_{15}	u_{25}	u_{35}	u_{45}	†
u_{16}	u_{26}	u_{36}	u_{46}	u_{56}

Stores that initially held A. After processing they hold W and most of the information for constructing the Householder orthogonal decomposition.

h:

u_{11}	u_{22}	u_{33}	u_{44}	u_{55}

Pivot scalars for the Householder transformations Q_i

g:

k_{11}	k_{22}	k_{33}		

Pivot scalars for the Householder transformations K_i.

p:

p_1	p_2	p_3	p_4	p_5

Interchange record.

†Indicates elements of R_{22} that are ignored.

Fig. 14.1

At the conclusion of the algorithm the solution vector x is stored in the storage array called x and the vector c is stored in the array called b. The algorithm is organized so that the names b and x could identify the same storage arrays. This avoids the need for an extra storage array for x. In this case the length of the b-array must be max (m, n).

Steps 1–13 of Algorithm HFTI accomplish forward triangularization with column interchanges. This is essentially the algorithm given in Golub and Businger (1965). The additional steps constitute the extension for rank-deficient cases as given in Hanson and Lawson (1969).

The coding details of Steps 3–10 are influenced by the code given in Björck and Golub (1967). The test at Step 6 is essentially equivalent to "If $h_\lambda > 10^3 \eta \bar{h}$." The form used in Step 6 avoids the explicit use of the machine-dependent relative precision parameter η.

(14.9) ALGORITHM HFTI($A, m, n, b, \tau, x, k, h, g, p$)

Step	Description
1	Set $\mu := \min (m, n)$.
2	For $j := 1, \ldots, \mu$, do Steps 3–12.
3	If $j = 1$, go to Step 7.
4	For $l := j, \ldots, n$, set $h_l := h_l - a_{j-1,l}^2$.
5	Determine λ such that $h_\lambda := \max \{h_l : j \le l \le n\}$.
6	If $(\bar{h} + 10^{-3} h_\lambda) > \bar{h}$, go to Step 9.
7	For $l := j, \ldots, n$, set $h_l := \sum_{i=j}^{m} a_{il}^2$.
8	Determine λ such that $h_\lambda := \max \{h_l : j \le l \le n\}$. Set $\bar{h} := h_\lambda$.
9	Set $p_j := \lambda$. If $p_j = j$, go to Step 11.
10	Interchange columns j and λ of A and set $h_\lambda := h_j$.
11	Execute Algorithm H1($j, j + 1, m, a_{1j}, h_j, a_{1,j+1}, n - j$).
12	Execute Algorithm H2($j, j + 1, m, a_{1j}, h_j, b, 1$).
13	*Comment:* The pseudorank k must now be determined. Note that the diagonal elements of R (stored in a_{11} through $a_{\mu\mu}$) are nonincreasing in magnitude. For example, the Fortran subroutine **HFTI** in Appendix C chooses k as the largest index j such that $|a_{jj}| > \tau$. If all $|a_{jj}| \le \tau$, the pseudorank k is set to zero, the solution vector x is set to zero, and the algorithm is terminated.
14	If $k = n$, go to Step 17.

Step	*Description*

15   *Comment:* Here $k < n$. Next, determine the orthogonal transformations K_i whose product constitutes K of Eq. (14.3).

16   For $i := k, k - 1, \ldots, 1$, execute Algorithm H1($i, k + 1$, $n, a_{i1}, g_i, a_{11}, i - 1$). (The parameters a_{i1} and a_{11} each identify the first element of a row vector that is to be referenced.)

17   Set $x_k := b_k/a_{kk}$. If $k \leq 1$, go to Step 19.

18   For $i := k - 1, k - 2, \ldots, 1$, $x_i := (b_i - \sum\limits_{j=i+1}^{k} a_{ij}x_j)/a_{ii}$ [see Eq. (14.4)].

19   If $k = n$, go to Step 22.

20   Define the $(n - k)$-vector y_2 of Eq. (14.5) and store its components in x_i, $i := k + 1, \ldots, n$. In particular, set $y_2 := 0$ if the minimal length solution is desired.

21   For $i := 1, \ldots, k$, execute Algorithm H2($i, k + 1, n, a_{i1}$, $g_i, x, 1$). [See Eq. (14.6). Here a_{i1} identifies the first element of a row vector.]

22   For $j := \mu, \mu - 1, \ldots, 1$, do Step 23.

23   If $p_j \neq j$, interchange the contents of x_j and x_{p_j} [see Eq. (14.6)].

A Fortran subroutine **HFTI**, implementing Algorithm HFTI, is given in Appendix C. In the subroutine the algorithm has been generalized to handle multiple right-side vectors. The subroutine also handles the (unlikely) case of the pseudorank k being zero by setting the solution vector x to zero.

15 ANALYSIS OF COMPUTING ERRORS FOR HOUSEHOLDER TRANSFORMATIONS

Throughout the preceding chapters we have treated primarily the mathematical aspects of various least squares problems. In Chapter 9 we studied the effect on the solution of uncertainty in the problem data. In practice, computer arithmetic will introduce errors into the computation. The remarkable conclusion of floating point error analyses [e.g., see Wilkinson (1965a)] is that in many classes of linear algebraic problems the computed solution of a problem is the exact solution to a problem that differs from the given problem by a relatively small perturbation of the given data.

Of great significance is the fact that these virtual perturbations of the data due to round-off are frequently smaller than the inherent uncertainty in the data so that, from a practical point of view, errors due to round-off can be ignored.

The purpose of Chapters 15, 16, and 17 is to present results of this type for the least squares algorithms given in Chapters 11, 13, and 14.

The set of real numbers that are exactly representable as normalized floating point numbers in a given computer will be called *machine numbers*.

For our purposes the floating point arithmetic capability of a given computer can be characterized by three positive numbers L, U, and η. The number L is the smallest positive number such that both L and $-L$ are machine numbers. The number U is the largest positive number such that both U and $-U$ are machine numbers.

In the analysis to follow we shall assume that all nonzero numbers x involved in a computation satisfy $L \leq |x| \leq U$. In the practical design of reliable general-purpose programs implementing these algorithms the condition $|x| \leq U$ can be achieved by appropriate scaling. It may not be feasible to guarantee simultaneously that all nonzero numbers satisfy $|x| \geq L$ but appropriate scaling can assure that numbers satisfying $|x| < L$ can safely be replaced by zero without affecting the accuracy of the results in the sense of vector norms.

The number η characterizes the relative precision of the normalized floating point arithmetic. Following Wilkinson we use the notation

$$\bar{z} = fl(x + y)$$

to indicate that $\bar{z}$ is the machine number produced by the computer when its addition operation is applied to the machine numbers x and y. If $z = x + y$, then even assuming $L \leq |z| \leq U$, it is frequently true that $\bar{z} - z \neq 0$. For convenience this difference will be called *round-off error* regardless of whether the particular computer uses some type of rounding or simply truncation.

The number η is the smallest positive number such that whenever the true result z of one of the five operations, add, subtract, multiply, divide, or square root, is zero or satisfies $L \leq |z| \leq U$, there exist numbers ϵ_i, bounded in magnitude by η, satisfying the corresponding one of the following five equations.

$$(15.1) \quad fl(x + y) = x(1 + \epsilon_1) + y(1 + \epsilon_2) = \frac{x}{1 + \epsilon_3} + \frac{y}{1 + \epsilon_4}$$

(When x and y have the same signs, $\epsilon_1 = \epsilon_2$ and $\epsilon_3 = \epsilon_4$.)

$$(15.2) \quad fl(x - y) = x(1 + \epsilon_5) - y(1 + \epsilon_6) = \frac{x}{1 + \epsilon_7} - \frac{y}{1 + \epsilon_8}$$

(When x and y have opposite signs, $\epsilon_5 = \epsilon_6$ and $\epsilon_7 = \epsilon_8$.)

$$(15.3) \quad fl(xy) = xy(1 + \epsilon_9) = \frac{xy}{1 + \epsilon_{10}}$$

$$(15.4) \quad fl\left(\frac{x}{y}\right) = \left(\frac{x}{y}\right)(1 + \epsilon_{11}) = \frac{x}{y(1 + \epsilon_{12})}$$

$$(15.5) \quad fl(x^{1/2}) = x^{1/2}(1 + \epsilon_{13}) = \frac{x^{1/2}}{1 + \epsilon_{14}}$$

The general approach to analyzing algorithms of linear algebra using assumptions of this type is well illustrated in the literature [e.g., Wilkinson (1960) and (1965a); Forsythe and Moler (1967); Björk (1967a and 1967b)]. Different techniques have been used to cope with the nuisance of keeping track of second- and higher-order terms in η that inevitably arise. Since the main useful information to be obtained is the coefficient of η in the bounds, we propose to determine only that coefficient in our analyses.

Due to grouping η^2 terms into a single $O(\eta^2)$ term, the "bounds" obtained in Chapters 15, 16, and 17 are not fully computable. For values of η corresponding to practical computers, however, the discrepancy from this source is negligible. Since maximum accumulation of errors is assumed throughout the analysis, the bounds obtained tend to be severe overestimates. The main practical purpose of these bounds is to give an indication of the dependence of the errors on the parameters m, n, and η and to establish the excellent numerical stability of the algorithms of Chapters 11, 13, and 14.

We first analyze the round-off error in Algorithm H1 (10.22). This algorithm constructs a single $m \times m$ Householder transformation that will zero all but the first element of a given m-vector v. Including scaling to avoid loss of accuracy from underflow, the algorithm is mathematically defined by

$$(15.6) \qquad \mu = \max_i |v_i|$$

$$(15.7) \qquad t = \sum_{i=1}^{m} \left(\frac{v_i}{\mu}\right)^2$$

$$(15.8) \qquad s = -(\operatorname{sgn} v_1)t^{1/2}\mu$$

$$(15.9) \qquad u_1 = v_1 - s$$

$$(15.10) \qquad u_i = v_i \qquad i = 2, \ldots, m$$

$$(15.11) \qquad b = su_1$$

$$(15.12) \qquad Q = I_m + b^{-1}uu^T$$

Using an overbar to denote computed machine numbers, the quantities actually computed using this algorithm may be written as

$$(15.13) \qquad \mu = \max_i |v_i|$$

$$(15.14) \qquad \bar{t} = fl\left[\sum_{i=1}^{m} \left(\frac{v_i}{\mu}\right)^2\right]$$

$$(15.15) \qquad \bar{s} = -fl[(\operatorname{sgn}(v_1))(\bar{t})^{1/2}\mu]$$

$$(15.16) \qquad \bar{u}_1 = fl(v_1 - \bar{s})$$

$$(15.17) \qquad \bar{u}_i = v_i \qquad i = 2, \ldots, m$$

$$(15.18) \qquad \bar{b} = fl(\bar{s}\bar{u}_1)$$

$$(15.19) \qquad \bar{Q} = fl(I_m + \bar{b}^{-1}\bar{u}\bar{u}^T)$$

The computed quantities defined in Eq. (15.14) to (15.19) are related to the true quantities of Eq. (15.7) to (15.12) as follows:

$$(15.20) \qquad \bar{t} = t(1 + \tau) \qquad |\tau| \leq (m + 2)\eta + O(\eta^2)$$

$$(15.21) \qquad \bar{s} = s(1 + \sigma) \qquad |\sigma| \leq \frac{(m + 6)\eta}{2} + O(\eta^2)$$

$$(15.22) \qquad \bar{u}_1 = u_1(1 + \kappa) \qquad |\kappa| \leq \frac{(m + 8)\eta}{2} + O(\eta^2)$$

$$(15.23) \qquad \bar{u}_i = u_i \qquad i = 2, \ldots, m$$

$$(15.24) \qquad \bar{b} = b(1 + \beta) \qquad |\beta| \leq (m + 8)\eta + O(\eta^2)$$

$$(15.25) \qquad \|\bar{Q} - Q\|_F \leq 4\kappa + 2\beta + O(\eta^2) \leq (4m + 32)\eta + O(\eta^2)$$

The details of the derivation of these bounds are omitted since they are somewhat lengthy and are very similar to the derivation of bounds given on pp. 152–156 of Wilkinson (1965a). Note, however, that the bounds given here are different from those in Wilkinson (1965a) primarily because we are assuming that all arithmetic is done with precision η, whereas it is assumed there that some operations are done with a precision of η^2. We defer discussion of such mixed precision arithmetic to Chapter 17.

We next consider the application of a Householder transformation to an $m \times n$ matrix C using Algorithm H2 or the optional capability of Algorithm H1 (10.22). Mathematically we wish to compute

$$(15.26) \qquad\qquad B = QC$$

In actual computation, however, we only have $\bar{Q}$ available in place of Q so the computed result will be

$$(15.27) \qquad\qquad \bar{B} = fl(\bar{Q}C)$$

The matrix $\bar{B}$ will satisfy

$$(15.28) \qquad \|\bar{B} - B\|_F = \|[fl(\bar{Q}C) - \bar{Q}C] + (\bar{Q}C - QC)\|_F$$
$$\leq \|fl(\bar{Q}C) - \bar{Q}C\|_F + \|\bar{Q} - Q\|_F \cdot \|C\|_F$$
$$\equiv \|E\|_F + \|\bar{Q} - Q\|_F \cdot \|C\|_F$$

Using an analysis similar to that on pp. 157–160 of Wilkinson (1965a), it can be shown that

$$(15.29) \qquad\qquad \|E\|_F \leq (2m + 5)\|C\|_F \eta + O(\eta^2)$$

Thus using Eq. (15.25), (15.28), and (15.29) gives

$$(15.30) \qquad\qquad \|\bar{B} - B\|_F \leq (6m + 37)\|C\|_F \eta + O(\eta^2)$$

We must consider the error associated with the application of k successive Householder transformations beginning with an $m \times n$ matrix A. The true mathematical operation desired is

$$(15.31) \qquad\qquad A_1 = A$$

$$(15.32) \qquad A_{i+1} = \hat{Q}_i A_i \qquad i = 1, \ldots, k$$

where the Householder matrix $\hat{Q}_i$ is constructed to produce zeros in positions $i + 1$ through m of column i of $\hat{Q}_i A_i$. This sequence of operations is basic to the various algorithms we are considering.

The computed quantities will be

$$(15.33) \qquad \bar{A}_1 = A$$

$$(15.34) \qquad \bar{A}_{i+1} = fl(\bar{Q}_i \bar{A}_i) \qquad i = 1, \ldots, k$$

where $\bar{Q}_i$ is the computed approximation to the true matrix Q_i that would have produced zeros in elements $i + 1$ through m of column i of the product matrix $Q_i \bar{A}_i$.

Define the error matrix

$$(15.35) \qquad F_i = Q_i \bar{A}_i - fl(\bar{Q}_i \bar{A}_i)$$
$$= Q_i \bar{A}_i - \bar{A}_{i+1} \qquad i = 1, \ldots, k$$

From Eq. (15.30) we have the bound

$$(15.36) \qquad \| F_i \|_F \leq \alpha_{m+1-i} \| \bar{A}_i \|_F + O(\eta^2)$$

where

$$(15.37) \qquad \alpha_j = (6j + 37)\eta$$

It follows by direct substitution and the orthogonality of the matrices Q_i that

$$(15.38) \quad \| \bar{A}_{k+1} - Q_k \cdots Q_1 A \|_F \leq \sum_{i=1}^{k} \alpha_{m+1-i} \| \bar{A}_i \|_F + O(\eta^2)$$
$$\leq \| A \|_F \sum_{i=1}^{k} \alpha_{m+1-i} + O(\eta^2)$$
$$\leq (6m - 3k + 40)k\eta \| A \|_F + O(\eta^2)$$

It is useful to define the error matrix

$$H_k = Q_1 \cdots Q_k \bar{A}_{k+1} - A$$

Then

$$(15.39) \qquad \bar{A}_{k+1} = Q_k \cdots Q_1 (A + H_k)$$

and from Eq. (15.38)

$$(15.40) \qquad \| H_k \|_F \leq (6m - 3k + 40)k\eta \| A \|_F + O(\eta^2)$$

Equation (15.39) and the bound in Eq. (15.40) can be interpreted as showing that the computed matrix $\bar{A}_{k+1}$ is the exact result of an orthogonal transformation of a matrix $A + H_k$ where $\|H_k\|_F$ is small relative to $\|A\|_F$.

Note that the result expressed by Eq. (15.39) and (15.40) is independent of the fact that the matrices $\bar{Q}_i$ were computed for the purpose of zeroing certain elements of $\bar{A}_i$. Equations (15.39) and (15.40) remain true for cases in which A (and thus $\bar{A}_{k+1}$) is replaced by a matrix or vector not having any special relation to the matrices $\bar{Q}_i$.

Furthermore if the computed matrices of Eq. (15.34) are used in the opposite order, say,

$$(15.41) \qquad \bar{Y} = fl(\bar{Q}_1 \bar{Q}_2 \cdots \bar{Q}_k X)$$

where X is an arbitrary matrix, then the computed matrix $\bar{Y}$ satisfies

$$(15.42) \qquad \bar{Y} = Q_1 \cdots Q_k (X + K)$$

where

$$(15.43) \qquad \|K\|_F \leq (6m - 3k + 40)k\eta \|X\|_F + O(\eta^2)$$

The orthogonal matrices Q_i in Eq. (15.42) are the same as those in Eq. (15.39). This latter observation will be needed in the proofs of Theorems (16.18), (16.36), (17.19), and (17.23).

Finally we shall need bounds on the error in solving a triangular system of equations. Let R be a $k \times k$ triangular matrix and c be a k-vector. Then the computed solution $\bar{x}$ of

$$(15.44) \qquad Rx = c$$

will satisfy

$$(15.45) \qquad (R + S)\bar{x} = c$$

where S is a triangular matrix of the same configuration as R and satisfies

$$(15.46) \qquad \|S\|_F \leq k\|R\|_F \eta + O(\eta^2)$$

For the derivation of this result, see Forsythe and Moler (1967), pp. 103–105.

In the course of proving Eq. (15.46) in the place cited, it is shown that

$$(15.47) \qquad |s_{ii}| \leq 2\eta |r_{ii}| + O(\eta^2) \qquad i = 1, \ldots, k$$

so that for reasonable values of η we are assured that nonsingularity of R implies nonsingularity of $R + S$.

EXERCISE

(15.48) Let G denote a Givens matrix computed by Algorithm G1 (10.25) using exact arithmetic, and let $\bar{G}$ denote the corresponding matrix computed using arithmetic of finite precision η. Let z be an arbitrary machine representable 2-vector. Derive a bound of the form

$$\| fl(\bar{G}z) - Gz \| \leq c\eta \| z \|$$

and determine a value for the constant c.

16 ANALYSIS OF COMPUTING ERRORS FOR THE PROBLEM LS

In this chapter the error analyses of Chapter 15 will be used to obtain bounds for the computational errors of the algorithms of Chapters 11, 13, and 14. These algorithms provide solution procedures for the six cases of Problem LS defined in Fig. 1.1.

It is assumed in this chapter that *all* arithmetic is done with the same precision η. This leads to error bounds involving n^2 and mn. Smaller bounds, involving only the first power of n and independent of m, are obtainable if higher precision arithmetic is used at certain critical points in these computations. The theorems of this chapter will be restated for the case of such mixed precision in Chapter 17.

(16.1) THEOREM (*The full rank overdetermined least squares problem*)

Let A *be an* m $\times$ n *matrix of pseudorank* n (*see Chapter 14 for definition of pseudorank*) *and* b *be an* m-*vector. If* $\bar{x}$ *is the solution of* $Ax \cong b$ *computed using Algorithms HFT (11.4) and HS1 (11.10) or HFTI (14.9) with computer arithmetic of relative precision* η, *then* $\bar{x}$ *is the exact least squares solution of a problem*

(16.2)
$$(A + E)x \cong b + f$$

with

(16.3)
$$\|E\|_F \le (6m - 3n + 41)n\eta\|A\|_F + O(\eta^2)$$

and

(16.4)
$$\|f\| \le (6m - 3n + 40)n\eta\|b\| + O(\eta^2)$$

90

Proof: The mathematical operations required may be written as

$$(16.5) \qquad \hat{Q}[A : b] = \begin{bmatrix} R & c \\ 0 & d \end{bmatrix}$$

$$(16.6) \qquad Rx = c$$

Using Eq. (15.39) and (15.40) with A there replaced by the matrix $[A : b]$ of Eq. (16.5), it follows that there exist an orthogonal matrix Q, a matrix G, and a vector f such that the computed quantities $\bar{R}$, $\bar{c}$, and $\bar{d}$ satisfy

$$(16.7) \qquad Q[A + G : b + f] = \begin{bmatrix} \bar{R} & \bar{c} \\ 0 & \bar{d} \end{bmatrix}$$

with

$$(16.8) \qquad \|G\|_F \le (6m - 3n + 40)n\eta\|A\|_F + O(\eta^2)$$

and

$$(16.9) \qquad \|f\| \le (6m - 3n + 40)n\eta\|b\| + O(\eta^2)$$

In place of Eq. (16.6) to be solved for x, the computer is presented with the problem

$$\bar{R}x = \bar{c}$$

From Eq. (15.45) and (15.46) the computed solution $\bar{x}$ will satisfy

$$(\bar{R} + S)\bar{x} = \bar{c}$$

with

$$(16.10) \qquad \|S\|_F \le n\eta\|\bar{R}\|_F + O(\eta^2) = n\eta\|A\|_F + O(\eta^2)$$

To relate this computed vector $\bar{x}$ to the original least squares problem, define the augmented matrix

$$\begin{bmatrix} \bar{R} + S & \bar{c} \\ 0 & \bar{d} \end{bmatrix}$$

Left multiplying this matrix by the matrix Q^T where Q is the orthogonal matrix of Eq. (16.7) gives

$$\left[A + G + Q^T \begin{bmatrix} S \\ 0 \end{bmatrix} : b + f \right]$$

Define

$$E = G + Q^T \begin{bmatrix} S \\ 0 \end{bmatrix}$$

Then $\tilde{x}$ is the least squares solution of Eq. (16.2). Using $\|E\| \leq \|G\| + \|S\|$ along with Eq. (16.8) to (16.10) establishes the bounds in Eq. (16.3) and (16.4). This completes the proof of Theorem (16.1).

(16.11) THEOREM (*The square nonsingular problem*)

Let A *be an* n × n *matrix of pseudorank* n *and* b *be an* n-*vector. If* $\tilde{x}$ *is the solution of* Ax = b *computed using Algorithms HFT (11.4) and HS1 (11.10) or HFTI (14.9) with computer arithmetic of relative precision* η, *then* $\tilde{x}$ *is the exact solution of a problem*

(16.12) $(A + E)\tilde{x} = b + f$

with

(16.13) $\|E\|_F \leq (3n^2 + 41n)\eta\|A\|_F + O(\eta^2)$

and

(16.14) $\|f\| \leq (3n^2 + 40n)\eta\|b\| + O(\eta^2)$

This theorem is simply a special case of Theorem (16.1) for the case $n = m$ so the bounds in Eq. (16.13) and (16.14) follow from the bounds in Eq. (16.3) and (16.4). We have stated Theorem (16.11) as a separate theorem because there is often independent interest in the square nonsingular case.

It is of interest to compare Theorem (16.11) with an analogous error analysis theorem for Gaussian elimination algorithms with pivoting [Forsythe and Moler (1967)].

In this case one has the inequality

(16.15) $\|E\|_\infty \leq 1.01(n^3 + 3n^2)\|A\|_\infty \eta \rho_n$

Here the symbol $\|\cdot\|_\infty$ denotes the max-row-sum norm defined for any $m \times n$ matrix B as

$$\|B\|_\infty = \max \left\{ \sum_{j=1}^{n} |b_{ij}| : 1 \leq i \leq m \right\}$$

The quantity ρ_n is a measure of the relative growth of elements in the course of Gaussian elimination (*loc. cit.*, p. 102).

If partial pivoting is used, one has the bound

(16.16) $\rho_n \leq 2^{n-1}$

If complete pivoting is used, one has the bound

(16.17) $p_n \leq 1.8n^{(\log n)/4}$

The usual implementations of Gaussian elimination employ the partial pivoting strategy. Thus the bound in Eq. (16.15) for the norm of the matrix E grows exponentially with n.

With the use of the Householder algorithms such as Algorithm HFTI (14.9), the situation is much more satisfactory and assuring. In this case the inequality in Eq. (16.13) does not involve a growth term p_n at all. Wilkinson (1965a) has also remarked on this but points out that roughly twice the number of operations must be performed with the Householder algorithm as with Gaussian elimination with partial pivoting. Furthermore the quantity p_n can easily be computed in the course of executing a Gaussian elimination algorithm and Wilkinson has stated that in practice p_n rarely exceeds 4 or 8 or 16.

Analogous remarks can be made about Theorem (17.15) where extended precision is used at certain points of the Householder algorithm.

In Theorems (16.1) and (16.11) it was shown that the computed solution is *the solution* of a perturbed problem. Theorems (16.18) and (16.36), in contrast, show that the computed solution is *close to the minimum length solution* of a perturbed problem. This more complicated result is due to the fact that Theorems (16.18) and (16.36) concern problems whose solution would be nonunique without imposition of the minimum length solution condition.

(16.18)   THEOREM   (*The minimum length solution of the full rank underdetermined problem*)

> *Let A be an* m × n *matrix of pseudorank* m *and* b *be an* m-*vector. If* x̄ *is the solution of Ax = b computed using Algorithms HBT (13.8) and HS2 (13.9) using computer arithmetic of relative precision* η, *then* x̄ *is close to the exact solution of a perturbed problem in the following sense: There exists a matrix* E *and a vector* x̂ *such that* x̄ *is the minimum length solution of*

(16.19) $(A + E)x = b$

> *with*

(16.20) $\|E\|_F \leq (6n - 3m + 41)m\eta\|A\|_F + O(\eta^2)$

> *and*

(16.21) $\|\bar{x} - \hat{x}\| \leq (6n - 3m + 40)m\eta\|\bar{x}\| + O(\eta^2)$

Proof: A brief mathematical statement of the algorithm of Chapter 13 is

(16.22) $$A\hat{Q} = [R : 0]$$

(16.23) $$Ry = b$$

(16.24) $$x = \hat{Q}\begin{bmatrix} y \\ 0 \end{bmatrix}$$

In place of Eq. (16.22) the computed lower triangular matrix $\bar{R}$ will satisfy

(16.25) $$(A + G)Q = [\bar{R} : 0]$$

where Q is orthogonal and, using the inequality in Eq. (15.40),

(16.26) $$\| G \|_F \leq (6n - 3m + 40)m\eta \| A \|_F + O(\eta^2)$$

Furthermore from Eq. (15.45) and (15.46) it follows that in place of Eq. (16.23) the computed vector $\bar{y}$ will satisfy

(16.27) $$(\bar{R} + S)\bar{y} = b$$

with

(16.28) $$\| S \| \leq m\eta \| \bar{R} \|_F + O(\eta^2) = m\eta \| A \|_F + O(\eta^2)$$

In place of Eq. (16.24) the computed vector $\bar{x}$ will be defined by

(16.29) $$\bar{x} = Q\left\{ \begin{bmatrix} \bar{y} \\ 0 \end{bmatrix} + h \right\}$$

where from Eq. (15.41) to (15.43) one obtains

(16.30) $$\| h \| \leq (6n - 3m + 40)m\eta \| \bar{y} \| + O(\eta^2)$$

The orthogonal matrix Q in Eq. (16.29) is the same as the matrix Q in Eq. (16.25) as noted in the remarks associated with Eq. (15.41) to (15.43).

Rewriting Eq. (16.27) as

(16.31) $$[\bar{R} + S : 0]\begin{bmatrix} \bar{y} \\ 0 \end{bmatrix} = b$$

then

(16.32) $$[\bar{R} + S : 0]Q^T Q \begin{bmatrix} \bar{y} \\ 0 \end{bmatrix} = b$$

and using Eq. (16.25) and (16.29) there follows

(16.33) $$\{A + G + [S:0]Q^T\}(\bar{x} - Qh) = b$$

Define

(16.34) $$E = G + [S:0]Q^T$$

and

(16.35) $$\hat{x} = \bar{x} - Qh \equiv Q\begin{bmatrix} \bar{y} \\ 0 \end{bmatrix}$$

From the inequality in Eq. (15.47) we may assume that $\bar{R} + S$ of Eq. (16.32) is nonsingular and thus that the row space of $A + E \equiv [\bar{R} + S:0]Q^T$ is the same as the space spanned by the first m columns of Q.

From Eq. (16.33) to (16.35), $\hat{x}$ satisfies Eq. (16.19), while again using Eq. (16.35), $\hat{x}$ lies in the space spanned by the first m columns of Q. Thus $\hat{x}$ is the unique minimum length solution of Eq. (16.19).

The matrix E and the vector $(\bar{x} - \hat{x})$ satisfy

$$\|E\|_F \le \|G\|_F + \|S\|_F \le (6n - 3m + 41)m\eta\|A\|_F + O(\eta^2)$$

and

$$\|\bar{x} - \hat{x}\| = \|h\| \le (6n - 3m + 40)m\eta\|\bar{y}\| + O(\eta^2)$$
$$= (6n - 3m + 40)m\eta\|\bar{x}\| + O(\eta^2)$$

which establish the inequalities in Eq. (16.20) and (16.21), completing the proof of the Theorem (16.18).

(16.36)   THEOREM   (*The rank-deficient Problem LS*)

Let A *be an* m $\times$ n *matrix and* b *be an* m-*vector. Let* $\bar{x}$ *be the solution of* Ax $\cong$ b *computed using Algorithm HFTI (14.9) with computer arithmetic of relative precision* η. *Let* k *be the pseudorank determined by the algorithm and let* $\bar{R}_{22}$ *be the computed matrix corresponding to* R$_{22}$ *of Eq. (14.1). Then* $\bar{x}$ *is close to the exact solution to a perturbed problem in the following sense: There exists a matrix* E *and vectors* f *and* $\hat{x}$ *such that* $\hat{x}$ *is the minimal length least squares solution of*

(16.37) $$(A + E)x \cong b + f$$

with

(16.38) $$\|E\|_F \le \|\bar{R}_{22}\|_F + (6m + 6n - 6k - 3\mu + 84)\mu\eta\|A\|_F$$
$$+ O(\eta^2) \qquad \mu = \min(m, n)$$

(16.39) $$\| f \| \leq (6m - 3k + 40)k\eta \| b \| + O(\eta^2)$$

(16.40) $$\| \bar{x} - \hat{x} \| \leq (6n - 6k + 43)k\eta \| \bar{x} \| + O(\eta^2)$$

Proof: Refer to Eq. (14.1) to (14.6) for the mathematical description of the algorithm.

Recall that the matrix Q of Eq. (14.1) is the product of n Householder transformations, $Q = Q_n \cdots Q_1$. Note that the matrices R_{11} and R_{12} of Eq. (14.1) and the vector c_1 of Eq. (14.2) are fully determined by just the first k of these transformations. Let

(16.41) $$\tilde{Q} = Q_k \cdots Q_1$$

Then

(16.42) $$\tilde{Q}AP = \begin{bmatrix} R_{11} & R_{12} \\ 0 & S \end{bmatrix}$$

and

(16.43) $$\tilde{Q}b = \begin{bmatrix} c_1 \\ d \end{bmatrix}$$

where

(16.44) $$\| S \|_F = \| R_{22} \|_F$$

and

(16.45) $$\| d \| = \| c_2 \|$$

In place of Eq. (16.42) the computed matrices $\bar{R}_{11}$, $\bar{R}_{12}$, and $\bar{S}$ will satisfy

(16.46) $$\hat{Q}(A + G)P = \begin{bmatrix} \bar{R}_{11} & \bar{R}_{12} \\ 0 & \bar{S} \end{bmatrix}$$

where, using the results of Chapter 15, $\hat{Q}_k$ is an exactly orthogonal matrix and

(16.47) $$\| G \|_F \leq \left(\sum_{l=1}^{k} \alpha_{m+1-l} \right) \| A \|_F + O(\eta^2)$$

and

$$\| \bar{S} \|_F \leq \| \bar{R}_{22} \|_F + \left(\sum_{l=k+1}^{\mu} \alpha_{m+1-l} \right) \| A \|_F + O(\eta^2)$$

Here α_j is defined by Eq. (15.37), $\mu = \min(m, n)$ as in Chapter 14, and $\bar{R}_{22}$ denotes the computed matrix corresponding to R_{22} of Eq. (14.1).

From Eq. (16.46) we may also write

(16.48)
$$\hat{Q}(A + H)P = \begin{bmatrix} \bar{R}_{11} & \bar{R}_{12} \\ 0 & 0 \end{bmatrix}$$

with

(16.49) $\|H\|_F \leq \|\bar{S}\|_F + \|G\|_F \leq \|\bar{R}_{22}\|_F + \left(\sum_{l=1}^{\mu} \alpha_{m+1-l}\right)\|A\|_F + O(\eta^2)$

$$\leq \|\bar{R}_{22}\|_F + (6m - 3\mu + 40)\mu\eta\|A\|_F + O(\eta^2)$$

In place of Eq. (14.3) the computed $\bar{W}$ satisfies

(16.50)
$$\{[\bar{R}_{11} : \bar{R}_{12}] + M\}\hat{K} = [\bar{W} : 0]$$

where $\hat{K}$ is an orthogonal matrix.

The bound for $\|M\|_F$ is somewhat different than is given by Eq. (15.40) since each of the k Householder transformations whose product constitutes K has only $n - k + 1$ columns that differ from columns of the identity matrix. Therefore, using α_j as defined in Eq. (15.37),

(16.51) $\|M\|_F \leq \alpha_{n-k+1}k\eta\|[\bar{R}_{11} : \bar{R}_{12}]\|_F + O(\eta^2)$

$$= (6n - 6k + 35)k\eta\|A\|_F + O(\eta^2)$$

In place of Eq. (16.43) the computed vectors $\bar{c}_1$ and $\bar{d}$ satisfy

(16.52)
$$\hat{Q}(b + f) = \begin{bmatrix} \bar{c}_1 \\ \bar{d} \end{bmatrix}$$

where $\hat{Q}$ is the same matrix as in Eq. (16.46) and

(16.53) $\|f\| \leq (6m - 3k + 40)k\eta\|b\| + O(\eta^2)$

which will establish Eq. (16.39) of Theorem (16.36).

Referring to Eq. (14.4) the computed vector $\bar{y}_1$ will satisfy

(16.54)
$$(\bar{W} + Z)\bar{y}_1 = \bar{c}_1$$

with $(\bar{W} + Z)$ nonsingular and

(16.55) $\|Z\|_F \leq k\eta\|\bar{W}\|_F + O(\eta^2)$

$$\leq k\eta\|A\|_F + O(\eta^2)$$

In Eq. (14.5) we shall consider only the case $y_2 = 0$. In place of Eq. (14.6) the computed $\bar{x}$ satisfies

$$(16.56) \qquad \bar{x} = P\hat{K} \begin{bmatrix} \bar{y}_1 \\ 0 \end{bmatrix} + h$$

with

$$(16.57) \qquad \|h\| \leq (6n - 6k + 43)k\eta \|\bar{y}_1\| + O(\eta^2)$$

The error coefficient in Eq. (16.57) is determined in the same way as that in Eq. (16.51).

Define

$$(16.58) \qquad \hat{x} = P\hat{K} \begin{bmatrix} \bar{y}_1 \\ 0 \end{bmatrix}$$

Then from Eq. (16.56) to (16.58)

$$(16.59) \qquad \|\bar{x} - \hat{x}\| \leq (6n - 6k + 43)k\eta \|\bar{x}\| + O(\eta^2)$$

which establishes Eq. (16.40) of the theorem.

It follows from Eq. (16.54) and (16.58), by arguments similar to those used in proving Theorem (16.18), that $\hat{x}$ is the minimal length solution of

$$(16.60) \qquad [\bar{W} + Z : 0]\hat{K}^T P^T x = \bar{c}_1$$

From Eq. (16.50)

$$(16.61) \qquad [\bar{W} + Z : 0]\hat{K}^T = [\bar{R}_{11} : \bar{R}_{12}] + X$$

where

$$(16.62) \qquad X = M + [Z : 0]\hat{K}^T$$

and from Eq. (16.51) and (16.55)

$$(16.63) \qquad \begin{aligned} \|X\|_F &\leq \|M\|_F + \|Z\|_F \\ &\leq (6n - 6k + 44)k\eta \|A\|_F + O(\eta^2) \end{aligned}$$

Substituting Eq. (16.61) into Eq. (16.60) and augmenting the system with additional rows shows that $\hat{x}$ is the minimal length least squares solution of

$$(16.64) \qquad \left\{ \begin{bmatrix} \bar{R}_{11} & \bar{R}_{12} \\ 0 & 0 \end{bmatrix} + \begin{bmatrix} X \\ 0 \end{bmatrix} \right\} P^T x \cong \begin{bmatrix} \bar{c}_1 \\ \bar{d} \end{bmatrix}$$

Left-multiplying Eq. (16.64) by $\hat{Q}^T$ and using Eq. (16.46) shows that $\hat{x}$ is the unique minimum length solution of the least squares problem, Eq. (16.37), with

$$(16.65) \qquad E = H + \hat{Q}^T \begin{bmatrix} X \\ 0 \end{bmatrix} P^T$$

Thus

$$(16.66) \qquad \|E\|_F \leq \|H\|_F + \|X\|_F$$

or using Eq. (16.49) and (16.63), and using $k \leq \mu$ to simplify the final expression,

$$(16.67) \qquad \|E\|_F \leq \|\bar{R}_{22}\|_F + (6m + 6n - 6k - 3\mu + 84)\mu\eta\|A\|_F + O(\eta^2)$$

With Eq. (16.53) and (16.59) this completes the proof of Theorem (16.36).

17

ANALYSIS OF COMPUTING ERRORS FOR THE PROBLEM LS USING MIXED PRECISION ARITHMETIC

In our analysis of computing errors we have assumed that all arithmetic was performed with the same precision characterized by the parameter η. If a particular computer has the capability of efficiently executing arithmetic in higher precision (say, with precision parameter $\omega < \eta$) and has efficient machine instructions for converting numbers from one precision to the other, then it is possible to use this extended precision at selected points in the computation to improve the computational accuracy with very small extra cost in execution time and essentially no extra storage requirements.

Generally, unless one decides to convert entirely from η-precision to ω-precision, one will limit the use of ω-precision arithmetic to those parts of the algorithm that will require only a small fixed number of ω-precision storage locations, the number of such locations being independent of the problem dimensions m and n.

Wilkinson has published extensive analyses [e.g., Wilkinson (1965a)] of computing errors for specific mixed precision versions of algorithms in linear algebra assuming $\omega = \eta^2$. We shall use similar methods of analysis to derive error bounds for specific mixed precision versions of the algorithms analyzed in the previous chapter.

For the computation of a Householder transformation we propose to modify Eq. (15.13) to (15.19) as follows. Equation (15.14) will be computed using ω-precision arithmetic throughout, with a final conversion to represent $\bar{t}$ as an η-precision number. Then the error expressions of Eq. (15.20) to (15.25) become

$$(17.1) \qquad \bar{t} = t(1 + \tau) \qquad |\tau| \leq \eta + O(\eta^2)$$

(17.2) $$\bar{s} = s(1 + \sigma) \qquad |\sigma| \leq \frac{5\eta}{2} + O(\eta^2)$$

(17.3) $$\bar{u}_1 = u_1(1 + \kappa) \qquad |\kappa| \leq \frac{7\eta}{2} + O(\eta^2)$$

(17.4) $$\bar{b} = b(1 + \beta) \qquad |\beta| \leq 7\eta + O(\eta^2)$$

(17.5) $$\| \bar{Q} - Q \|_F \leq 28\eta + O(\eta^2)$$

In place of Eq. (15.27) the matrix

(17.6) $$\bar{B} = fl[C + \bar{u}(\bar{u}^T C)\bar{b}^{-1}]$$

will be computed using ω-precision arithmetic with a final conversion of the elements of $\bar{B}$ to η-precision form. Then in place of the error bounds in Eq. (15.29) and (15.30) we have, respectively,

(17.7) $$\| E \|_F \leq \| C \|_F \eta + O(\eta^2)$$

and, using Eq. (17.5),

(17.8) $$\| \bar{B} - B \|_F \leq 29 \| C \|_F \eta + O(\eta^2)$$

Then the bound in Eq. (15.40) on the error associated with multiplication by k Householder transformations becomes

(17.9) $$\| H_k \|_F \leq 29k \| A \|_F \eta + O(\eta^2)$$

Finally, ω-precision arithmetic could be used in solving the triangular system of Eq. (15.44) in which case the bound in Eq. (15.46) would become

(17.10) $$\| S \|_F \leq \| R \|_F \eta + O(\eta^2)$$

The error bound in Eq. (15.46) is already small relative to the error bounds from other sources in the complete solution of Problem LS and thus the additional reduction of the overall error bounds due to the use of ω-precision arithmetic in solving Eq. (15.44) is very slight. For this reason we shall assume that η-precision rather than ω-precision arithmetic is used in solving Eq. (15.44).

With this mixed precision version of the algorithms, Theorems (16.1), (16.11), (16.18) and (16.36) are replaced by the following four theorems. Proofs are omitted due to the close analogy with those of Chapter 16.

(17.11) THEOREM (*The full rank overdetermined least squares problem*)

Let A be an m $\times$ n *matrix of pseudorank* n *and* b *be an* m-*vector.*

If $\bar{x}$ is the solution of $Ax \cong b$ *computed using Algorithms HFT (11.4) and HS1 (11.10) or HFTI (14.9), with mixed precision arithmetic of precisions η and $\omega \leq \eta^2$ as described above, then $\bar{x}$ is the exact least squares solution of a problem*

(17.12)
$$(A + E)x \cong b + f$$

where

(17.13)
$$\|E\|_F \leq 30n \|A\|_F \eta + O(\eta^2)$$

and

(17.14)
$$\|f\| \leq 29n \|b\| \eta + O(\eta^2)$$

(17.15) THEOREM (*The square nonsingular problem*)

Let A be an n × n *matrix of pseudorank* n *and* b *be an* n-*vector. If $\bar{x}$ is the solution of* $Ax = b$ *computed using Algorithms HFT (11.4) and HS1 (11.10) or HFTI (14.9) with mixed precision arithmetic of precisions η and $\omega \leq \eta^2$ as described above, then $\bar{x}$ is the exact solution of a problem*

(17.16)
$$(A + E)x = b + f$$

with

(17.17)
$$\|E\|_F \leq 30n \|A\|_F \eta + O(\eta^2)$$

and

(17.18)
$$\|f\|_F \leq 29n \|b\| \eta + O(\eta^2)$$

(17.19) THEOREM (*The minimum length solution of the full rank underdetermined problem*)

Let A be an m × n *matrix of pseudorank* m *and* b *be an* m-*vector. If $\bar{x}$ is the solution of* $Ax = b$ *computed using the Algorithms HBT (13.8) and HS2 (13.9) with mixed precision arithmetic of precisions η and $\omega \leq \eta^2$ as described above, then $\bar{x}$ is close to the exact solution of a perturbed problem in the following sense: There exists a matrix E and a vector $\hat{x}$ such that $\hat{x}$ is the minimum length solution of*

(17.20)
$$(A + E)x = b$$

with

(17.21)
$$\|E\|_F \leq 30m \|A\|_F \eta + O(\eta^2)$$

and

(17.22)
$$\|\bar{x} - \hat{x}\| \leq 29m \|\bar{x}\| \eta + O(\eta^2)$$

(17.23) THEOREM (*The rank-deficient Problem LS*)

Let A be an m × n *matrix and* b *be an* m-*vector. Let* x̄ *be the solution of* Ax ≅ b *computed using Algorithm HFTI (14.9) with mixed precision arithmetic of precisions* η *and* $\omega \leq \eta^2$ *as described above. Let* k *be the pseudorank determined by the algorithm and let* R̄$_{22}$ *be the computed matrix corresponding to* R$_{22}$ *of Eq. (14.1). Then* x̄ *is close to the exact solution to a perturbed problem in the following sense: There exists a matrix* E *and vectors* f *and* x̂ *such that* x̄ *is the minimal length least squares solution of*

(17.24) $$(A + E)x \cong b + f$$

with

(17.25) $$\|E\|_F \leq \|\bar{R}_{22}\|_F + 59k\,\|A\|_F\eta + O(\eta^2)$$

(17.26) $$\|f\| \leq 29k\,\|b\|\,\eta + O(\eta^2)$$

(17.27) $$\|\bar{x} - \hat{x}\| \leq 29k\,\|\bar{x}\|\,\eta + O(\eta^2)$$

A different type of round-off error analysis of the Householder transformation algorithm as applied to the problem $A_{m \times n}x \cong b$, $m \geq n$, Rank $(A) = n$, has been carried out by Powell and Reid (1968a and 1968b). This analysis was motivated by consideration of the disparately row-weighted least squares problem which arises in the approach to the equality-constrained least squares problem described in Chapter 22.

In Chapter 22 we shall wish to solve a least squares problem in which some rows of the data matrix $[A : b]$ have the property that all elements of the row are very small in magnitude relative to the elements of some other rows but the relative accuracy of these small elements is nevertheless essential to the problem. By this it is meant that perturbations of these elements that are significant relative to the size of the elements will cause significant changes in the solution vector.

Clearly, Theorem (17.11) does not assure that the Householder algorithm will produce a satisfactory solution to such a problem since Eq. (17.13) and (17.14) allow perturbations to all elements of A and b which, although "small" relative to the largest elements of A and b, respectively, may be significantly large relative to the small elements.

Indeed Powell and Reid found experimentally that the performance of the Householder algorithm [such as Algorithm HFTI (14.9)] on disparately row-weighted problems depended critically on the *ordering of the rows* of the matrix $[A : b]$. They observed that if a row of small elements was used as the pivotal row while some of the following rows contained large elements in the pivotal column, then in Eq. (10.7) the quantity v_p would be small relative to s. In the extreme case of $|v_p|/|s| < \eta$ the quantity v_p makes no contribution in Eq. (10.7). Furthermore in such cases it may also be true that the pth component of c_j will be much smaller in magnitude than the pth component of t_ju in Eq. (10.21) so that the pth component of c_j makes no contribution in Eq. (10.21). Thus if the elements of the pivotal row are very small relative to the elements of some of the following rows, the effect can be essentially the same as replacing the data of the pivotal row with zeros. If the solution is sensitive to this loss of data, then this situation must be avoided if possible.

Powell and Reid observed satisfactory results in such problems after adding a row-interchange strategy, which we shall describe preceding Theorem (17.37).

We shall state three theorems given in Powell and Reid (1968a) that provide insight into the performance of Householder transformations in disparately row-weighted problems.

For notational convenience, assume the columns of A are ordered a priori so that no column interchanges occur during the execution of Algorithm HFTI. Let $[\bar{A}^{(1)} : \bar{b}^{(1)}] = [A : b]$ and let $[\bar{A}^{(i+1)} : \bar{b}^{(i+1)}]$ denote the computed matrix resulting from application of the ith Householder transformation to $[\bar{A}^{(i)} : \bar{b}^{(i)}]$ for $i = 1, \ldots, n$.

Define

(17.28) $$\gamma_i = \max_{j,k} |\bar{a}_{ij}^{(k)}| \qquad i = 1, \ldots, m$$

where $\bar{a}_{ij}^{(k)}$ is the (i, j)-element of $\bar{A}^{(k)}$.

Let $Q^{(k)}$ denote the exact Householder orthogonal matrix determined by the condition that elements in positions (i, k), $i = k + 1, \ldots, m$, in the product matrix $Q^{(k)}\bar{A}^{(k)}$ must be zero.

(17.29) THEOREM [*Powell and Reid (1968a)*]

> Let A be an m × n *matrix of pseudorank* n. Let A be reduced to an upper triangular m × n *matrix* $\bar{A}^{(n+1)}$ *by Algorithm HFTI (14.9) using mixed precision arithmetic of precisions* η *and* $\omega \leq \eta^2$. *Then*

(17.30) $$\bar{A}^{(n+1)} = Q^{(n)}Q^{(n-1)} \cdots Q^{(1)}(A + E)$$

> *where the elements* e_{ij} *of* E *satisfy*

(17.31) $$|e_{ij}| \leq (20n^2 - 17n + 44)\gamma_i\eta + O(\eta^2)$$

For the statement of Theorem (17.32) we need the following additional definitions to account for the fact that the right-side vector b cannot participate in the column interchanges:

$$\mu_i = \max_k |\bar{b}_i^{(k)}| \qquad i = 1, \ldots, m$$

$$\nu_k = \left\{ \sum_{i=k}^{m} [\bar{b}_i^{(k)}]^2 \right\}^{1/2} \qquad k = 1, \ldots, n$$

$$\sigma_k = \left\{ \sum_{i=k}^{m} [\bar{a}_{ik}^{(k)}]^2 \right\}^{1/2} \qquad k = 1, \ldots, n$$

$$\rho = \max \left(\max_i \frac{\mu_i}{\gamma_i}, \ \max_k \frac{\nu_k}{\sigma_k} \right)$$

Note that because of our assumptions on the a priori ordering of the columns of A, the quantity σ_k satisfies

$$\sigma_k = \max_{k \leq j \leq n} \left(\sum_{i=k}^{m} [\bar{a}_{ij}^{(k)}]^2 \right)^{1/2} \qquad k = 1, \ldots, n$$

(17.32)   THEOREM   [*Powell and Reid (1968a)*]

> Let A *be an* m $\times$ n *matrix of pseudorank* n *and let* b *be an* m-*vector. Let* $\bar{x}$ *be the solution of* Ax $\cong$ b *computed using Algorithm HFTI (14.9) with mixed precision arithmetic of precisions* η *and* $\omega \leq \eta^2$. *Then* $\bar{x}$ *is the exact least squares solution of a problem*

(17.33) $$(A + E)x \cong b + f$$

where the elements e_{ij} of E satisfy

(17.34) $$|e_{ij}| \leq (20n^2 - 15n + 44)\gamma_i \eta + O(\eta^2)$$

and the elements f_i of f satisfy

(17.35) $$|f_i| \leq (20n^2 + 20n + 44)\rho \gamma_i \eta + O(\eta^2)$$

Using the general fact that $|e_{ij}| \leq \|E\|_F$, one can obtain the inequality

(17.36) $$|e_{ij}| \leq 30n \|A\|_F \eta + O(\eta^2)$$

from Eq. (17.13). Comparing inequalities in Eq. (17.34) and (17.36) we note that inequality in Eq. (17.34) will be most useful in those cases in which γ_i is substantially smaller than $\|A\|_F$. Analogous remarks hold regarding Eq. (17.14) and (17.35).

Recall that γ_i denotes the magnitude of the largest element occurring in row i of any of the matrices $A^{(1)}, \ldots, A^{(n+1)}$. Thus for γ_i to be small (relative to $\|A\|_F$) requires that the elements of row i of the original matrix $\bar{A}^{(1)} = A$ must be small and that elements in row i of the successive transforming matrices $\bar{A}^{(k)}$ must not grow substantially in magnitude.

Powell and Reid have given an example showing that if no restriction is placed on the ordering of the rows of A, the growth of initially small elements in some rows can be very large. They recommend the following row-interchange strategy: Suppose that a pivot column has been selected and moved to the kth column of the storage array in preparation for the kth Householder transformation. Determine a row index l such that

$$| a_{lk} | = \max \{ | a_{ik} | : k \leq i \leq m \}$$

Interchange rows l and k. The kth Householder transformation is then constructed and applied as usual. This process is executed for $k = 1, \ldots, n$.

(17.37) THEOREM [*Powell and Reid (1968a)*]

> *With γ_i defined by Eq. (17.28) and using the column- and row-interchange strategy just described, the quantities γ_i satisfy*
>
> $$\gamma_i \leq (1 + 2^{1/2})^{n-1} m^{1/2} \max_j | a_{ij}^{(1)} |$$
>
> *for* $1 \leq i \leq m.$

Powell and Reid report that using this combined column- and row-interchange strategy they have obtained satisfactory solutions to certain disparately row-weighted problems for which results were unsatisfactory when rows of small elements were permitted to be used as pivotal rows.

18

COMPUTATION OF THE SINGULAR VALUE DECOMPOSITION AND THE SOLUTION OF PROBLEM LS

Section 1. INTRODUCTION

A review of the singular value decomposition of a matrix was given in Golub and Kahan (1965). Included was a bibliography dealing with applications and algorithms, and some work toward a new algorithm. This work was essentially completed in Businger and Golub (1967). An improved version of the algorithm is given by Golub and Reinsch in Wilkinson and Reinsch (1971). This algorithm is a special adaptation of the QR algorithm [due to Francis (1961)] for computing the eigenvalues and eigenvectors of a symmetric matrix.

In this chapter we present the symmetric matrix QR algorithm and the Golub and Reinsch algorithm (slightly enhanced) for computing the singular value decomposition.

$$A_{m \times n} = U_{m \times m} \begin{bmatrix} S_{n \times n} \\ 0 \end{bmatrix} V_{n \times n}^T$$

Included are additional details relating to the use of the singular value decomposition in the analysis and solution of Problem LS.

For notational convenience and because it is the case in which one is most likely to apply singular value analysis, we shall assume that $m \geq n$ throughout this chapter. The case of $m < n$ can be converted to this case by adjoining $n - m$ rows of zeros to the matrix A or, in the case of Problem

LS, to the augmented data matrix $[A : b]$. This adjoining of zero rows can be handled implicitly in a computer program so that it does not actually require the storing of these zero rows or arithmetic operations involving these zero elements. Such an implementation of singular value decomposition with analysis and solution of Problem LS is provided by the set of Fortran subroutines **SVA, SVDRS,** and **QRBD** in Appendix C.

Section 2.   THE SYMMETRIC MATRIX QR ALGORITHM

The symmetric matrix QR algorithm of Francis, incorporating origin shifts, is one of the most satisfactory in all numerical analysis. Its properties and rate of convergence are particularly well understood. The following discussion and related theorems define this algorithm and form the basis for its convergence proof.

An $n \times n$ symmetric matrix A can initially be transformed to tridiagonal form by the use of $(n - 2)$ Householder or $[(n - 1)(n - 2)/2]$ Givens orthogonal similarity transformations. Thus we may assume, without loss of generality, that the symmetric matrix A whose eigenvalues are to be computed is tridiagonal.

The QR *algorithm with origin shifts for computing the eigenvalue decomposition of an* n $\times$ n *symmetric tridiagonal matrix* A *can be written in the form*

$$(18.1) \qquad A_1 = A$$

$$(18.2) \qquad Q_k(A_k - \sigma_k I_n) = R_k \left.\right\}$$
$$(18.3) \qquad R_k Q_k^T + \sigma_k I_n = A_{k+1} \left.\right\} \quad k = 1, 2, \ldots$$

where for k = 1, 2, 3, . . . ,

   (a) Q_k *is orthogonal*
   (b) R_k *is upper triangular*
   (c) σ_k *is the* k*th shift parameter*

Before stating the method of computing the shift parameters σ_k, notation will be introduced for the elements of the matrices A_k. It can be verified [see Exercise (18.46)] that A_1 being tridiagonal implies that all the matrices A_k, $k = 2, 3, \ldots$, will be tridiagonal also. We shall write the diagonal terms of each tridiagonal matrix A_k as $a_i^{(k)}$, $i = 1, \ldots, n$, and the superdiagonal and subdiagonal terms as $b_i^{(k)}$, $i = 2, \ldots, n$, for reference in this chapter and in Appendix B.

The shift parameters σ_k may now be defined as follows:

(18.4)   *Each σ_k is the eigenvalue of the lower right 2×2 submatrix of A_k,*

$$\begin{pmatrix} a_{n-1}^{(k)} & b_n^{(k)} \\ b_n^{(k)} & a_n^{(k)} \end{pmatrix}$$

which is closest to $a_n^{(k)}$.

From Eq. (18.1) to (18.3) we have

$$A_{k+1} = Q_k A_k Q_k^T$$

so that all A_k have the same eigenvalues. We denote them as $\lambda_1, \ldots, \lambda_n$.

If for any k some of the $b_i^{(k)}$ are zero, the matrix A_k may be split into independent submatrices in which all superdiagonal and subdiagonal elements are nonzero. Thus we need only give our attention to the problem for which all $b_i^{(k)}$ are nonzero.

Furthermore, as is stated in Lemma (B.1) of Appendix B, all subdiagonal elements being nonzero implies that all eigenvalues are distinct. Thus we need only concern ourselves with computing eigenvalue–eigenvector decompositions of symmetric tridiagonal matrices whose eigenvalues are distinct.

The convergence of the QR algorithm is characterized by the following theorem of Wilkinson (1968a and 1968b).

(18.5)   THEOREM   (*Global Quadratic Convergence of the Shifted QR Algorithm*)

> Let $A = A_1$ be an $n \times n$ *symmetric tridiagonal matrix with nonzero subdiagonal terms. Let matrices A_k orthogonally similar to A be generated with the QR algorithm using origin shifts as described in Eq. (18.1) to (18.3) and (18.4). Then*
>
> (a) *Each A_k is tridiagonal and symmetric.*
> (b) *The entry $b_n^{(k)}$ of A_k at the intersection of the nth row and $(n-1)$st column tends to zero as $k \to \infty$.*
> (c) *The convergence of $b_n^{(k)}$ to zero is ultimately quadratic: There exists $e > 0$, depending on A, such that for all k*
>
> $$|b_n^{(k+1)}| \le e \, |b_n^{(k)}|^2$$

The proof of this theorem is given in Appendix B.

In practice the algorithm appears to be cubically convergent. However, quadratic convergence is the best that has been proved. Further remarks on this point are given near the end of Appendix B.

Section 3. COMPUTING THE SINGULAR VALUE DECOMPOSITION

We now consider the construction of the singular value decomposition of an $m \times n$ matrix A. We assume $m \geq n$. See Section 1 of this chapter for remarks on the case $m < n$.

The singular value decomposition (SVD) will be computed in two stages. In the first stage A is transformed to an upper bidiagonal matrix $\begin{bmatrix} B \\ 0 \end{bmatrix}$ by a sequence of at most $2n - 1$ Householder transformations:

$$(18.6) \qquad \begin{bmatrix} B \\ 0 \end{bmatrix} = Q_n(\cdots ((Q_1 A)H_2) \cdots H_n) \equiv Q^T A H$$

where

$$(18.7) \qquad B = \begin{bmatrix} q_1 & e_2 & & & \\ & q_2 & \cdot & & \\ & & \cdot & \cdot & \\ & & & \cdot & e_n \\ & & & & q_n \end{bmatrix}$$

The transformation matrix Q_i is selected to zero elements in rows $i + 1$ through m of column i, whereas the matrix H_i is selected to zero elements in columns $i + 1$ through n of row $i - 1$.

Note that H_n is simply an identity matrix. It is included in Eq. (18.6) for notational convenience. Furthermore, Q_n is an identity matrix if $m = n$ but is generally nontrivial if $m > n$.

The second stage of the SVD computation is the application of a specially adapted QR algorithm to compute the singular value decomposition of B,

$$(18.8) \qquad B = \hat{U} S \hat{V}^T$$

where $\hat{U}$ and $\hat{V}$ are orthogonal and S is diagonal. Then the singular value decomposition of A is

$$(18.9) \qquad A = (Q\hat{U}) \begin{bmatrix} S \\ 0 \end{bmatrix} (H\hat{V})^T$$

$$\equiv U \begin{bmatrix} S \\ 0 \end{bmatrix} V^T$$

We now treat the computation represented by Eq. (18.8). First note that if any e_i is zero, the matrix B separates into blocks that can be treated independently. Next we will show that if any q_i is zero, this permits the application of certain transformations which also produce a partition of the matrix.

Suppose that $q_k = 0$, with $q_j \neq 0$, and $e_j \neq 0$, $j = k + 1, \ldots, n$. Premultiplying B by $(n - k)$ Givens rotations T_j we will have

$$(18.10) \qquad T_n \cdots T_{k+1} B = B' \equiv \begin{bmatrix} q_1 & e_2 & & & & & & \\ & \cdot & \cdot & & & & & \\ & & \cdot & \cdot & & & & \\ & & & \cdot & \cdot & & & \\ & & & q_{k-1} & e_k & & & \\ & & & & 0 & 0 & & \\ & & & & & q'_{k+1} & e'_{k+2} & \\ & & & & & & \cdot & \cdot \\ & & & & & & & \cdot & \cdot \\ & & & & & & & & \cdot & \cdot \\ & & & & & & & & & q'_n & e'_n \end{bmatrix}$$

with $e'_j \neq 0$, $j = k + 2, \ldots, n$, and $q'_j \neq 0$, $j = k + 1, \ldots, n$.

The rotation T_j is constructed to zero the entry at the intersection of row k and column j, $j = k + 1, \ldots, n$, and is applied to rows k and j, $j = k + 1$, $\ldots, n$.

The matrix B' of Eq. (18.10) is of the form

$$B' = \begin{bmatrix} B'_1 & 0 \\ 0 & B'_2 \end{bmatrix}$$

where all diagonal and superdiagonal elements of B'_2 are nonzero. Before we discuss B'_2 further, note that B'_1 has at least one zero singular value since its lower right corner element [position (k, k)] is zero.

When the algorithm later returns to operate on B'_1 this fact can be used to eliminate e_k, with the following sequence of rotations:

$$B''_1 = B'_1 R_{k-1} \cdots R_1$$

Here R_i operates on columns i and k to zero position (i, k). For $i > 1$, the application of this rotation creates a nonzero entry in position $(i - 1, k)$.

We return to consideration of B'_2 and for convenience revert to the symbol B to denote this bidiagonal matrix, all of whose diagonal and superdiagonal elements are nonzero. The symbol n continues to denote the order of B.

The singular value decomposition of B [Eq. (18.8)] will be obtained by an iterative procedure of the form

$$B_1 = B$$
$$B_{k+1} = U_k^T B_k V_k \qquad k = 1, 2, \ldots$$

where U_k and V_k are orthogonal and B_k is upper bidiagonal for all k. The choice of U_k and V_k is such that the matrix

$$\tilde{S} = \lim_{k \to \infty} B_k$$

exists and is diagonal.

Note that the diagonal elements of the matrix $\tilde{S}$ which results directly from this iterative procedure are not generally positive or ordered. These conditions are obtained by appropriate postprocessing steps.

This iterative procedure is the QR algorithm of Francis as organized for the singular value problem by Golub and Reinsch (1970). One step of the algorithm proceeds as follows. Given B_k, the algorithm determines the eigenvalues λ_1 and λ_2 of the lower right 2×2 submatrix of $B_k^T B_k$ and sets the shift parameter σ_k equal to the λ_i nearest the value of the lower right element of $B_k^T B_k$.

The orthogonal matrix V_k is determined so that the product

$$(18.11) \qquad V_k^T(B_k^T B_k - \sigma_k I_n)$$

is upper triangular.

The orthogonal matrix U_k is determined so that the product matrix

$$(18.12) \qquad B_{k+1} = U_k^T B_k V_k$$

is upper bidiagonal.

The computational details differ from what would be suggested by the formulas above. In particular the matrix $B_k^T B_k$ is never formed and the shift by σ_k is accomplished implicitly.

To simplify the notation we shall drop the subscripts indicating the iteration stage, k.

Using the notation of (18.7) the lower right 2×2 submatrix of $B^T B$ is

$$\begin{bmatrix} q_{n-1}^2 + e_{n-1}^2 & e_n q_{n-1} \\ e_n q_{n-1} & q_n^2 + e_n^2 \end{bmatrix}$$

Its characteristic equation is

$$(18.13) \qquad (q_{n-1}^2 + e_{n-1}^2 - \lambda)(q_n^2 + e_n^2 - \lambda) - (e_n q_{n-1})^2 = 0$$

Since we seek the root of Eq. (18.13) that is closest to $q_n^2 + e_n^2$, it is convenient to make the substitution

$$(18.14) \qquad \delta = q_n^2 + e_n^2 - \lambda$$

Then δ satisfies

$$(18.15) \qquad \delta^2 + (q_{n-1}^2 - q_n^2 + e_{n-1}^2 - e_n^2)\delta - (e_n q_{n-1})^2 = 0$$

To aid in solving Eq. (18.15), define

(18.16) $$f = \frac{q_n^2 - q_{n-1}^2 + e_n^2 - e_{n-1}^2}{2e_n q_{n-1}}$$

and

(18.17) $$\gamma = \frac{\delta}{e_n q_{n-1}}$$

Then γ satisfies

(18.18) $$\gamma^2 - 2f\gamma - 1 = 0$$

The root $\hat{\gamma}$ of Eq. (18.18) having smallest magnitude is given by

(18.19) $$\hat{\gamma} = \frac{1}{t}$$

where

(18.20) $$t = \begin{cases} [f + (1 + f^2)^{1/2}] & \text{if } f \geq 0 \\ [f - (1 + f^2)^{1/2}] & \text{if } f < 0 \end{cases}$$

Thus, using Eq. (18.14) and (18.17), the root $\hat{\lambda}$ of Eq. (18.13) closest to $q_n^2 + e_n^2$ is

(18.21) $$\hat{\lambda} = q_n^2 + e_n^2 - \hat{\gamma}e_n q_{n-1} = q_n^2 + e_n\left(e_n - \frac{q_{n-1}}{t}\right)$$

and the shift parameter is defined as

(18.22) $$\sigma = \hat{\lambda}$$

Next we must determine V so that $V^T(B^T B - \sigma I)$ is upper triangular [see Eq. (18.11)]. Note that since $B^T B$ is tridiagonal, it will follow that both $V^T(B^T B - \sigma I)V$ and $V^T(B^T B)V$ are tridiagonal [see Exercise (18.46)].

There is a partial converse to these assertions which leads to the algorithm actually used to compute V.

(18.23) THEOREM [*Paraphrased from Francis (1961)*]

If $B^T B$ *is tridiagonal with all subdiagonal elements nonzero,* V *is orthogonal,* σ *is an arbitrary scalar,*

(18.24) $$V^T(B^T B)V \text{ is tridiagonal}$$

and

(18.25) *the first column of* $V^T(B^T B - \sigma I)$ *is zero below the first element, then* $V^T(B^T B - \sigma I)$ *is upper triangular.*

Using this theorem the matrices V and U of Eq. (18.11) and (18.12) will be computed as products of Givens rotations,

$$(18.26) \qquad V = R_1 R_2 \cdots R_{n-1}$$

and

$$(18.27) \qquad U^T = T_{n-1} \cdots T_2 T_1$$

where R_i operates on columns i and $i + 1$ of B and T_i operates on rows i and $i + 1$ of B.

The first rotation R_1 is determined to satisfy condition (18.25) of Theorem (18.23). The remaining rotations R_i and T_i are determined to satisfy condition (18.24) without disturbing condition (18.25).

Note that it is in the use of the matrices T_i that this algorithm departs from the standard implicit QR algorithm for a symmetric matrix [Martin and Wilkinson in Wilkinson and Reinsch (1971)]. In the symmetric matrix problem one has a tridiagonal symmetric matrix, say, Y, and wishes to produce

$$(18.28) \qquad \tilde{Y} = V^T Y V$$

In the present algorithm one has Y in the factored form $B^T B$ where B is bidiagonal. One wishes to produce $\tilde{Y}$ in a factored form, $\tilde{Y} = \tilde{B}^T \tilde{B}$. Thus $\tilde{B}$ must be of the form $\tilde{B} = U^T B V$ where V is the same orthogonal matrix as in Eq. (18.28) and U is also orthogonal.

Turning now to the determination of the rotations R_i and T_i of Eq. (18.26) and (18.27), note that the first column of $(B^T B - \sigma I)$ is

$$(18.29) \qquad \begin{bmatrix} q_1^2 - \sigma \\ q_1 e_2 \\ 0 \\ . \\ . \\ . \\ 0 \end{bmatrix}$$

with σ given by Eq. (18.22). The first rotation R_1 is determined by the requirement that the second element in the first column of $R_1^T (B^T B - \sigma I)$ must be zero.

The remaining rotations are determined in the order $T_1, R_2, T_2, \ldots,$ R_{n-1}, T_{n-1} and applied in the order indicated by the parentheses in the following expression:

$$(18.30) \qquad T_{n-1}(\cdots T_2((T_1(BR_1))R_2) \cdots R_{n-1}$$

The rotation $T_i, i = 1, \ldots, n - 1$, operates upon rows i and $i + 1$ and zeros the element in position $(i + 1, i)$. The rotation R_i, $i = 1, \ldots, n - 1$, operates on columns i and $i + 1$ and, for $i \geq 2$, zeros the element in position $(i - 1, i + 1)$.

This is an instance of an algorithmic process sometimes called *chasing*. The pattern of appearance and disappearance of elements outside the bidiagonal positions is illustrated for the case of $n = 6$ by Fig. 18.1.

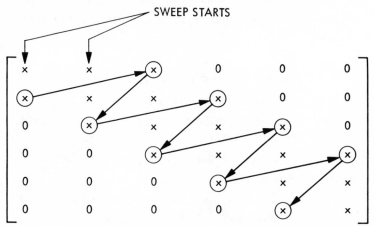

Fig. 18.1   The chasing pattern for one QR sweep for the case of $n = 6$.

From the preceding discussion it is clear that an SVD algorithm for a general $m \times n$ matrix can be based upon an algorithm for the more specialized problem of the singular value decomposition of a nonsingular bidiagonal matrix B with elements identified as in Eq. (18.7).

The algorithm QRBD will be stated for performing one sweep of this fundamental part of the SVD computation. This algorithm first tests the off-diagonal elements e_i against a given tolerance parameter ϵ. If $|e_i| < \epsilon$ for any i, $2 \leq i \leq n$, the largest index l for which $|e_i| < \epsilon$ is stored in the location named l, and the algorithm terminates. Otherwise the algorithm sets $l = 1$ to indicate that $|e_i| \geq \epsilon$ for $2 \leq i \leq n$. The algorithm proceeds to compute the shift parameter [see Eq. (18.22)]. Next the algorithm determines and applies the rotations R_i and T_i, $i = 1, \ldots, n - 1$ [see Eq. (18.30)]. The algorithm terminates with the elements of the transformed matrix $\tilde{B} = U^T B V$ replacing those of B in storage.

For some purposes, such as solving Problem LS, it is necessary to multiply certain other vectors or matrices by the matrices U^T and V. Algorithm QRBD has provision for the user to supply a $k \times n$ matrix $W = \{w_{ij}\}$ and an $n \times p$ matrix $G = \{g_{ij}\}$ which the algorithm will replace by the $k \times n$ product matrix WV and the $n \times p$ product matrix $U^T G$, respectively.

(18.31)   ALGORITHM   QRBD($q, e, n, \epsilon, l, W, k, G, p$)

Step	Action
1	*Comment:* Steps 2 and 3 scan for small off-diagonal elements.
2	For $i := n, n - 1, \ldots, 2$, do Step 3.
3	If $|e_i| \leq \epsilon$, set $l := i$ and go to Step 14.
4	Set $l := 1$ and compute σ by means of Eq. (18.16) to (18.22).
5	Set $i := 2$.
6	Set $e_1 := q_1 - \sigma/q_1$ and $z := e_2$ [see Eq. (18.29)].
7	Execute Algorithm G1($e_{i-1}, z, c, s, e_{i-1}$).
8	Execute Algorithm G2(c, s, q_{i-1}, e_i) and for $j := 1, \ldots, k$ execute Algorithm G2($c, s, w_{j,i-1}, w_{j,i}$).
9	Set $z := sq_i$ and $q_i := cq_i$.
10	Execute Algorithm G1($q_{i-1}, z, c, s, q_{i-1}$).
11	Execute Algorithm G2(c, s, e_i, q_i) and for $j := 1, \ldots, p$ execute Algorithm G2($c, s, g_{i-1,j}, g_{i,j}$).
12	If $i = n$, go to Step 14.
13	Set $z := se_{i+1}, e_{i+1} := ce_{i+1}$, and $i := i + 1$ and go to Step 7.
14	*Comment:* If $l = 1$, one full QR sweep on the bidiagonal matrix has been performed. If $l > 1$, the element e_l is small, so the matrix can be split at this point.

By repeated application of Algorithm QRBD to a nonsingular upper bidiagonal matrix B, one constructs a sequence of bidiagonal matrices B_k with diagonal terms $q_1^{(k)}, \ldots, q_n^{(k)}$ and superdiagonal terms $e_2^{(k)}, \ldots, e_n^{(k)}$. From Theorem (18.5) we know that the product $e_n^{(k)} q_n^{(k)}$ converges quadratically to zero as $k \to \infty$. The assumption that B is nonsingular implies that $q_n^{(k)}$ is bounded away from zero. It follows that $e_n^{(k)}$ converges quadratically to zero. In practice the iterations will be stopped when $|e_n^{(k)}| < \epsilon$.

Having accepted $|e_n^{(k)}|$ as being sufficiently small, one next constructs the SVD of one or more bidiagonal matrices of order $n - 1$ or lower; this process is continued as above with n replaced by $n - 1$ until $n = 1$. With no more than $n - 1$ repetitions of the procedures above, the decomposition $B = \tilde{U}\tilde{S}\tilde{V}^T$ is produced.

With the precision parameter η in the range of about 10^{-18} to 10^{-8} and $\epsilon = \eta \|B\|$, experience shows that generally a total of about $2n$ executions of

Algorithm QRBD are required to reduce all superdiagonal elements to less than ϵ in magnitude.

The diagonal elements of $\tilde{S}$ are not generally nonnegative or nonincreasing. To remedy this, choose a diagonal matrix D whose diagonal terms are $+1$ or -1 with the signs chosen so that

$$(18.32) \qquad\qquad \hat{S} = \tilde{S}D$$

has nonnegative diagonal terms. Next choose a permutation matrix P such that the diagonal terms of

$$(18.33) \qquad\qquad S = P^T\hat{S}P$$

are nonincreasing. It is easily seen that

$$B = (\tilde{U}P)(P^T\hat{S}P)(\tilde{V}DP)^T \equiv \hat{U}S\hat{V}^T$$

is a singular value decomposition of B with the diagonal elements of S nonnegative and nonincreasing.

The matrix V of Eq. (18.9) will be produced as a product of the form

$$(18.34) \qquad\qquad V = H_2 \cdots H_n R_1 \cdots R_v DP$$

where the matrices H_i are the Householder transformations of Eq. (18.6), the matrices R_i are all the postmultiplying Givens rotations produced and used preceeding and during QR sweeps [see Eq. (18.30) and the text following Eq. (18.10)], and the matrices D and P are given by Eq. (18.32) and (18.33), respectively.

Similarly the matrix U^T of Eq. (18.9) is generated as

$$(18.35) \qquad\qquad U^T = PT_\mu \cdots T_1 Q_n \cdots Q_1$$

where P is given by Eq. (18.33), the rotations T_i are all the premultiplying Givens rotations that arise preceeding and during the QR sweeps [see Eq. (18.10) and (18.30)], and the matrices Q_i are the Householder transformations of Eq. (18.6).

Section 4.   SOLVING PROBLEM LS USING SVD

For application to Problem LS the singular value decomposition,

$$(18.36) \qquad\qquad A_{m \times n} = U_{m \times m} \begin{bmatrix} S_{n \times n} \\ 0_{(m-n) \times n} \end{bmatrix} V_{n \times n}^T$$

can be used to replace the problem $Ax \cong b$ by the equivalent problem

$$(18.37) \qquad \begin{bmatrix} S \\ 0 \end{bmatrix} p \cong g$$

where

$$(18.38) \qquad g \equiv \begin{bmatrix} g^{(1)} \\ g^{(2)} \end{bmatrix} \begin{matrix} \}n \\ \}m-n \end{matrix} = \begin{bmatrix} U_1^T \\ U_2^T \end{bmatrix} b = U^T b$$

and

$$(18.39) \qquad x = Vp$$

One may wish to consider a sequence of *candidate solutions* $x^{(k)}$ defined by

$$(18.40) \qquad x^{(k)} = \sum_{j=1}^{k} p_j v_j$$

where v_j is the jth column vector of V.

Note that it is convenient to compute the vectors $x^{(k)}$ by the formulas

$$(18.41) \qquad x^{(0)} = 0$$

$$(18.42) \qquad x^{(k)} = x^{(k-1)} + p_k v_k \qquad k = 1, \ldots, n$$

The residual norm ρ_k associated with the candidate solution $x^{(k)}$, defined by

$$\rho_k = \| A x^{(k)} - b \|$$

also satisfies

$$(18.43) \qquad \rho_k^2 = \| g^{(2)} \|^2 + \sum_{i=k+1}^{n} (g_i^{(1)})^2$$

These numbers can be computed in the following manner:

$$(18.44) \qquad \rho_n^2 = \| g^{(2)} \|^2$$

$$(18.45) \qquad \rho_k^2 = \rho_{k+1}^2 + (g_{k+1}^{(1)})^2 \qquad k = n-1, \ldots, 1, 0$$

where $g^{(1)}$ and $g^{(2)}$ are the subvectors of g defined by Eq. (18.38).

The columns of V associated with small singular values may be interpreted as indicating near linear dependencies among the columns of A. This has been noted in Chapter 12. Further remarks on the practical use of the singular values and the quantities ρ_k are given in Chapter 25, Section 6.

Section 5.   ORGANIZATION OF A COMPUTER PROGRAM FOR SVD

We shall describe a way in which these various quantities can be computed with economical use of storage. This description assumes $m \geq n$. See Section 1 of this chapter for remarks on the case $m < n$.

First, if mn is large and $m \gg n$, we could arrange that the data matrix $[A : b]$ be first reduced to an $(n + 1)$ upper triangular matrix by sequential Householder processing (see Chapter 27). The matrix A is reduced to bidiagonal form $\begin{bmatrix} B \\ 0 \end{bmatrix}$, as indicated in Eq. (18.6). The nonzero elements of B replace the corresponding elements of A in the storage array called A. The transformations Q_i of Eq. (18.6) may be applied to b as they are formed and need not be saved. The vector resulting from this transformation replaces b in the storage array called b. The transformations H_i of Eq. (18.6) must be saved in the storage space that becomes available in the upper triangle of the A array plus an n-array, called h, say.

After the bidiagonalization the nonzero elements of B are copied into the two n-arrays q and e, occupying locations $q_i, i = 1, \ldots, n$, and $e_i, i = 2, \ldots, n$, with location e_1 available for use as working storage by the QR algorithm.

The computation of the matrix V of Eq. (18.36) is initiated by explicitly forming the product $(H_2 \cdots H_n)$ required in Eq. (18.34). This computation can be organized (and is so organized in the Fortran subroutine **SVDRS** of Appendix C) so that the resulting product matrix occupies the first n rows of the storage array A, and no auxiliary arrays of storage are required.

The QR algorithm is applied to B, as represented by the data stored in arrays q and e. As each rotation R_i is produced during the QR algorithm it is multiplied into the partial product matrix stored in the A array for the purpose of forming V [see Eq. (18.34)]. Similarly, each rotation T_i is multiplied times the vector stored in the array b in order to form the vector $U^T b$ [see Eq. (18.35) and (18.38)].

At the termination of the QR iterations the numbers stored in the locations $e_i, i = 2, \ldots, n$ will be small. The numbers stored in $q_i, i = 1, \ldots, n$, must next be set to be nonnegative and sorted. Application of these sign changes and permutations to the storage array A completes the computation of the matrix V [see Eq. (18.34)]. Application of the permutations to the storage array b completes the computation of $g = U^T b$ [see Eq. (18.35) and (18.38)].

If desired the candidate solutions of Eq. (18.41) and (18.42) can be computed and stored as column vectors in the upper $n \times n$ submatrix of the array A, overwriting the matrix V.

The Fortran subroutine **SVDRS** (Appendix C) contains the additional feature of initially scanning the matrix A for any entirely zero columns. If l such columns are found, the columns are permuted so that the first $n - l$ columns are nonzero. Such a matrix has at least l of the computed singular values exactly zero.

This feature was introduced into the subroutine so that a user can delete a variable from a problem by the simple process of zeroing the corresponding column of the matrix A. Without this procedure some of the computed singular values that should be zero would have values of approximately $\eta \| A \|$. In many circumstances this would be satisfactory, but in some contexts it is desirable to produce the exact zero singular values.

The subroutine **SVDRS** also tests the rows of A, and if necessary permutes the rows of the augmented matrix $[A : b]$ so that if A contains l nonzero rows the first $\mu = \min (l, n)$ rows of the permuted matrix are nonzero. This is only significant when $l < n$, in which case A has at least $n - l$ zero singular values and this process assures that at least $n - l$ of the computed singular values will be exactly zero.

When the singular value decomposition is computed for the purpose of reaching a better understanding of a particular least squares problem, it is desirable to have a program that prints various quantities derived from the singular value decomposition in a convenient format. Such a Fortran subroutine, **SVA**, is described and included in Appendix C. An example of the use of **SVA** is given in Chapter 26.

EXERCISES

(18.46)   Prove that if A_1 is symmetric and tridiagonal, then so are the A_k, $k = 2, \ldots$, of Eq. (18.3).

(18.47)   Prove that the special QR algorithm [Eq. (18.11), (18.12), and (18.22)] converges in one iteration for 2×2 bidiagonal matrices.

19 OTHER METHODS FOR LEAST SQUARES PROBLEMS

Section 1. Normal Equations with Cholesky Decomposition
Section 2. Modified Gram–Schmidt Orthogonalization

As a computational approach to least squares problems we have stressed the use of Householder orthogonal transformations. This approach has the numerical stability characteristic of orthogonal transformations (Chapters 15 to 17) along with ready adaptability to special requirements such as sequential accumulation of data (see Chapter 27). We have also suggested the use of singular value analysis as a technique for reaching a better understanding of a poorly conditioned problem (Chapters 18 and 25).

In this chapter we shall discuss some of the other computational approaches to the least squares problem. For convenience Table 19.1 gives the high-order terms in the count of operations for the solution of the least squares problem $A_{m \times n} x \cong b, m > n$, using various methods.

One class of methods to be discussed is based on the mathematical equivalence of the least squares problem $Ax \cong b$ and the system of linear equations $(A^T A)x = (A^T b)$. This system is referred to as the *normal equations* for the problem. Methods based on forming and solving the normal equations typically require only about half as many operations as the Householder algorithm, but these operations must be performed using precision η^2 to satisfactorily encompass the same class of problems as can be treated using the Householder algorithm with precision η.

Most special processes, such as sequential accumulation of data (see Chapter 27), for example, can be organized to require about the same *number* of units of storage in either a normal equations algorithm or a Householder algorithm. Note, however, that if the units of storage are taken to be computer words of the same length for both methods, the Householder algorithm will successfully process a larger class of problems than the normal equations algorithm.

121

The second method to be discussed in this chapter is *modified Gram–Schmidt orthogonalization* (MGS). The numerical properties of this method are very similar to those of the Householder algorithm. The factored representation of the Q matrix which is used in the Householder algorithm occupies about $n^2/2$ fewer storage locations than the explicit representation used in MGS. Largely because of this feature, we have found the Householder algorithm more conveniently adaptable than MGS to various special applications.

Table 19.1   OPERATION COUNTS FOR VARIOUS LEAST SQUARES
COMPUTATIONAL METHODS

Method	Asymptotic Number of Operations Where an Operation Is a Multiply or Divide Plus an Add
Householder triangularization	$mn^2 - n^3/3$
Singular value analysis	
Direct application to A	$2mn^2 + \sigma(n)$†
Householder reduction of A to triangular R	
plus singular value analysis of R	$mn^2 + 5n^3/3 + \sigma(n)$†
Form normal equations	$mn^2/2$
Cholesky solution of normal equations	$n^3/6$
Gauss–Jordan solution of normal equations	
(for stepwise regression)	$n^3/3$
Eigenvalue analysis of normal equations	$4n^3/3 + \sigma(n)$†
Gram–Schmidt (either classical or modified)	mn^2

†The term $\sigma(n)$ accounts for the iterative phase of the singular value or eigenvalue computation. Assuming convergence of the QR algorithm in about $2n$ sweeps and implementation as in the Fortran subroutine **QRBD** (Appendix C), $\sigma(n)$ would be approximately $4n^3$.

As a specific example of this compact storage feature of the Householder transformation, note that one can compute the solution $\hat{x}$ of $Ax \cong b$ and the residual vector $r = b - A\hat{x}$ using a total of $m(n + 1) + 2n + m$ storage locations. Either the use of normal equations or the MGS algorithm requires an additional $n(n + 1)/2$ storage locations. Furthermore, as mentioned before, the assessment of storage requirements must take into consideration the fact that the normal equations method will require η^2-precision storage locations to handle the class of problems that can be handled by Householder or MGS using η-precision storage locations.

Section 1.   NORMAL EQUATIONS WITH CHOLESKY DECOMPOSITION

The method most commonly discussed in books giving brief treatment of least squares computation is the use of normal equations. Given the problem $A_{m \times n}x \cong b$ one premultiplies by A^T, obtaining the system

(19.1)
$$P_{n \times n} x = d_{n \times 1}$$

which is called the system of *normal equations* for the problem. Here

(19.2)
$$P = A^T A$$

and

(19.3)
$$d = A^T b$$

Equation (19.1) arises directly from the condition that the residual vector $b - Ax$ for the least squares solution must be orthogonal to the column space of A. This condition is expressed by the equation

(19.4)
$$A^T(b - Ax) = 0$$

which is seen to be equivalent to Eq. (19.1) using Eq. (19.2) and (19.3).

While P has the same rank as A and thus can be singular, it is nevertheless true that Eq. (19.1) is always consistent. To verify this note that, from Eq. (19.3), the vector d is in the row space of A but from Eq. (19.2), the row space of A is identical with the column space of P.

The system (19.1) could be solved by any method that purports to solve a square consistent system of linear equations. From Eq. (19.2), however, the matrix P is symmetric and nonnegative definite, which permits the use of Cholesky elimination with its excellent properties of economy and stability [Wilkinson (1965a), pp. 231–232].

The Cholesky method, which we will now describe, is based on the fact that there exists an $n \times n$ upper triangular (real) matrix U such that

(19.5)
$$U^T U = P$$

and thus the solution x of Eq. (19.1) can be obtained by solving the two triangular systems

(19.6)
$$U^T y = d$$

and

(19.7)
$$Ux = y$$

Alternatively, the process of solving Eq. (19.6) for y can be accomplished as a part of the Cholesky decomposition of an appropriate augmented matrix. For this purpose define

$$\tilde{A} = [A:b]$$

and

(19.8)
$$\tilde{P} = \tilde{A}^T \tilde{A} \equiv \begin{bmatrix} P & d \\ d^T & \omega^2 \end{bmatrix}$$

Note that P and d of Eq. (19.8) denote the same matrix and vector as in Eq. (19.2) and (19.3) and ω satisfies

$$\omega^2 = b^T b$$

The upper triangular Cholesky factorization of $\tilde{P}$ is then of the form

(19.9) $$\tilde{P} = \tilde{U}^T \tilde{U}$$

with

(19.10) $$\tilde{U} = \begin{bmatrix} U & y \\ 0 & \rho \end{bmatrix} \begin{matrix} \}n \\ \}1 \end{matrix}$$
$$\underbrace{}_{n} \ \underbrace{}_{1}$$

where U and y of Eq. (19.10) satisfy Eq. (19.5) and (19.6). Furthermore it is easily verified that ρ of Eq. (19.10) satisfies

$$|\rho| = \| b - A\hat{x} \|$$

where $\hat{x}$ is any vector minimizing $\| b - Ax \|$.

For convenience the details of computing the Cholesky decomposition will be described using the notation of Eq. (19.5). Obviously the algorithm also applies directly to Eq. (19.9).

From Eq. (19.5) we obtain the equations

(19.11) $$\sum_{k=1}^{i} u_{ki} u_{kj} = p_{ij} \quad \text{for } i = 1, \ldots, n; j = i, \ldots, n$$

Solving for u_{ij} in the equation involving p_{ij} leads to the following equations, which constitute the Cholesky (also called square root or Banachiewicz) factorization algorithm:

(19.12) $$v_i = p_{ii} - \sum_{k=1}^{i-1} u_{ki}^2$$

(19.13) $$u_{ii} = v_i^{1/2} \qquad\qquad\qquad\qquad \left. \begin{matrix} \\ \\ \\ \\ \\ \\ \end{matrix} \right\} i = 1, \ldots, n$$

(19.14) $$u_{ij} = \frac{p_{ij} - \sum_{k=1}^{i-1} u_{ki} u_{kj}}{u_{ii}} \quad j = i + 1, \ldots, n$$

In Eq. (19.12) and (19.14) the summation term is taken to be zero when $i = 1$.

Theoretically if Rank $(A) = n$, then $v_i > 0$, $i = 1, \ldots, n$. Equations (19.12) to (19.14) then define unique values for each u_{ij}.

If Rank $(A) = k < n$ however, there will be a first value of i, say, $i = t$, such that $v_t = 0$. Then for $i = t$ the numerator in the right side of Eq. (19.14) must also be zero for $j = t + 1, \ldots, n$, since solutions exist for all the equations in (19.11). In this case there is some freedom in the assignment of values to $u_{tj}, j = t + 1, \ldots, n$. One set of values that is always admissible is $u_{tj} = 0, j = t + 1, \ldots, n$ [see Exercise (19.37)].

It follows that the theoretical algorithm of Eq. (19.12) to (19.14) provides a solution for Eq. (19.5) regardless of the rank of A if it is modified so that Eq. (19.14) is replaced by

$$(19.15) \qquad u_{ij} = 0 \qquad j = i + 1, \ldots, n$$

for any value of i for which $u_{ii} = 0$.

Note that the upper triangular matrix R of the Householder decomposition $QA = \begin{bmatrix} R \\ 0 \end{bmatrix}$ satisfies $R^T R = P$. If Rank $(A) = n$, the solution of Eq. (19.5) is unique to within the signs of rows of U [see Exercise (2.17)]. Thus if Rank $(A) = n$, the matrix R of the Householder decomposition is identical with the Cholesky matrix U to within signs of rows.

The Cholesky decomposition can also be computed in the form

$$(19.16) \qquad L^T L = P$$

where L is an $n \times n$ *lower* triangular matrix. The formulas for elements of L are

$$(19.17) \qquad l_{ii} = \left(p_{ii} - \sum_{k=i+1}^{n} l_{ki}^2 \right)^{1/2} $$

$$(19.18) \qquad l_{ij} = \frac{p_{ij} - \sum_{k=i+1}^{n} l_{ki} l_{kj}}{l_{ii}} \qquad j = 1, \ldots, i-1 $$

$$i = n, \ldots, 1$$

In actual computation there arises the possibility that the value of v_i computed using Eq. (19.12) for the Cholesky decomposition of the (theoretically) nonnegative definite matrix P of Eq. (19.2) may be negative due to round-off error. Such errors can arise cumulatively in the computation associated with Eq. (19.2), (19.12), (19.13), and (19.14).

If a computed v_i is negative for some i, one method of continuing the computation is to set $v_i = 0$, $u_{ii} = 0$, and apply Eq. (19.15). A Fortran subroutine using essentially this idea is given by Healy (1968).

Another approach is to perform and record a symmetric interchange of rows and columns to maximize the value of v_i at the ith step of the algorithm. This has the effect of postponing the occurrence of nonpositive values of v_i, if any, until all positive values have been processed. When all remaining values of v_i are nonpositive, then the corresponding remaining rows of U can be set to zero.

The implementation of such an interchange strategy requires some reordering of the operations expressed in Eq. (19.12) to (19.14) so that partially computed values of v_i will be available when needed to make the choice of pivital element. When interchanges are used on an augmented matrix of the form occurring in Eq. (19.8), the last column and row must not take part in the interchanges.

A Cholesky decomposition for $\tilde{P}$ of Eq. (19.8) gives rise to an upper triangular matrix $\tilde{U}$ of Eq. (19.10) having the same relationship to Problem LS as does the triangular matrix obtained by Householder triangularization of $\tilde{A}$. This can be a useful observation in case one already has data in the form of normal equations and then wishes to obtain a singular value analysis of the problem.

Ignoring the effects of round-off error, the same information obtainable from singular value analysis of A (see Chapter 18 and Chapter 25, Section 5) can be obtained from an eigenvalue analysis of the matrix P of Eq. (19.2). Thus if the eigenvalue–eigenvector decomposition of P is

$$P = VS^2V^T$$

with eigenvalues, $s_{11}^2 \geq s_{22}^2 \geq \cdots \geq s_{nn}^2$, one can compute

$$g^{(1)} = S^{-1}V^Td$$
$$\gamma = (\omega^2 - \|g^{(1)}\|^2)^{1/2}$$

where d and ω^2 are defined by Eq. (19.8). Then with the change of variables

$$x = Vp$$

Problem LS is equivalent to the problem

(19.19)
$$\begin{bmatrix} S \\ 0 \end{bmatrix}p \cong \begin{bmatrix} g^{(1)} \\ \gamma \end{bmatrix}$$

This least squares problem is equivalent to problem (18.37) obtained from singular value analysis since the diagonal matrix S and the n-vector $g^{(1)}$ of Eq. (19.19) are the same as in Eq. (18.37) and γ of Eq. (19.19) and $g^{(2)}$ of Eq. (18.37) satisfy $\gamma = \|g^{(2)}\|$.

For a given arithmetic precision η, however, the quantities $S, g^{(1)}$, and $\gamma = \|g^{(2)}\|$ will be determined more accurately by singular value analysis of $[A : b]$ or of the triangular matrix $R = Q[A : b]$ obtained by Householder triangularization than by an eigenvalue analysis of $[A : b]^T[A : b]$.

As mentioned in the introduction of this chapter, using arithmetic of a fixed relative precision η, a larger class of problems can be solved using Householder transformations than can be solved by forming normal equations and using the Cholesky decomposition. This can be illustrated by considering the following example.

Define the 3×2 matrix

$$A = \begin{bmatrix} 1 & 1 \\ 1 & 1 \\ 1 & 1 - \epsilon \end{bmatrix}$$

Suppose the value of ϵ is such that it is significant to the problem, $\epsilon > 100\eta$, say, but $\epsilon^2 < \eta$, so that when computing with relative precision η we have $1 - \epsilon \neq 1$ but $3 + \epsilon^2$ is computed as 3. Thus instead of computing

$$A^T A = \begin{bmatrix} 3 & 3 - \epsilon \\ 3 - \epsilon & 3 - 2\epsilon + \epsilon^2 \end{bmatrix}$$

we shall compute the matrix

$$\begin{bmatrix} 3 & 3 - \epsilon \\ 3 - \epsilon & 3 - 2\epsilon \end{bmatrix}$$

plus some random errors of order of magnitude 3η.

Further computation of the upper triangular matrix

$$R = \begin{bmatrix} r_{11} & r_{12} \\ 0 & r_{22} \end{bmatrix}$$

that satisfies $R^T R = A^T A$ yields, using the Cholesky method [Eq. (19.12) to (19.14)],

$$r_{11} = \sqrt{3}$$

$$r_{12} = \frac{3 - \epsilon}{\sqrt{3}}$$

$$r_{22}^2 = p_{22} - r_{12}^2$$

This final equation is computed as

$$r_{22}^2 = 3 - 2\epsilon - \frac{9 - 6\epsilon}{3} = 0$$

plus random errors of the order of magnitude 3η and of arbitrary sign. The correct result at this point would be of course

$$r_{22}^2 = 3 - 2\epsilon + \epsilon^2 - \frac{9 - 6\epsilon + \epsilon^2}{3} = \frac{2\epsilon^2}{3}$$

Thus using arithmetic of relative precision η, no significant digits are obtained in the element r_{22}. Consequently the matrix R is not distinguishable from a singular matrix. This difficulty is not a defect of the Cholesky decomposition algorithm but rather reflects the fact that the column vectors of $A^T A$ are so nearly parallel (linearly dependent) that the fact that they are not parallel cannot be established using arithmetic of relative precision η.

Contrast this situation with the use of Householder transformations to triangularize A directly without forming $A^T A$. Using Eq. (10.5) to (10.11) one would compute

$$s = -\sqrt{3}$$

$$u = \begin{bmatrix} 1 + \sqrt{3} \\ 1 \\ 1 \end{bmatrix}$$

$$b = -\sqrt{3}(1 + \sqrt{3}) = -(3 + \sqrt{3})$$

Then to premultiply the second column, a_2, of A by $I + b^{-1} uu^T$ one computes

$$t = b^{-1}(u^T a_2) = -\frac{3 + \sqrt{3} - \epsilon}{3 + \sqrt{3}} = \frac{\epsilon(3 - \sqrt{3}) - 6}{6}$$

and

$$\tilde{a}_2 = a_2 + tu = \begin{bmatrix} 1 \\ 1 \\ 1 - \epsilon \end{bmatrix} + \frac{\epsilon(3 - \sqrt{3}) - 6}{6} \begin{bmatrix} 1 + \sqrt{3} \\ 1 \\ 1 \end{bmatrix}$$

$$= \begin{bmatrix} \dfrac{\epsilon - 3}{\sqrt{3}} \\[2mm] \dfrac{\epsilon(3 - \sqrt{3})}{6} \\[2mm] \dfrac{-\epsilon(3 + \sqrt{3})}{6} \end{bmatrix}$$

The important point to note is that the second and third components of $\tilde{a}_2$, which are of order of magnitude ϵ, were computed as the difference of quantities of order of magnitude unity. Thus, using η-precision arithmetic, these components are not lost in the round-off error.

The final step of Householder triangularization would be to replace the second components of $\tilde{a}_2$ by the negative euclidean norm of the second and third components, and this step involves no numerical difficulties.

$$r_{22} = -\left\{ \left[\frac{\epsilon(3 - \sqrt{3})}{6} \right]^2 + \left[\frac{-\epsilon(3 + \sqrt{3})}{6} \right]^2 \right\}^{1/2}$$

$$= \frac{-\epsilon\sqrt{6}}{3}$$

Collecting results we obtain

$$R = \begin{bmatrix} -\sqrt{3} & \dfrac{\epsilon - 3}{\sqrt{3}} \\ 0 & \dfrac{-\epsilon\sqrt{6}}{3} \end{bmatrix}$$

plus *absolute* errors of the order of magnitude 3η.

In concluding this example we note that, using η-precision arithmetic, applying Householder transformations directly to A produces a triangular matrix that is clearly nonsingular, whereas the formation of normal equations followed by Cholesky decomposition does not.

Section 2. MODIFIED GRAM–SCHMIDT ORTHOGONALIZATION

Gram–Schmidt orthogonalization is a classical mathematical technique for solving the following problem: Given a linearly independent set of vectors $\{a_1, \ldots, a_n\}$, produce a mutually orthogonal set of vectors $\{q_1, \ldots, q_n\}$ such that for $k = 1, \ldots, n$, the set $\{q_1, \ldots, q_k\}$ spans the same k-dimensional subspace as the set $\{a_1, \ldots, a_k\}$. The classical set of mathematical formulas expressing q_j in terms of a_j and the previously determined vectors $q_1, \ldots, q_{j-1}$ appear as follows:

$$(19.20) \qquad\qquad q_1 = a_1$$

$$(19.21) \qquad q_j = a_j - \sum_{i=1}^{j-1} r_{ij} q_i \qquad j = 2, \ldots, n$$

where

$$(19.22) \qquad\qquad r_{ij} = \frac{a_j^T q_i}{q_i^T q_i}$$

To convert to matrix notation, define A to be the matrix with columns a_j, Q to be the matrix with columns q_j, and R to be the upper triangular matrix with unit diagonal elements with the strictly upper triangular elements given by Eq. (19.22). Then Eq. (19.20) and (19.21) can be written as

$$(19.23) \qquad\qquad A = QR$$

It is apparent that Gram–Schmidt orthogonalization can be regarded as being another method for decomposing a matrix into the product of a matrix with orthogonal columns and a triangular matrix.

Experimental evidence [Rice (1966)] indicated that Eq. (19.20) to (19.22) have significantly less numerical stability than a certain mathematically equivalent variant. The stability of this modified Gram–Schmidt method was established by Björck (1967a), who obtained the expressions given below as Eq. (19.35) and (19.36).

Note that the value of the inner product $(a_j^T q_i)$ will not change if a_j is replaced by any vector of the form

$$(19.24) \qquad a_j - \sum_{k=1}^{i-1} \alpha_k q_k$$

since $q_k^T q_i = 0$ for $k \neq i$. In particular, it is recommended that Eq. (19.22) be replaced by

$$(19.25) \qquad r_{ij} = \frac{a_j^{(i)T} q_i}{q_i^T q_i}$$

where

$$(19.26) \qquad a_j^{(i)} = a_j - \sum_{k=1}^{i-1} r_{kj} q_k$$

In fact, if one poses the problem of choosing the numbers α_k to minimize the norm of the vector defined by expression (19.24), it is easily verified that the minimal length vector is given by Eq. (19.26). Thus the vector a_j of Eq. (19.22) and the vector $a_j^{(i)}$ of Eq. (19.25) are related by the inequality $\| a_j^{(i)} \| \leq \| a_j \|$. The superior numerical stability of this modified algorithm derives from this fact.

The amount of computation and storage required is not increased by this modification since the vectors $a_j^{(i)}$ can be computed recursively rather than using Eq. (19.26) explicitly. Furthermore, the vector $a_j^{(i+1)}$ can replace the vector $a_j^{(i)}$ in storage.

This algorithm, which has become known as the *modified Gram–Schmidt* (*MGS*) algorithm [Rice (1966)], is described by the following equations:

$$(19.27) \qquad a_j^{(1)} = a_j \qquad j = 1, \ldots, n$$

$$(19.28) \qquad q_i = a_i^{(i)}$$

$$(19.29) \qquad d_i^2 = q_i^T q_i$$

$$(19.30) \qquad r_{ij} = \frac{a_j^{(i)T} q_i}{d_i^2} \left. \begin{array}{c} \\ \\ \end{array} \right\} \quad \left. \begin{array}{c} \\ j = i+1, \ldots, n \\ \end{array} \right\} \quad i = 1, \ldots, n$$

$$(19.31) \qquad a_j^{(i+1)} = a_j^{(i)} - r_{ij} q_i$$

To use MGS in the solution of Problem LS one can form the augmented matrix

$$(19.32) \qquad \tilde{A} = [A : b]$$

and apply MGS to the $m \times (n + 1)$ matrix $\tilde{A}$ to obtain

$$(19.33) \qquad \tilde{A} = \tilde{Q} \tilde{R}$$

where the matrix $\tilde{R}$ is upper triangular with unit diagonal elements. The strictly upper triangular elements of $\tilde{R}$ are given by Eq. (19.30). The vectors q_i given by Eq. (19.28) constitute the column vectors of the $m \times (n + 1)$ matrix $\tilde{Q}$. One also obtains the $(n + 1) \times (n + 1)$ diagonal matrix $\tilde{D}$ with diagonal elements d_i, $i = 1, \ldots, n + 1$, given by Eq. (19.29).

For the purpose of mathematical discussion leading to the solution of Problem LS, let Q be an $m \times m$ orthogonal matrix satisfying

$$Q_{m \times m} \begin{bmatrix} \tilde{D} \\ 0 \end{bmatrix} \begin{matrix} \}n + 1 \\ \}m - n - 1 \end{matrix} = \tilde{Q}_{m \times (n+1)}$$
$$\underbrace{}_{n + 1}$$

Then

(19.34)
$$\tilde{A} = Q \begin{bmatrix} \tilde{D} \\ 0 \end{bmatrix} \tilde{R}$$

$$= Q \begin{bmatrix} DR & Dc \\ 0 & d_{n+1} \\ 0 & 0 \end{bmatrix}$$

where $\tilde{D}$ and $\tilde{R}$ have been partitioned as

$$\tilde{R} = \begin{bmatrix} R & c \\ 0 & 1 \end{bmatrix} \begin{matrix} \}n \\ \}1 \end{matrix}$$
$$\underbrace{}_{n} \; \underbrace{}_{1}$$

and

$$\tilde{D} = \begin{bmatrix} D & 0 \\ 0 & d_{n+1} \end{bmatrix} \begin{matrix} \}n \\ \}1 \end{matrix}$$
$$\underbrace{}_{n} \; \underbrace{}_{1}$$

Then for arbitrary x we have

$$\| Ax - b \|^2 = \| Q^T (Ax - b) \|^2$$
$$= \| D(Rx - c) \|^2 + d_{n+1}^2$$

Therefore the minimum value of the quantity $\| Ax - b \|$ is $|d_{n+1}|$ and this value is attained by the vector $\hat{x}$, which satisfies the triangular system

$$Rx = c$$

For a given precision of arithmetic MGS has about the same numerical stability as Householder triangularization. The error analysis given in Björck (1967a) obtains the result that the solution to $Ax \cong b$ computed using MGS with mixed precision (η, η^2) is the exact solution to a perturbed problem $(A + E)x \cong b + f$ with

$$(19.35) \qquad \qquad \|E\|_F \leq 2n^{3/2} \|A\|_F \eta$$

and

$$(19.36) \qquad \qquad \|f\| \leq 2n^{3/2} \|b\| \eta$$

Published tests of computer programs [e.g., Wampler (1969)] have found MGS and Householder programs to be of essentially equivalent accuracy.

The number of arithmetic operations required is somewhat larger for MGS than for Householder (see Table 19.1) because in MGS one is always dealing with column vectors of length m, whereas in Householder triangularization one deals with successively shorter columns. This fact also means that programs for MGS usually require more storage than Householder programs, since it is not as convenient in MGS to produce the R matrix in the storage initially occupied by the matrix A. With some extra work of moving quantities in storage, however, MGS can be organized to produce R^T in the storage initially occupied by A if one is not saving the q_i vectors.

The MGS method can be organized to accumulate rows or blocks of rows of $[A : b]$ sequentially to handle cases in which the product mn is very large and $m \gg n$. This possibility rests on the fact that the $(n + 1) \times (n + 1)$ matrix $(\tilde{D}\tilde{R})$ [see Eq. (19.34)] defines the same least squares problem as the $m \times (n + 1)$ matrix $[A : b]$. This fact can be used to develop a sequential MGS algorithm in the same way as the corresponding property of the Householder triangularization is used in Chapter 27 to develop a sequential Householder method.

Specialized adaptations of Gram–Schmidt orthogonalization include the conjugate gradient method of solving a positive definite set of linear equations [Hestenes and Stiefel (1952); Beckman, appearing in pp. 62–72 of Ralston and Wilf (1960); Kammerer and Nashed (1972)] that has some attractive features for large sparse problems [Reid (1970), (1971a), and (1971b)] and the method of Forsythe (1957) for polynomial curve fitting using orthogonalized polynomials.

EXERCISES

(19.37) Let R be an $n \times n$ upper triangular matrix with some diagonal elements $r_{ii} = 0$. Show that there exists an $n \times n$ upper triangular matrix U such that $U^T U = R^T R$ and $u_{ij} = 0$, $j = i, \ldots, n$, for all values of i for which $r_{ii} = 0$. *Hint:* Construct U in the form $U = QR$, where Q is the product of a finite sequence of appropriately chosen Givens rotation matrices.

(19.38) [Peters and Wilkinson (1970)] Assume Rank $(A_{m \times n}) = n$ and $m > n$. By Gaussian elimination (assuming partial pivoting) the decomposition $A = PLR$ can be obtained where $L_{m \times n}$ is lower triangular with unit diagonal elements, $R_{n \times n}$ is upper triangular, and P is a permutation matrix accounting for the row interchanges.

(a) Count operations needed to compute this decomposition.

(b) Problem LS can be solved by solving $L^T L y = L^T P^T b$ by the Cholesky method and then solving $Rx = y$. Count the operations needed to form $L^T L$ and to compute its Cholesky factorization.

(c) Show that the total operation count of this method is the same as for the Householder method (see Table 19.1). (We suggest determining only the coefficients of mn^2 and n^3 in these operation counts.)

(19.39) Let A be an $m \times n$ matrix of rank n. Let $A = Q \begin{bmatrix} R \\ 0 \end{bmatrix}$ be the decomposition of A obtained by the (exact) application of the modified Gram–Schmidt algorithm to A. Assume Q and R are adjusted so that the columns of Q have unit euclidean length. Consider the (exact) application of the algorithm HFT (11.4) to the $(m + n) \times n$ matrix $\begin{bmatrix} 0_{n \times n} \\ A_{m \times n} \end{bmatrix}$ and let $\begin{bmatrix} B_{n \times n} \\ C_{m \times n} \end{bmatrix}$ denote the data that replace $\begin{bmatrix} 0 \\ A \end{bmatrix}$ in storage after execution of HFT. Show that B is the same as R to within signs of rows and C is the same as Q to within signs of columns.

(19.40) (Cholesky Method Without Square Roots) In place of Eq. (19.5) consider the decomposition $W^T D W = P$, where W is upper triangular with unit diagonal elements and d is diagonal. Derive formulas analogous to Eq. (19.12) to (19.14) for computing the diagonal elements of D and the strictly upper triangular elements of W. Show that using this decomposition of $\tilde{P}$ [Eq. (19.8)] problem (19.1) can be solved without computing any square roots.

20

LINEAR LEAST SQUARES WITH LINEAR EQUALITY CONSTRAINTS USING A BASIS OF THE NULL SPACE

In this chapter we begin the consideration of least squares problems in which the variables are required to satisfy specified linear equality or inequality constraints. Such problems arise in a variety of applications. For example, in fitting curves to data, equality constraints may arise from the need to interpolate some data or from a requirement for adjacent fitted curves to match with continuity of the curves and possibly of some derivatives. Inequality constraints may arise from requirements such as positivity, monotonicity, and convexity.

We shall refer to the linear equality constrained least squares problem as Problem LSE. The problem with linear inequality, as well as possibly linear equality constraints, will be called Problem LSI. Three distinct algorithms for solving Problem LSE will be given here and in Chapters 21 and 22, respectively. Problem LSI will be treated in Chapter 23.

To establish our notation we state Problem LSE as follows:

(20.1)   PROBLEM LSE   (*Least Squares with Equality Constraints*)

Given an $m_1 \times n$ *matrix* C *of rank* k_1, *an* m_1-*vector* d, *an* $m_2 \times n$ *matrix* E, *and an* m_2-*vector* f, *among all* n-*vectors* x *that satisfy*

(20.2) $$Cx = d$$

find one that minimizes

(20.3) $$\| Ex - f \|$$

Clearly Problem LSE has a solution if and only if Eq. (20.2) is consistent. We shall assume the consistency of Eq. (20.2) throughout Chapters 20, 21, 22, and 23. In the usual practical cases of this problem one would have

134

$n > m_1 = k_1$, which would assure that Eq. (20.2) is consistent and has more than one solution.

It will subsequently be shown that, if a solution for Problem LSE exists, it is unique if and only if the augmented matrix $\begin{bmatrix} C \\ E \end{bmatrix}$ is of rank n. In the case of nonuniqueness there is a unique solution of minimal length.

Clearly Problem LSE could be generalized to the case in which Eq. (20.2) is inconsistent but is interpreted in a least squares sense. We have not seen this situation arise in practice; however, our discussion of Problem LSE would apply to that case with little modification.

The three algorithms to be given for Problem LSE are each compact in the sense that no two-dimensional arrays of computer storage are needed beyond that needed to store the problem data. Each of the three methods can be interpreted as having three stages:

1. Derive a lower-dimensional unconstrained least squares problem from the given problem.
2. Solve the derived problem.
3. Transform the solution of the derived problem to obtain the solution of the original constrained problem.

The first method [Hanson and Lawson (1969)], which will be described in this chapter, makes use of an orthogonal basis for the null space of the matrix of the constraint equations. This method uses both postmultiplication and premultiplication of the given matrix $\begin{bmatrix} C \\ E \end{bmatrix}$ by orthogonal transformation matrices.

If the problem does not have a unique solution, this method has the property that the (unique) minimum length solution of the derived unconstrained problem gives rise to the (unique) minimum length solution of the original constrained problem. Thus this first method is suggested for use if it is expected that the problem may have deficient pseudorank and if stabilization methods (see Chapter 25) based on the notion of limiting the length of the solution vector are to be used.

This method is amenable to numerically stable updating techniques for solving a sequence of LSE problems in which the matrix C is successively changed by adding or deleting rows. The method is used in this way by Stoer (1971) in his algorithm for Problem LSI.

Furthermore, this first method is of theoretical interest. In Theorem (20.31) this method is used to show the existence of an unconstrained least squares problem which is equivalent to Problem LSE in the sense of having the same set of solutions for any given right-side vectors d and f. This permits the application to Problem LSE of certain computational procedures, such as singular value analysis.

The second method, presented in Chapter 21, uses direct elimination by premultiplication using both orthogonal and nonorthogonal transformation matrices.

Either of the first two methods is suitable for use in removing equality constraints (if any) as a first step in solving Problem LSI. Also, either method is adaptable to sequential data processing in which new rows of data are adjoined to either the constraint system $[C: d]$ or the least squares system $[E: f]$ after the original data have already been triangularized.

The third method for Problem LSE, presented in Chapter 22, illuminates some significant properties of the Householder transformation applied to a least squares problem with disparately weighted rows. From a practical point of view the main value of this third method is the fact that it provides a way of solving Problem LSE in the event that one has access to a subroutine implementing the Householder algorithm for the unconstrained least squares problem but one does not have and does not choose to produce code implementing one of the specific algorithms for Problem LSE.

We turn now to our first method for Problem LSE. This method makes use of an explicit parametric representation of the members of the linear flat

$$(20.4) \qquad\qquad X = \{x : Cx = d\}$$

If

$$(20.5) \qquad\qquad C = HRK^T$$

is any orthogonal decomposition (see Chapter 2 for definition) of C, and K is partitioned as

$$(20.6) \qquad\qquad K = [\underset{k_1}{\underbrace{K_1,}} \quad \underset{n-k_1}{\underbrace{K_2]}}\}n$$

then from Theorem (2.3), Eq. (2.8), it follows that X may be represented as

$$(20.7) \qquad\qquad X = \{x : x = \bar{x} + K_2 y_2\}$$

where

$$(20.8) \qquad\qquad \bar{x} = C^+ d$$

and y_2 ranges over the space of all vectors of dimension $n - k_1$.

(20.9)   THEOREM

> *Assuming Eq. (20.2) is consistent, Problem LSE has a unique minimal length solution given by*

(20.10) $$\hat{x} = C^+d + (EZ)^+(f - EC^+d)$$

where

$$Z = I_n - C^+C$$

or equivalently

(20.11) $$\hat{x} = C^+d + K_2(EK_2)^+(f - EC^+d)$$

where K_2 is defined by Eq. (20.6).

The vector $\hat{x}$ is the unique solution vector for Problem LSE if and only if Eq. (20.2) is consistent and the rank of $\begin{bmatrix} C \\ E \end{bmatrix}$ is n.

Proof: To establish the equivalence of Eq. (20.10) and (20.11) note first that

$$Z = I_n - C^+C = KK^T - KR^+RK^T$$

$$= KK^T - K\begin{bmatrix} I_{k_1} & 0 \\ 0 & 0 \end{bmatrix}K^T = KK^T - K_1K_1^T$$

$$= K_2K_2^T$$

and thus

$$EZ = EK_2K_2^T$$

The fact that

$$(EK_2K_2^T)^+ = K_2(EK_2)^+$$

can be verified by directly checking the four Penrose conditions [Theorem (7.9)] and recalling that

$$K_2^TK_2 = I_{n-k_1}$$

Thus Eq. (20.10) and (20.11) are equivalent.

Using Eq. (20.7) it is seen that the problem of minimizing the expression of Eq. (20.3) over all $x \in X$ is equivalent to finding an $(n - k_1)$-vector y_2 that minimizes

$$\| E(\bar{x} + K_2y_2) - f \|$$

or equivalently

(20.12) $$\|(EK_2)y_2 - (f - E\bar{x})\|$$

From Eq. (7.5) the unique minimal length solution vector for this problem is given by

(20.13) $$\hat{y}_2 = (EK_2^T)^+(f - E\bar{x})$$

Thus using Eq. (20.7) a solution vector for Problem LSE is given by

$$(20.14) \qquad \hat{x} = \bar{x} + K_2 \hat{y}_2$$

which is equivalent to Eq. (20.11) and thus to Eq. (20.10) also.

Since the column vectors of K_2 are mutually orthogonal, and also orthogonal to $\bar{x}$, the norm of any vector $x \in X$ satisfies

$$(20.15) \qquad \|x\|^2 = \|\bar{x}\|^2 + \|y_2\|^2$$

If $\tilde{y}_2 \neq \hat{y}_2$ also minimizes Eq. (20.10), then $\|\tilde{y}_2\| > \|\hat{y}_2\|$. Thus $\tilde{x} = \bar{x} + K_2 \tilde{y}_2$ is a solution to LSE but

$$(20.16) \qquad \|\tilde{x}\|^2 = \|\bar{x}\|^2 + \|\tilde{y}_2\|^2 > \|\bar{x}\|^2 + \|\hat{y}_2\|^2 = \|\hat{x}\|^2$$

It follows that $\hat{x}$ is the unique minimum length solution vector for Problem LSE.

It remains to relate the uniqueness of the solution of Problem LSE to $k = \text{Rank}\left(\begin{bmatrix} C \\ E \end{bmatrix}\right)$. Clearly if $k < n$ there exists an n-vector $w \neq 0$ satisfying

$$(20.17) \qquad \begin{bmatrix} C \\ E \end{bmatrix} w = 0$$

and if $\hat{x}$ is a solution of Problem LSE, so is $\hat{x} + w$.

On the other hand, if $k = n$, consider the $(m_1 + m_2) \times n$ matrix

$$(20.18) \qquad M = \begin{bmatrix} C \\ E \end{bmatrix} K = \begin{bmatrix} CK_1 & CK_2 \\ EK_1 & EK_2 \end{bmatrix} \begin{matrix} \}m_1 \\ \}m_2 \end{matrix}$$
$$\underbrace{}_{k_1} \underbrace{}_{n-k_1}$$

whose rank is also n. Having only n columns, all columns must be independent. Since $CK_2 = 0$, it follows that EK_2 must be of rank $n - k_1$. Thus $\hat{y}$ of Eq. (20.13) is the unique vector minimizing Eq. (20.12), and consequently $\hat{x}$ of Eq. (20.14) is the unique solution of Problem LSE.

This completes the proof of Theorem (20.9).

A computational procedure can be based on Eq. (20.11). We shall assume $k_1 = m_1$ since this is the usual case in practice. Then the matrix H of Eq. (20.5) can be taken as the identity matrix.

Let K be an $n \times n$ orthogonal matrix that when postmultiplied times C transforms C to a lower triangular matrix. Postmultiply C and E by K; thus

(20.19)
$$\begin{bmatrix} C \\ E \end{bmatrix} K = \begin{bmatrix} \tilde{C}_1 & 0 \\ \tilde{E}_1 & \tilde{E}_2 \end{bmatrix} \begin{matrix} \}m_1 \\ \}m_2 \end{matrix}$$
$$\underbrace{}_{m_1} \underbrace{}_{n - m_1}$$

where $\tilde{C}_1$ is $m_1 \times m_1$ lower triangular and nonsingular.
Solve the lower triangular system

(20.20)
$$\tilde{C}_1 y_1 = d$$

for the m_1-vector $\hat{y}_1$. Compute

(20.21)
$$\tilde{f} = f - \tilde{E}_1 \hat{y}_1$$

Solve the least squares problem

(20.22)
$$\tilde{E}_2 y_2 \cong \tilde{f}$$

for the $(n - m_1)$-vector $\hat{y}_2$. Finally, compute the solution vector,

(20.23)
$$x = K \begin{bmatrix} \hat{y}_1 \\ \hat{y}_2 \end{bmatrix}$$

The actual computational steps are described by means of the following algorithm. Initially the matrices and vectors C, d, E, and f are stored in arrays of the same names. The array g is of length m_1. The arrays h, u, and p are each of length $n - m_1$. The solution of minimal length is returned in the storage array, x, of length n. The parameter τ is a user provided tolerance parameter needed at Step 6.

(20.24) ALGORITHM LSE(C, d, E, f, m_1, m_2, n, g, h, u, p, x, τ)

Step	Description
1	For $i := 1, \ldots, m_1$, execute Algorithm H1(i, $i + 1$, n, c_{i1}, g_i, $c_{i+1,1}$, $m_1 - i$). [See Eq. (20.19). Here and in Step 2 Algorithms H1 and H2 operate on *rows* of the arrays C and E.]
2	For $i := 1, \ldots, m_1$, execute Algorithm H2(i, $i + 1$, n, c_{i1}, g_i, e_{11}, m_2). [See Eq. (20.19). If the matrices C and E were stored in a single $(m_1 + m_2) \times n$ array, say W, then Steps 1 and 2 could be combined as: For $i := 1, \ldots, m_1$ execute Algorithm H1(i, $i + 1$, n, w_{i1}, g_i, $w_{i+1,1}$, $m_1 + m_2 - i$).]
3	Set $x_1 := d_1 / c_{11}$. If $m_1 = 1$, go to Step 5.

Step	Description
4	For $i := 2, \ldots, m_1$, set

$$x_i := \left(d_i - \sum_{j=1}^{i-1} c_{ij}x_j \right) \bigg/ c_{ii}.$$

[See Eq. (20.20).]

| 5 | For $i := 1, \ldots, m_2$, set |

$$f_i := f_i - \sum_{j=1}^{m_1} e_{ij}x_j.$$

[See Eq. (20.21).]

| 6 | Execute Algorithm HFTI(e_{1,m_1+1}, m_2, $n - m_1$, f, τ, x_{m_1+1}, h, u, p) of Eq. (14.9). [The solution $\hat{y}_2$ of Eq. (20.22) has been computed and stored in locations x_i, $i := m_1 + 1$, $\ldots, n$.] |

| 7 | For $i := m_1, m_1 - 1, \ldots, 1$, execute Algorithm H2($i$, $i + 1$, n, c_{i1}, g_i, x, 1). [See Eq. (20.23). Here Algorithm H2 refers to *rows* of the array C and applies transformations to the singly subscripted array, x.] |

| 8 | *Comment:* The array named x now contains the minimal length solution of Problem LSE. |

As an example of Problem LSE, consider the minimization of $\| Ex - f \|$ subject to $Cx = d$ where

$$(20.25) \qquad C = [0.4087 \quad 0.1593]$$

$$(20.26) \qquad E = \begin{bmatrix} 0.4302 & 0.3516 \\ 0.6246 & 0.3384 \end{bmatrix}$$

$$(20.27) \qquad d = 0.1376$$

$$(20.28) \qquad f = \begin{bmatrix} 0.6593 \\ 0.9666 \end{bmatrix}$$

The Householder orthogonal matrix, K, which triangularizes C from the right is

$$K = \begin{bmatrix} -0.9317 & -0.3632 \\ -0.3632 & 0.9317 \end{bmatrix}$$

Of course this matrix would not usually be computed explicitly in executing Algorithm LSE (20.24). Using Eq. (20.19) to (20.23) we compute

$$\begin{bmatrix} \tilde{C}_1 & 0 \\ \tilde{E}_1 & \tilde{E}_2 \end{bmatrix} = \begin{bmatrix} -0.4386 & 0.0 \\ -0.5285 & 0.1714 \\ -0.7049 & 0.0885 \end{bmatrix}$$

$$\hat{y}_1 = \frac{0.1376}{-0.4386} = -0.3137$$

$$\tilde{f} = \begin{bmatrix} 0.4935 \\ 0.7455 \end{bmatrix}$$

$$\hat{y}_2 = 4.0472$$

(20.29)
$$\hat{x} = \begin{bmatrix} -1.1775 \\ 3.8848 \end{bmatrix}$$

We turn now to some theoretical consequences of Theorem (20.9). Equation (20.10) can be rewritten as

$$\hat{x} = \hat{A}^+ b$$

if we define

(20.30)
$$\hat{A}^+ = [C^+ - (EZ)^+ EC^+ : (EZ)^+]$$
$$= [C^+ - K_2(EK_2)^+ EC^+ : K_2(EK_2)^+]$$

and

$$b = \begin{bmatrix} d \\ f \end{bmatrix}$$

But Eq. (20.30) implies that $\hat{x}$ is the unique minimum length solution to the problem of minimizing

$$\| \hat{A}x - b \|$$

where $\hat{A}$ is the pseudoinverse of the matrix $\hat{A}^+$ defined in Eq. (20.30).

(20.31) THEOREM

The pseudoinverse of the matrix $\hat{A}^+$ *defined in Eq. (20.30) is*

(20.32)
$$\hat{A} = \begin{bmatrix} C \\ \hat{E} \end{bmatrix}$$

where

(20.33) $\hat{E} = (EZ)(EZ)^+ E = E(EZ)^+ E = (EK_2)(EK_2)^+ E$

and Z *and* K_2 *are defined as in Theorem (20.9).*

Proof: Recall that

$$CK_2 = 0$$
$$K_2^T C^+ = 0$$
$$K_2^T K_2 = I_{n-m_1}$$
$$Z = I_n - C^+C = K_2 K_2^T$$

and

$$(EZ)^+ = (EK_2 K_2^T)^+ = K_2 (EK_2)^+$$

Other relations that follow directly from these and will be used in the proof are

$$EZC^+ = EK_2 K_2^T C^+ = 0$$
$$C(EZ)^+ = CK_2 (EK_2)^+ = 0$$

and

(20.34)
$$E(EZ)^+ = EK_2 (EK_2)^+$$
$$= EK_2 K_2^T K_2 (EK_2)^+$$
$$= (EZ)(EZ)^+$$

Equation (20.34) establishes the equality of the different expressions for $\hat{E}$ in Eq. (20.33).

Next define

$$X = \begin{bmatrix} C \\ (EZ)(EZ)^+ E \end{bmatrix}$$

and verify the four Penrose equations [Theorem (7.9)] as follows:

$$\hat{A}^+ X = C^+C - (EZ)^+ EC^+C + (EZ)^+ (EZ)(EZ)^+ E$$
$$= C^+C - (EZ)^+ E(C^+C - I)$$
$$= C^+C + (EZ)^+ (EZ)$$

which is symmetric.

$$X\hat{A}^+ = \begin{bmatrix} CC^+ - C(EZ)^+ EC^+ & C(EZ)^+ \\ (EZ)(EZ)^+ E[I - (EZ)^+ E]C^+ & (EZ)(EZ)^+ E(EZ)^+ \end{bmatrix}$$
$$= \begin{bmatrix} CC^+ & 0 \\ 0 & (EZ)(EZ)^+ \end{bmatrix}$$

which is symmetric.

$$\hat{A}^+X\hat{A}^+ = [C^+C + (EZ)^+(EZ)]\cdot[C^+ - (EZ)^+EC^+:(EZ)^+]$$
$$= \{C^+ - (EZ)^+(EZ)[I - (EZ)^+E]C^+:(EZ)^+(EZ)(EZ)^+\}$$
$$= [C^+ - (EZ)^+(EZ)(EZ)^+EC^+:(EZ)^+]$$
$$= [C^+ - (EZ)^+EC^+:(EZ)^+] = \hat{A}^+$$

and

$$XÂ^+X = \begin{bmatrix} C \\ (EZ)(EZ)^+E \end{bmatrix} \cdot [C^+C + (EZ)^+(EZ)]$$

$$= \begin{bmatrix} C \\ (EZ)(EZ)^+EC^+C + (EZ)(EZ)^+E(EZ)^+EZ \end{bmatrix}$$

$$= \begin{bmatrix} C \\ (EZ)(EZ)^+E[C^+C + Z] \end{bmatrix} = \begin{bmatrix} C \\ (EZ)(EZ)^+E \end{bmatrix} = X$$

This completes the proof of Theorem (20.31).

As a numerical illustration of Theorem (20.31) we evaluate Eq. (20.32) and (20.33) for the data (20.25) to (20.28). From Eq. (20.33) we have

$$\hat{E} = (EK_2)(EK_2)^+E$$
$$= \begin{bmatrix} 0.1714 \\ 0.0885 \end{bmatrix}[4.6076 \quad 2.3787]\begin{bmatrix} 0.4302 & 0.3516 \\ 0.6246 & 0.3384 \end{bmatrix}$$
$$= \begin{bmatrix} 0.5943 & 0.4155 \\ 0.3068 & 0.2145 \end{bmatrix}$$

and from Eq. (20.32)

$$\hat{A} = \begin{bmatrix} C \\ \hat{E} \end{bmatrix} = \begin{bmatrix} 0.4087 & 0.1593 \\ 0.5943 & 0.4155 \\ 0.3068 & 0.2145 \end{bmatrix}$$

According to Theorem (20.31) this matrix has the property that for an arbitrary one-dimensional vector, $\bar{d}$, and two-dimensional vector, $\bar{f}$, the solution of the least squares problem

$$\hat{A}x \cong \begin{bmatrix} \bar{d} \\ \bar{f} \end{bmatrix}$$

is also the solution of the following Problem LSE: Minimize $\|Ex - \bar{f}\|$ subject to $Cx = \bar{d}$ where C and E are defined by Eq. (20.25) and (20.26).

It should be noted, however, that the residual norms for these two least squares problems are not in general the same.

21

LINEAR LEAST SQUARES WITH LINEAR EQUALITY CONSTRAINTS BY DIRECT ELIMINATION

In this chapter Problem LSE, which was introduced in the preceding chapter, will be treated using a method of direct elimination by premultiplication. It will be assumed that the rank k_1 of C is m_1 and the rank of $\begin{bmatrix} C \\ E \end{bmatrix}$ is n. From Theorem (20.9) this assures that Problem LSE has a unique solution for any right-side vectors d and f.

It will be further assumed that the columns of $\begin{bmatrix} C \\ E \end{bmatrix}$ are ordered so that the first m_1 columns of C are linearly independent. *Column interchanges to achieve such an ordering would be a necessary part of any computational procedure of the type to be discussed in this chapter.*

Introduce the partitioning

$$(21.1) \qquad \begin{bmatrix} C \\ E \end{bmatrix} = \begin{bmatrix} C_1 & C_2 \\ E_1 & E_2 \end{bmatrix} \begin{matrix} \}m_1 \\ \}m_2 \end{matrix}$$
$$\qquad\qquad\qquad\quad \underbrace{}_{m_1} \underbrace{}_{n-m_1}$$

and

$$(21.2) \qquad x = \begin{bmatrix} x_1 \\ x_2 \end{bmatrix} \begin{matrix} \}m_1 \\ \}n-m_1 \end{matrix}$$

The constraint equation $Cx = d$ can be solved for x_1 and will give

$$x_1 = C_1^{-1}(d - C_2 x_2)$$

Substituting this expression for x_1 in $\| Ex - f \|$ gives

(21.3)
$$\| Ex - f \| = \| E_1 C_1^{-1}(d - C_2 x_2) + E_2 x_2 - f \|$$
$$= \| (E_2 - E_1 C_1^{-1} C_2) x_2 - (f - E_1 C_1^{-1} d) \|$$
$$\equiv \| \tilde{E}_2 x_2 - \tilde{f} \|$$

as the expression to be minimized to determine x_2.

Conceptually this leads to the following solution procedure: Compute

(21.4)
$$\tilde{E}_2 = E_2 - E_1 C_1^{-1} C_2$$

and

(21.5)
$$\tilde{f} = f - E_1 C_1^{-1} d$$

Solve the least squares problem

(21.6)
$$\tilde{E}_2 x_2 \cong \tilde{f}$$

and finally compute

(21.7)
$$x_1 = C_1^{-1}(d - C_2 x_2)$$

There are a variety of ways in which these steps can be accomplished. For example, following Björck and Golub (1967), suppose a QR decomposition of C_1 is computed so that

(21.8)
$$C_1 = Q_1^T \tilde{C}_1$$

where Q_1 is orthogonal and $\tilde{C}_1$ is upper triangular. Then Eq. (21.4) and (21.5) can be written as

(21.9)
$$\tilde{E}_2 = E_2 - (E_1 \tilde{C}_1^{-1})(Q_1 C_2) \equiv E_2 - \tilde{E}_1 \tilde{C}_2$$

and

(21.10)
$$\tilde{f} = f - (E_1 \tilde{C}_1^{-1})(Q_1 d) \equiv f - \tilde{E}_1 \tilde{d}$$

The resulting algorithm may be described as follows. Compute Householder transformations to triangularize C_1 and also apply these transformations to C_2 and d:

(21.11)
$$Q_1[C_1 : C_2 : d] = [\tilde{C}_1 : \tilde{C}_2 : \tilde{d}]$$

Compute the $m_2 \times m_1$ matrix $\tilde{E}_1$ as the solution of the triangular system

(21.12)
$$\tilde{E}_1 \tilde{C}_1 = E_1$$

Compute

(21.13)
$$\tilde{E}_2 = E_2 - \tilde{E}_1 \tilde{C}_2$$

and

(21.14) $$\tilde{f} = f - \tilde{E}_1 \tilde{d}$$

Compute Householder transformations to triangularize $\tilde{E}_2$ and also apply these transformations to $\tilde{f}$:

(21.15) $$Q_2[\tilde{E}_2 : \tilde{f}] = \begin{bmatrix} \hat{E}_2 & \hat{f} \\ 0 & \varphi \\ 0 & 0 \end{bmatrix} \begin{matrix} \}n - m_1 \\ \}1 \\ \}m_1 + m_2 - n - 1 \end{matrix}$$

Finally compute the solution vector $\begin{bmatrix} \hat{x}_1 \\ \hat{x}_2 \end{bmatrix}$ by solving the triangular system

(21.16) $$\begin{bmatrix} \tilde{C}_1 & \tilde{C}_2 \\ 0 & \tilde{E}_2 \end{bmatrix} \cdot \begin{bmatrix} x_1 \\ x_2 \end{bmatrix} = \begin{bmatrix} \tilde{d} \\ \hat{f} \end{bmatrix}$$

This algorithm requires no two-dimensional arrays of computer storage other than those required to store the initial data since quantities written with a tilde can overwrite the corresponding quantities written without a tilde and quantities written with a circumflex can overwrite the corresponding quantities written with a tilde.

The coding of this algorithm can also be kept remarkably compact, as is illustrated by the ALGOL procedure *decompose* given in Björck and Golub (1967). Suppose the data matrix $\begin{bmatrix} C \\ E \end{bmatrix}$ is stored in an $m \times n$ array, A, with $m = m_1 + m_2$ and the data vector $\begin{bmatrix} d \\ f \end{bmatrix}$ is stored in an m-array, b. Steps (21.11) and (21.15) can be accomplished by the same code, essentially Steps 2–12 of Algorithm HFTI (14.9). When executing Eq. (21.11) arithmetic operations are performed only on the first m_1 rows of $[A : b]$ but any interchanges of columns that are required must be done on the full m-dimensional columns of A.

Steps (21.12) to (21.14), which may be interpreted as Gaussian elimination, are accomplished by the operations

(21.17) $$a_{i1} := \frac{a_{i1}}{a_{11}}$$

(21.18) $$a_{ij} := \frac{a_{ij} - \sum_{k=1}^{j-1} a_{ik} a_{kj}}{a_{jj}} \quad j := 2, \ldots, m_1 \qquad \left.\begin{matrix} \\ \\ \end{matrix}\right\} \quad i := m_1 + 1, \\ \ldots, m$$

(21.19) $$a_{ij} := a_{ij} - \sum_{k=1}^{m_1} a_{ik} a_{kj} \quad j := m_1 + 1, \ldots, n$$

(21.20) $$b_i := b_i - \sum_{k=1}^{m_1} a_{ik} b_k$$

As a numerical example of this procedure, let us again consider Problem LSE with the data given in Eq. (20.25) to (20.28). For these data the matrix Q_1 of Eq. (21.11) can be taken to be the 1×1 identity matrix. Using Eq. (21.11) to (21.16) we compute

$$\tilde{E}_2 = \begin{bmatrix} 0.1839 \\ 0.0949 \end{bmatrix}$$

$$\tilde{f} = \begin{bmatrix} 0.5145 \\ 0.7563 \end{bmatrix}$$

$$\hat{x} = \begin{bmatrix} -1.1775 \\ 3.8848 \end{bmatrix}$$

22 LINEAR LEAST SQUARES WITH LINEAR EQUALITY CONSTRAINTS BY WEIGHTING

An observation that many statisticians and engineers have made is the following. Suppose one has a linear least squares problem where one would like some of the equations to be exactly satisfied. This can be accomplished approximately by weighting these constraining equations heavily and solving the resulting least squares system. Equivalently, those equations which one does not necessarily want to be exactly satisfied can be downweighted and the resulting least squares system can then be solved.

In this chapter we shall analyze this formal computational procedure for solving Problem LSE (20.1). Basically the idea is simple. *Compute the solution of the least squares problem:*

$$(22.1) \qquad \begin{bmatrix} C \\ \epsilon E \end{bmatrix} x \cong \begin{bmatrix} d \\ \epsilon f \end{bmatrix}$$

(using Householder's method, for example) for a "small" but nonzero value of ϵ. Then the solution $\tilde{x}(\epsilon)$ of the least squares problem (22.1) is "close" [in a sense made precise by inequality (22.38) and Eq. (22.37)] to the solution $\hat{x}$ of Problem LSE.

This general approach is of practical interest since some existing linear least squares subroutines or programs can effectively solve Problem LSE by means of solving the system of Eq. (22.1).

An apparent practical drawback of this idea is the fact that the matrix of problem (22.1) becomes arbitrarily poorly conditioned [assuming Rank (C) $< n$] as the "downweighting parameter" ϵ is made smaller. This observation certainly does limit the practicality of the idea if problem (22.1) is solved by the method of normal equations [see Eq. (19.1)]. Thus the normal equation

148

for problem (22.1) is

$$(C^T C + \epsilon^2 E^T E)x = C^T d + \epsilon^2 E^T f$$

and unless the matrices involved have very special structure, the computer representation of the matrix $C^T C + \epsilon^2 E^T E$ will be indistinguishable from the matrix $C^T C$ (and $C^T d + \epsilon^2 E^T f$ will be indistinguishable from $C^T d$) when ϵ is sufficiently small.

Nevertheless, Powell and Reid (1968a and 1968b) have found experimentally that a satisfactory solution to a disparately weighted problem such as Eq. (22.1) can be computed by Householder transformations if care is taken to introduce *appropriate row interchanges*. The reader is referred to Chapter 17, in which the principal theoretical results of Powell and Reid (1968a) are presented in the form of three theorems.

In Powell and Reid (1968a) the interchange rules preceding the construction of the kth Householder transformation are as follows. First, do the usual column interchange as in Algorithm HFTI (14.9), moving the column whose euclidean length from row k through m is greatest into column k. Next scan the elements in column k from row k through m. If the element of largest magnitude is in row l, interchange rows l and k.

This row interchange rule would generally be somewhat overcautious for use in problem (22.1). The main point underlying the analysis of Powell and Reid (1968a) is that a very damaging loss of numerical information occurs if the pivot element v_p [using the notation of Eq. (10.5) to (10.11)] is significant for the problem but of insignificant magnitude relative to some elements v_i, $i > p$. By "significant for the problem" we mean that the solution would be significantly changed if v_p were replaced by zero. But if $|v_p|$ is sufficiently small relative to some of the numbers $|v_i|$, $i > p$, then v_p will be computationally indistinguishable from zero in the calculation of s and u_p by Eq. (10.5) and (10.7), respectively.

Suppose the nonzero elements of C and E in problem (22.1) are of the same order of magnitude so that any large disparity in sizes of nonzero elements of the coefficient matrix of Eq. (22.1) is due to the small multiplier ϵ. Then the disastrous loss of numerical information mentioned above would generally occur only if the pivot element v_p [of Eq. (10.5)] were chosen from a row of ϵE (say, $v_p = \epsilon e_{ij}$) while some row of C (say, row l) has not yet been used as a pivot row and contains an element (say, c_{lj}) such that $|c_{lj}| \gg \epsilon |e_{ij}|$. This situation can be avoided by keeping the rows in the order indicated in Eq. (22.1) so that all rows of C are used as pivot rows before any row of ϵE is used. Here we have used the symbols ϵE and C to denote those matrices or their successors in the algorithm.

We now turn to an analysis of convergence of the solution vector of problem (22.1) to the solution vector of Problem LSE as $\epsilon \rightarrow 0$. First consider the special case of Problem LSE in which the matrix C is diagonal and E is an identity matrix. This case is easily understood, and it will be shown subsequently that the analysis of more general cases is reducible to this special case.

Define

$$(22.2) \qquad C = \begin{bmatrix} s_1 & & 0 & 0 & \cdots & 0 \\ & \cdot & & & & \\ & & \cdot & & & \\ 0 & & s_{m_1} & 0 & \cdots & 0 \end{bmatrix}_{m_1 \times n} \qquad s_1 \geq \cdots \geq s_{m_1} > 0$$

and

$$(22.3) \qquad\qquad\qquad E = I_n$$

Together the following two lemmas show that the solution of the down-weighted least squares problem (22.1) converges to the solution of Problem LSE, as $\epsilon \to 0$, for the special case of C and E as given in Eq. (22.2) and (22.3).

(22.4)   LEMMA

Suppose that C *and* E *are given in Eq. (22.2) and (22.3). Then the solution of Problem LSE using the notation of Eq. (20.2) and (20.3) is given by*

$$(22.5) \qquad\qquad \hat{x}_i = \begin{cases} \dfrac{d_i}{s_i} & i = 1, \ldots, m_1 \\ f_i & i = m_1 + 1, \ldots, n \end{cases}$$

The length of $E\hat{x} - f$ *is*

$$(22.6) \qquad \| E\hat{x} - f \| = \left[\left(f_1 - \frac{d_1}{s_1} \right)^2 + \cdots + \left(f_{m_1} - \frac{d_{m_1}}{s_{m_1}} \right)^2 \right]^{1/2}$$

(22.7)   LEMMA

Let C *and* E *be as in Eq. (22.2) and (22.3). The unique least squares solution of problem (22.1) is given by*

$$(22.8) \qquad \tilde{x}_i = \tilde{x}_i(\epsilon) = \begin{cases} \dfrac{d_i}{s_i} + \epsilon^2 \left(f_i - \dfrac{d_i}{s_i} \right) \left\{ s_i^2 \left[1 + \left(\dfrac{\epsilon}{s_i} \right)^2 \right] \right\}^{-1} & \\ & i = 1, \ldots, m_1 \\ f_i \qquad i = m_1 + 1, \ldots, n \end{cases}$$

Lemma (22.7) can be easily verified by forming and solving the normal equations $(C^T C + \epsilon^2 E^T E)x = (C^T d + \epsilon^2 E^T f)$. With the hypotheses present on C and E this system has a diagonal coefficient matrix from which Eq. (22.8) easily follows.

For convenience in expressing the difference between $\tilde{x}(\epsilon)$ and $\hat{x}$ define the vector-valued function $h(\epsilon)$ by

$$(22.9) \qquad \tilde{x}_i(\epsilon) - \hat{x}_i = \epsilon^2 h_i(\epsilon) \qquad i = 1, \ldots, n$$

Thus, using Eq. (22.5) and (22.8),

$$(22.10) \qquad h_i(\epsilon) = \begin{cases} \dfrac{s_i^{-2}(f_i - d_i/s_i)}{1 + (\epsilon/s_i)^2} & i = 1, \ldots, m_1 \\ 0 & i = m_1 + 1, \ldots, n \end{cases}$$

and

$$(22.11) \qquad |h_i(\epsilon)| \leq \begin{cases} s_i^{-2}\left|f_i - \dfrac{d_i}{s_i}\right| \leq s_{m_1}^{-2}\left|f_i - \dfrac{d_i}{s_i}\right| = s_{m_1}^{-2}|f_i - \hat{x}_i| \\ \qquad\qquad\qquad\qquad\qquad\qquad i = 1, \ldots, m_1 \\ 0 \quad i = m_1 + 1, \ldots, n \end{cases}$$

The residual vector for problem (22.1) is

$$(22.12) \qquad \begin{pmatrix} d \\ \epsilon f \end{pmatrix} - \begin{bmatrix} C \\ \epsilon E \end{bmatrix} \tilde{x}(\epsilon) = \begin{pmatrix} d \\ \epsilon f \end{pmatrix} - \begin{bmatrix} C \\ \epsilon E \end{bmatrix} [\hat{x} + \epsilon^2 h(\epsilon)]$$

$$= \begin{pmatrix} -\epsilon^2 C h(\epsilon) \\ \epsilon(f - E\hat{x}) - \epsilon^3 E h(\epsilon) \end{pmatrix}$$

Introduce the weighted euclidean norm

$$\overbrace{\qquad\qquad}^{m_1 \text{ terms}} \qquad \overbrace{\qquad\qquad}^{n \text{ terms}}$$
$$(22.13) \qquad \|\cdot\|_\epsilon = [(\cdot)^2 + \cdots + (\cdot)^2 + (\cdot/\epsilon)^2 + \cdots + (\cdot/\epsilon)^2]^{1/2}$$

then

$$(22.14) \qquad \left\| \begin{pmatrix} d \\ \epsilon f \end{pmatrix} - \begin{pmatrix} C \\ \epsilon E \end{pmatrix} \tilde{x}(\epsilon) \right\|_\epsilon^2 = \epsilon^4 \|C h(\epsilon)\|^2$$
$$+ \| f - E\hat{x} + \epsilon^2 E h(\epsilon) \|^2$$

From Eq. (22.10) we see that

$$\tilde{x}(\epsilon) \longrightarrow \hat{x} \quad \text{as } \epsilon \longrightarrow 0$$

while from Eq. (22.14) we see that

$$\left\| \begin{pmatrix} d \\ \epsilon f \end{pmatrix} - \begin{pmatrix} C \\ \epsilon E \end{pmatrix} \tilde{x}(\epsilon) \right\|_\epsilon \longrightarrow \| f - E\hat{x} \| \quad \text{as } \epsilon \longrightarrow 0$$

For practical computation one is interested in the question of how small ϵ must be to assure that $\tilde{x}(\epsilon)$ and $\hat{x}$ are indistinguishable when their components are represented as computer words with a relative precision of η. If all components of the vector $\hat{x}$ are nonzero, this condition is achieved by requiring that

$$(22.15) \qquad |\epsilon^2 h_i(\epsilon)| \le \eta |\hat{x}_i| \qquad i = 1, \ldots, n$$

This will be satisfied for all $|\epsilon| \le \epsilon_0$, where

$$(22.16) \qquad \epsilon_0^2 = \min_i \left\{ \frac{\eta s_i^2 |d_i/s_i|}{|f_i - d_i/s_i|} : f_i - d_i/s_i \ne 0 \right\}$$

Using Eq. (22.9) and (22.11) the vector norm of the difference between $\tilde{x}(\epsilon)$ and $\hat{x}$ is bounded as follows:

$$(22.17) \qquad \| \tilde{x}(\epsilon) - \hat{x} \| \le \frac{\epsilon^2 \| f - \hat{x} \|}{s_{m_1}^2}$$

Thus in order to have $\| \tilde{x}(\epsilon) - \hat{x} \| \le \eta \| \hat{x} \|$ it suffices to choose $|\epsilon| \le \epsilon_0$, where

$$(22.18) \qquad \epsilon_0^2 = \frac{\eta s_{m_1}^2 \| \hat{x} \|}{\| f - \hat{x} \|}$$

Note that Eq. (22.18) does not apply if $\| f - \hat{x} \| = 0$. In this case, however, Eq. (22.1) is consistent and has the same solution for all $\epsilon \ne 0$.

We now consider Problem LSE for the more general case in which C is $m_1 \times n$ of rank $k_1 \le m_1 < n$, E is $m_2 \times n$, and the matrix $[C^T : E^T]$ is of rank n. It is further assumed that the constraint equations $Cx = d$ are consistent.

By appropriate changes of variables we shall reduce this problem to the special one previously considered. First note that for $|\epsilon| < 1$ the least squares problem (22.1) is exactly equivalent to the least squares problem

$$(22.19) \qquad \begin{bmatrix} (1 - \epsilon^2)^{1/2} C \\ \epsilon E' \end{bmatrix} x \cong \begin{bmatrix} (1 - \epsilon^2)^{1/2} d \\ \epsilon f' \end{bmatrix}$$

where

$$(22.20) \qquad E' = \begin{bmatrix} C \\ E \end{bmatrix} \quad \text{and} \quad f' = \begin{bmatrix} d \\ f \end{bmatrix}$$

Next we introduce a uniform rescaling of problem (22.19), multiplying all coefficients and right-side components by $(1 - \epsilon^2)^{-1/2}$ to obtain the problem

$$(22.21) \qquad \begin{bmatrix} C \\ \rho E' \end{bmatrix} x \cong \begin{bmatrix} d \\ \rho f' \end{bmatrix}$$

where

$$p = \epsilon(1 - \epsilon^2)^{-1/2}$$

Introduce a QR decomposition of E':

$$(22.22) \qquad E' = Q^T \begin{bmatrix} R \\ 0 \end{bmatrix}$$

where Q is $(m_1 + m_2) \times (m_1 + m_2)$ and orthogonal and R is $n \times n$ and nonsingular. With

$$(22.23) \qquad g = Qf', \qquad Rx = y$$

we have the equivalent least squares problem

$$(22.24) \qquad \begin{matrix} m_1 \{ \\ n \{ \\ m_1 + m_2 - n \{ \end{matrix} \begin{bmatrix} CR^{-1} \\ pI_n \\ 0 \end{bmatrix} y \cong \begin{bmatrix} d \\ pg \end{bmatrix} \begin{matrix} \} m_1 \\ \} m_1 + m_2 \end{matrix}$$

Let

$$(22.25) \qquad CR^{-1} = USV^T$$

be a singular value decomposition of CR^{-1} [see Theorem (4.1)]. Since C is of rank k_1, CR^{-1} is also, and we have

$$S_{m_1 \times n} = \begin{bmatrix} s_1 & & & \\ & \cdot & & 0 \\ & & \cdot & \\ & & & s_{k_1} \\ & 0 & & 0 \end{bmatrix} \begin{matrix} \} k_1 \\ \\ \\ \} m_1 - k_1 \end{matrix}$$
$$\underbrace{}_{k_1} \quad \underbrace{}_{n - k_1}$$

where $s_1 \geq \cdots \geq s_{k_1} > 0$.

Partition g into two segments,

$$g = \begin{bmatrix} g_1 \\ g_2 \end{bmatrix} \begin{matrix} \} n \\ \} m_1 + m_2 - n \end{matrix}$$

Then with $V^T y = z$ the least squares problem of Eq. (22.24) is equivalent to the least squares problem

$$(22.26) \qquad \begin{bmatrix} S \\ pI_n \\ 0 \end{bmatrix} z \cong \begin{bmatrix} U^T d \\ pV^T g_1 \\ pg_2 \end{bmatrix}$$

Since $Cx = d$ is consistent,

$$(22.27) \qquad U^T d = \begin{bmatrix} d'_1 \\ \cdot \\ \cdot \\ \cdot \\ d'_{k_1} \\ 0 \\ \cdot \\ \cdot \\ \cdot \\ 0 \end{bmatrix}$$

Thus problem (22.1) has been transformed to problem (22.26), where the coefficient matrix consists of a zero matrix and two diagonal matrices, one of which is a scalar multiple of the identity matrix.

From Lemmas (22.4) and (22.7) it follows that as p tends to zero the solution $\tilde{z}(p)$ of problem (22.26) converges to the (unique) solution $\hat{z}$ of the problem

$$(22.28) \qquad \text{minimize} \quad \|z - V^T g_1\|^2 + \|g_2\|^2$$
$$\text{subject to} \quad Sz = U^T d$$

On the other hand, by means of the substitutions

$$(22.29) \qquad Rx = y$$

and

$$(22.30) \qquad V^T y = z$$

where R and V are defined by Eq. (22.22) and (22.25), respectively, Problem LSE (20.1) is converted to problem (22.28).

Thus using Eq. (22.29) and (22.30) we obtain

$$(22.31) \qquad \tilde{x}(p) = R^{-1} V \tilde{z}(p)$$

as the (unique) solution of problem (22.1) with $p = \epsilon(1 - \epsilon^2)^{-1/2}$ and

$$(22.32) \qquad \hat{x} = R^{-1} V \hat{z}$$

as the (unique) solution of Problem LSE (20.1).

To express the difference between $\tilde{x}(p)$ and $\hat{x}$, note that $\tilde{z}(p)$ and $\hat{z}$ satisfy

$$(22.33) \qquad \tilde{z}(p) - \hat{z} = p^2 h(p)$$

with the vector-valued function h given by Eq. (22.10) with appropriate changes of notation, and thus

$$(22.34) \qquad \tilde{x}(\rho) - \hat{x} = \rho^2 R^{-1} V h(\rho)$$

In order to apply the bounds in Eq. (22.11) using the quantities appearing in problem (22.26), define

$$(22.35) \qquad d' = U^T d$$

and

$$(22.36) \qquad g' = V^T g_1$$

Then, using Eq. (22.11) and (22.34),

$$\| \tilde{x}(\rho) - \hat{x} \| \le \rho^2 \| R^{-1} \| s_{k_1}^{-2} \left[\sum_{i=1}^{k_1} \left(g_i' - \frac{d_i'}{s_i} \right)^2 \right]^{1/2}$$

where g_i' and d_i' are the ith components of the vectors of Eq. (22.35) and (22.36), respectively.

For $\hat{x} \ne 0$, one obtains the relative precision $\| \tilde{x}(\rho) - \hat{x} \| \le \eta \| \hat{x} \|$ by choosing $0 < \rho \le \rho_0$ where

$$\rho_0^2 = \eta \| \hat{x} \| s_{k_1}^2 \left(\| R^{-1} \| \left[\sum_{i=1}^{k_1} \left(g_i' - \frac{d_i'}{s_i} \right)^2 \right]^{1/2} \right)^{-1}$$

Then defining

$$(22.37) \qquad \epsilon_0 = \frac{\rho_0}{(1 + \rho_0^2)^{1/2}}$$

the solution $\tilde{x}(\epsilon)$ of Eq. (22.1) satisfies

$$(22.38) \qquad \| \tilde{x}(\epsilon) - \hat{x} \| \le \eta \| \hat{x} \|$$

for ϵ satisfying $0 < \epsilon \le \epsilon_0$.

It should be emphasized that the artifact of considering the various extended problems (22.19), (22.21), (22.24), (22.26), and (22.28) and the associated coordinate changes is only for the purpose of *proving* that the solution of problem (22.1) converges to the solution of Problem LSE (20.1) as ϵ tends to zero. One should not lose sight of the fact that in any actual computational procedure based on these ideas, one can simply solve problem (22.1) directly, say, by Householder transformations.

Although the value of ϵ_0 given by Eq. (22.37) is an upper bound for $|\epsilon|$ in order to achieve full relative precision η, it involves quantities that one would not generally compute in solving problem (22.1). Note that there is *no positive lower bound* on permissible values of $|\epsilon|$ imposed by the mathematical analysis of the problem or the numerical stability of the Householder method as analyzed in Powell and Reid (1968a). The only practical lower limit is set by the computer's floating point underflow value, L, referred to in Chapter 15.

For example, consider a machine with $\eta = 10^{-8}$ and $L = 10^{-38}$. Suppose all nonzero data in the matrices C and E and the vectors d and f are approximately of unit magnitude.

In practice one would probably need to have $\epsilon < \eta^{1/2} = 10^{-4}$ and $\epsilon > L = 10^{-38}$. With this wide range available one might, for example, decide to set $\epsilon = 10^{-12}$.

As a numerical illustration of this weighted approach to solving Problem LSE, consider the data given in Eq. (20.25) to (20.28). We now formulate this as the weighted least squares problem

$$(22.39) \qquad \begin{bmatrix} 0.4087 & 0.1593 \\ 0.4302\epsilon & 0.3516\epsilon \\ 0.6246\epsilon & 0.3384\epsilon \end{bmatrix} \begin{bmatrix} x_1 \\ x_2 \end{bmatrix} \cong \begin{bmatrix} 0.1376 \\ 0.6593\epsilon \\ 0.9666\epsilon \end{bmatrix}$$

Problem (22.39) was solved on a UNIVAC 1108 computer using subroutine **HFTI** (Appendix C) with mixed precision (precisions: $\eta = 2^{-27} \doteq 10^{-8.1}$ and $\omega = 2^{-58} \doteq 10^{-17.5}$) for 40 values of ϵ ($\epsilon = 10^{-r}, r = 1, \ldots, 40$). Let $x^{(r)}$ denote the solution obtained when $\epsilon = 10^{-r}$.

Table 22.1

r	$\delta^{(r)}$
1	3.7×10^{-2}
2	3.7×10^{-4}
3	3.6×10^{-6}
4	6.3×10^{-8}
5	2.6×10^{-7}
6	9.1×10^{-8}
7	9.1×10^{-8}
8	1.2×10^{-7}
9	4.5×10^{-8}
10	5.8×10^{-8}
.	.
.	.
.	.
36	8.7×10^{-8}
37	3.6×10^{-9}
38 39 40	ϵ too small. Numbers multiplied by ϵ underflow to zero.

A "true" solution vector x was computed by the method of Chapter 20 using full double precision (10^{-18}) arithmetic. To 12 figures this vector was

$$\hat{x} = \begin{bmatrix} -1.17749898217 \\ 3.88476983058 \end{bmatrix}$$

The relative error of each solution $x^{(r)}$ was computed as

$$\delta^{(r)} = \frac{\| x^{(r)} - \hat{x} \|}{\| \hat{x} \|}$$

A selection of values of $\delta^{(r)}$ is given in Table 22.1.

From Table 22.1 we see that any value of ϵ from 10^{-4} to 10^{-37} gave reasonable single precision $(\eta \doteq 10^{-8.1})$ agreement with the "true" answer, $\hat{x}$.

EXERCISE

(22.40)   In Eq. (22.1) the matrix C is $m_1 \times n$, E is $m_2 \times n$, and $m = m_1 + m_2$. Let H_ϵ denote the first Householder matrix that would be constructed when triangularizing $\begin{bmatrix} C \\ \epsilon E \end{bmatrix}$. Define

$$J_\epsilon = \begin{bmatrix} I_{m_1} & 0 \\ 0 & \epsilon I_{m_2} \end{bmatrix}$$

Show that $\tilde{H} = \lim_{\epsilon \to 0} J_\epsilon^{-1} H_\epsilon J_\epsilon$ exists and give formulas for computing vectors u_1 and u_2 of dimensions m_1 and m_2, respectively, and a scalar b (as functions of the first column of $\begin{bmatrix} C \\ E \end{bmatrix}$) such that

$$\tilde{H} = I_m + b^{-1} \begin{bmatrix} u_1 \\ u_2 \end{bmatrix} [u_1^T, 0]$$

Show that the product matrix $\tilde{H} \begin{bmatrix} C \\ E \end{bmatrix}$ has zeros below the first element in the first column and thus that an appropriately constructed sequence of m_1 matrices like $\tilde{H}$ can be used to zero all elements below the diagonal in the first m_1 columns of $\begin{bmatrix} C \\ E \end{bmatrix}$. Show that this sequence of operations has the same effect as Eq. (21.11) to (21.14).

23

LINEAR LEAST SQUARES WITH LINEAR INEQUALITY CONSTRAINTS

Section 1. INTRODUCTION

There are many applications in applied mathematics, physics, statistics, mathematical programming, economics, control theory, social science, and other fields where the usual least squares problem must be reformulated by the introduction of certain inequality constraints. These constraints constitute additional information about a problem.

We shall concern ourselves with linear inequality constraints only. A variety of methods have been presented in the literature. We mention particularly papers that have given serious attention to the numerical stability of their methods: Golub and Saunders (1970), Gill and Murray (1970), and Stoer (1971).

The ability to consider least squares problems with linear inequality constraints allows us, in particular, to have such constraints on the solution as nonnegativity, or that each variable is to have independent upper and lower bounds, or that the sum of all the variables cannot exceed a specified value or that a fitted curve is to be monotone or convex.

Let E be an $m_2 \times n$ matrix, f an m_2-vector, G an $m \times n$ matrix, and h an m-vector. The least squares problem with linear inequality constraints will be stated as follows:

(23.1)   PROBLEM LSI

Minimize $\| Ex - f \|$ *subject to* $Gx \geq h$.

158

The following important special cases of Problem LSI will also be treated in detail:

(23.2)   PROBLEM NNLS (*Nonnegative Least Squares*)

   Minimize $\| Ex - f \|$ *subject to* $x \geq 0$.

(23.3)   PROBLEM LDP (*Least Distance Programming*)

   Minimize $\| x \|$ *subject to* $Gx \geq h$.

   Conditions characterizing a solution for Problem LSI are the subject of the Kuhn–Tucker theorem. This theorem is stated and discussed in Section 2 of this chapter.

   In Section 3 Problem NNLS is treated. A solution algorithm, also called NNLS, is presented. This algorithm is fundamental for the subsequent algorithms to be discussed in this chapter. A Fortran implementation of Algorithm NNLS is given in Appendix C as subroutine **NNLS**.

   In Section 4 it is shown that the availability of an algorithm for Problem NNLS makes possible an elegantly simple algorithm for Problem LDP. Besides stating Algorithm LDP in Section 4, a Fortran implementation, subroutine **LDP**, is given in Appendix C.

   The problem of determining whether or not a set of linear inequalities $Gx \geq h$ is consistent and if consistent finding some feasible vector arises in various contexts. Algorithm LDP can of course be used for this purpose. The fact that no assumptions need be made regarding the rank of G or the relative row and column dimensions of G may make this method particularly useful for some feasibility problems.

   In Section 5 the general problem LSI is treated by transforming it to Problem LDP. Problem LSI with equality constraints is treated in Section 6.

   Finally in Section 7 a numerical example of curve fitting with inequality constraints is presented as an application of these methods for handling constrained least squares problems. A Fortran program, **PROG6**, which carries out this example, is given in Appendix C.

Section 2.   CHARACTERIZATION OF A SOLUTION

   The following theorem characterizes the solution vector for Problem LSI:

(23.4)   THEOREM (*Kuhn–Tucker Conditions for Problem LSI*)

   An n-*vector* $\hat{x}$ *is a solution for Problem LSI* (*23.1*) *if and only if there exists an* m-*vector* $\hat{y}$ *and a partitioning of the integers 1 through* m *into subsets* $\mathcal{E}$ *and* $\mathcal{S}$ *such that*

(23.5) $$G^T \hat{y} = E^T(E\hat{x} - f)$$

(23.6) $\qquad \hat{r}_i = 0$ for $i \in \mathcal{E}$, $\qquad \hat{r}_i > 0$ for $i \in \mathcal{S}$

(23.7) $\qquad \hat{y}_i \geq 0$ for $i \in \mathcal{E}$, $\qquad \hat{y}_i = 0$ for $i \in \mathcal{S}$

   where

(23.8) $\qquad \qquad \qquad \hat{r} = G\hat{x} - h$

This theorem may be interpreted as follows. Let g_i^T denote the ith row vector of the matrix G. The ith constraint, $g_i^T x \geq h_i$, defines a feasible halfspace, $\{x : g_i^T x \geq h_i\}$. The vector g_i is orthogonal (normal) to the bounding hyperplane of this halfspace and is directed into the feasible halfspace. The point $\hat{x}$ is interior to the halfspaces indexed in $\mathcal{S}$ ($\mathcal{S}$ for *slack*) and on the boundary of the halfspaces indexed in $\mathcal{E}$ ($\mathcal{E}$ for *equality*).

The vector

$$p = E^T(E\hat{x} - f)$$

is the gradient vector of $\varphi(x) = \frac{1}{2}\| Ex - f \|^2$ at $x = \hat{x}$. Since $\hat{y}_i = 0$ for $i \notin \mathcal{E}$ Eq. (23.5) can be written as

(23.9) $\qquad \qquad \qquad \sum_{i \in \mathcal{E}} \hat{y}_i(-g_i) = -p$

which states that the negative gradient vector of φ at $\hat{x}$ is expressible as a nonnegative ($\hat{y}_i \geq 0$) linear combination of outward-pointing normals ($-g_i$) to the constraint hyperplanes on which $\hat{x}$ lies ($i \in \mathcal{E}$). Geometrically this means that the negative gradient vector $-p$ lies in the convex cone based at the point $\hat{x}$ and generated by the outward-pointing normals $-g_i$.

Any perturbation u of $\hat{x}$ such that $\hat{x} + u$ remains feasible must satisfy $u^T g_i \geq 0$ for all $i \in \mathcal{E}$. Multiplying both sides of Eq. (23.9) by such a vector u^T and using the fact that the $\hat{y}_i \geq 0$, it follows that u also satisfies $u^T p \geq 0$. From the identity $\varphi(\hat{x} + u) = \varphi(\hat{x}) + u^T p + \| Eu \|^2 / 2$, it follows that no feasible perturbation of $\hat{x}$ can reduce the value of φ.

The vector $\hat{y}$ (or the negative of this vector) which occurs in the Kuhn–Tucker theorem is sometimes called the *dual* vector for the problem. Further discussion of this theorem, including its proof, will be found in the literature on constrained optimization [see, e.g., Fiacco and McCormick (1968), p. 256].

Section 3. PROBLEM NNLS

Problem NNLS is defined by statement (23.2). We shall state Algorithm NNLS for solving Problem NNLS. The finite convergence of this algorithm will be proved.

We are initially given the $m_2 \times n$ matrix E, the integers m_2 and n, and the m_2-vector f. The n-vectors w and z provide working space. Index sets $\mathcal{P}$

and Z will be defined and modified in the course of execution of the algorithm. Variables indexed in the set Z will be held at the value zero. Variables indexed in the set $\mathcal{P}$ will be free to take values different from zero. If such a variable takes a nonpositive value, the algorithm will either move the variable to a positive value or else set the variable to zero and move its index from set $\mathcal{P}$ to set Z.

On termination x will be the solution vector and w will be the dual vector.

(23.10)   ALGORITHM NNLS($E, m_2, n, f, x, w, z, \mathcal{P}, Z$)

Step	Description
1	Set $\mathcal{P} :=$ NULL, $Z := \{1, 2, \ldots, n\}$, and $x := 0$.
2	Compute the n-vector $w := E^T(f - Ex)$.
3	If the set Z is empty or if $w_j \leq 0$ for all $j \in Z$, go to Step 12.
4	Find an index $t \in Z$ such that $w_t = \max\{w_j : j \in Z\}$.
5	Move the index t from set Z to set $\mathcal{P}$.
6	Let $E_\mathcal{P}$ denote the $m_2 \times n$ matrix defined by

$$\text{Column } j \text{ of } E_\mathcal{P} := \begin{cases} \text{column } j \text{ of } E \text{ if } j \in \mathcal{P} \\ 0 \quad \text{if } j \in Z \end{cases}$$

Compute the n-vector z as a solution of the least squares problem $E_\mathcal{P}z \cong f$. Note that only the components $z_j, j \in \mathcal{P}$, are determined by this problem. Define $z_j := 0$ for $j \in Z$.

7	If $z_j > 0$ for all $j \in \mathcal{P}$, set $x := z$ and go to Step 2.
8	Find an index $q \in \mathcal{P}$ such that $x_q/(x_q - z_q) = \min \{x_j/(x_j - z_j): z_j \leq 0, j \in \mathcal{P}\}$.
9	Set $\alpha := x_q/(x_q - z_q)$.
10	Set $x := x + \alpha(z - x)$.
11	Move from set $\mathcal{P}$ to set Z all indices $j \in \mathcal{P}$ for which $x_j = 0$. Go to Step 6.
12	*Comment:* The computation is completed.

On termination the solution vector x satisfies

(23.11) $x_j > 0 \quad j \in \mathcal{P}$

(23.12) $x_j = 0 \quad j \in Z$

and is a solution vector for the least squares problem

(23.13) $E_\mathcal{P}x \cong f$

The dual vector w satisfies

(23.14) $w_j = 0 \qquad j \in \mathcal{P}$

(23.15) $w_j \leq 0 \qquad j \in \mathbb{Z}$

and

(23.16) $w = E^T(f - Ex)$

Equations (23.11), (23.12), (23.14), (23.15), and (23.16) constitute the Kuhn–Tucker conditions [see Theorem (23.4)] characterizing a solution vector x for Problem NNLS. Equation (23.13) is a consequence of Eq. (23.12), (23.14), and (23.16).

Before discussing the convergence of Algorithm NNLS it will be convenient to establish the following lemma:

(23.17) LEMMA

Let A be an $m \times n$ matrix of rank n and let b be an m-vector satisfying

(23.18)
$$A^T b = \begin{bmatrix} 0 \\ \cdot \\ \cdot \\ \cdot \\ 0 \\ \omega \end{bmatrix} \begin{matrix} \left.\right\} n-1 \\ \\ \left.\right\} 1 \end{matrix}$$

with

(23.19) $\omega > 0$

If $\hat{x}$ is the least squares solution of $Ax \cong b$, then

(23.20) $\hat{x}_n > 0$

where $\hat{x}_n$ denotes the nth component of $\hat{x}$.

Proof: Let Q be an $m \times m$ orthogonal matrix that zeros the subdiagonal elements in the first $n - 1$ columns of A, thus

(23.21)
$$Q[\underset{n}{\underbrace{A}} : \underset{1}{\underbrace{b}}] = \begin{bmatrix} R & s & u \\ 0 & t & v \end{bmatrix} \begin{matrix} \}n-1 \\ \}m-n+1 \end{matrix}$$
$$\underset{n-1 \quad 1 \quad 1}{}$$

where R is upper triangular and nonsingular.

Since Q is orthogonal the conditions (23.18) imply

(23.22) $R^T u = 0$

and

(23.23) $s^T u + t^T v = \omega > 0$

Since R is nonsingular, Eq. (23.22) implies that $u = 0$. Thus Eq. (23.23) reduces to

(23.24) $t^T v = \omega > 0$

From Eq. (23.21) it follows that the nth component $\hat{x}_n$ of the solution vector $\hat{x}$ is the least squares solution of the reduced problem

(23.25) $t x_n \cong v$

Since the pseudoinverse of the column vector t is $t^T/(t^T t)$, the solution of problem (23.25) can be immediately written as

(23.26) $$\hat{x}_n = \frac{t^T v}{t^T t} = \frac{\omega}{t^T t} > 0$$

which completes the proof of Lemma (23.17).

Algorithm NNLS may be regarded as consisting of a main loop, Loop A, and an inner loop, Loop B. Loop B consists of Steps 6–11 and has a single entry point at Step 6 and a single exit point at Step 7.

Loop A consists of Steps 2–5 and Loop B. Loop A begins at Step 2 and exits from Step 3.

At Step 2 of Loop A the set $\mathcal{P}$ identifies the components of the current vector x that are positive. The components of x indexed in Z are zero at this point.

In Loop A the index t selected at Step 4 selects a coefficient not presently in set $\mathcal{P}$ that will be positive [by Lemma (23.17)] if introduced into the solution. This coefficient is brought into the tentative solution vector z at Step 6 in Loop B. If all other components of z indexed in set $\mathcal{P}$ remain positive, then at Step 7 the algorithm sets $x := z$ and returns to the beginning of Loop A. In this process set $\mathcal{P}$ is augmented and set Z is diminished by the transfer of the index t.

In many examples this sequence of events simply repeats with the addition of one more positive coefficient on each iteration of Loop A until the termination test at Step 3 is eventually satisfied.

However, if some coefficient indexed in set $\mathcal{P}$ becomes zero or negative in the vector z at Step 6, then Step 7 causes the algorithm to remain in Loop B performing a move that replaces x by $x + \alpha(z - x)$, $0 < \alpha \leq 1$, where α is chosen as large as possible subject to keeping the new x nonnegative. Loop B is repeated until it eventually exits successfully at Step 7.

The finiteness of Loop B can be proved by showing that all operations within Loop B are well defined, that at least one more index, the index called q at that point, is removed from set $\mathcal{P}$ each time Step 11 is executed, and that z_t is always positive [Lemma (23.17) applies here]. Thus exit from Loop B at Step 7 must occur after not more than $\pi - 1$ iterations within Loop B, where π denotes the number of indices in set $\mathcal{P}$ when Loop B was entered. In practice Loop B usually exits immediately on reaching Step 7 and does not reach Steps 8—11 at all.

Finiteness of Loop A can be proved by showing that the residual norm function

$$\rho(x) = \| f - Ex \|$$

has a strictly smaller value each time Step 2 is reached and thus that at Step 2 the vector x and its associated set $\mathcal{P} = \{i: x_i > 0\}$ are distinct from all previous instances of x and $\mathcal{P}$ at Step 2. Since $\mathcal{P}$ is a subset of the set $\{1, 2, \ldots, n\}$ and there are only a finite number of such subsets, Loop A must terminate after a finite number of iterations. In a set of small test cases it was observed that Loop A typically required about $\frac{1}{2}n$ iterations.

Updating the QR Decomposition of E

The least squares problem being solved at Step 6 differs from the problem previously solved at Step 6 either due to the addition of one more column of E into the problem at Step 5 or the deletion of one or more columns of E at Step 11. Updating techniques can be used to compute the QR decomposition for the new problem based upon retaining the QR decomposition of the previous problem. Three updating methods are described in Chapter 24. The third of these methods has been used in the Fortran subroutine **NNLS** (Appendix C).

Coping with Finite Precision Arithmetic

When Step 6 is executed immediately after Step 5 the component z_t computed during Step 6 will theoretically be positive. If z_t is not positive, as can happen due to round-off error, the algorithm may attempt to divide by zero at Step 8 or may incorrectly compute $\alpha = 0$ at Step 9.

This can be avoided by testing z_t following Step 6 whenever Step 6 has been entered directly from Step 5. If $z_t \leq 0$ at this point, it can be interpreted to mean that the number w_t computed at Step 2 and tested at Steps 3 and 4 should be taken to be zero rather than positive. Thus one can set $w_t := 0$ and loop back to Step 2. This will result either in termination at Step 3 or the assignment of a new value to t at Step 4.

At Step 11 any x_i whose computed value is negative (which can only be due to round-off error) should be treated as being zero by moving its index from set $\mathcal{P}$ to set Z.

The sign tests on z_i, $i \in \mathcal{P}$, at Steps 7 and 8 do not appear to be critical. The consequences of a possible misclassification here do not seem to be damaging.

A Fortran subroutine **NNLS** implementing Algorithm NNLS and using these ideas for enhancing the numerical reliability appears in Appendix C.

Section 4. PROBLEM LDP

The solution vector for Problem LDP (23.3) can be obtained by an appropriate normalization of the residual vector in a related Problem NNLS (23.2). This method of solving Problem LDP and its verification was brought to the authors' attention by Alan Cline.

We are given the $m \times n$ matrix G, the integers m and n, and the $\tilde{m}$-vector h. If the inequalities $Gx \geq h$ are compatible, then the algorithm will set the logical variable $\varphi = $ TRUE and compute the vector $\hat{x}$ of minimal norm satisfying these inequalities. If the inequalities are incompatible, the algorithm will set $\varphi = $ FALSE and no value will be assigned to $\hat{x}$. Arrays of working space needed by this algorithm are not explicitly indicated in the parameter list.

(23.27)   ALGORITHM LDP(G, m, n, h, $\hat{x}$, φ)

Step	Description
1	Define the $(n + 1) \times m$ matrix E and the $(n + 1)$-vector f as $E := \begin{bmatrix} G^T \\ h^T \end{bmatrix}$ and $f := [\overbrace{0, \ldots, 0}^{n}, 1]^T$. Use Algorithm NNLS to compute an m-vector $\hat{u}$ solving Problem NNLS: Minimize $\| Eu - f \|$ subject to $u \geq 0$.
2	Compute the $(n + 1)$-vector $r := E\hat{u} - f$.
3	If $\| r \| = 0$, set $\varphi := $ FALSE and go to Step 6.
4	Set $\varphi := $ TRUE.
5	For $j := 1, \ldots, n$, compute $\hat{x}_j := -r_j/r_{n+1}$.
6	The computation is completed.

Proof of Validity of Algorithm LDP

First consider the Problem NNLS, which is solved in Step 1 of Algorithm LDP. The gradient vector for the objective function $\frac{1}{2}\| Eu - f \|^2$ at the solution point $\hat{u}$ is

(23.28) $p = E^T r$

From the Kuhn–Tucker conditions [Theorem (23.4)] for this Problem NNLS there exist disjoint index sets $\mathcal{E}$ and $\mathcal{S}$ such that

(23.29) $$\mathcal{E} \cup \mathcal{S} = \{1, 2, \ldots, m\}$$

(23.30) $$\hat{u}_i = 0 \text{ for } i \in \mathcal{E}, \qquad \hat{u}_i > 0 \text{ for } i \in \mathcal{S}$$

and

(23.31) $$p_i \geq 0 \text{ for } i \in \mathcal{E}, \qquad p_i = 0 \text{ for } i \in \mathcal{S}$$

Using Eq. (23.28) to (23.31) we obtain

(23.32) $$\|r\|^2 = r^T r = r^T[E\hat{u} - f]$$
$$= p^T \hat{u} - r_{n+1} = -r_{n+1}$$

Consider the case in which $\|r\| > 0$ at Step 3. From Eq. (23.32) this implies that $r_{n+1} < 0$, so division by r_{n+1} at Step 5 is valid. Using Eq. (23.31) and (23.32) and the equations of Steps 2 and 5, we establish the feasibility of $\hat{x}$ as follows:

$$0 \leq p = E^T r$$

(23.33) $$= [G : h] \begin{bmatrix} \hat{x} \\ -1 \end{bmatrix}(-r_{n+1})$$

$$= (G\hat{x} - h)\|r\|^2$$

Therefore,

(23.34) $$G\hat{x} \geq h$$

From Eq. (23.31) and (23.33) it follows that the rows of the system of inequalities of Eq. (23.34) indexed in set $\mathcal{S}$ are satisfied with equality. The gradient vector for the objective function $\frac{1}{2}\|x\|^2$ of Problem LDP is simply x. The Kuhn–Tucker conditions for $\hat{x}$ to minimize $\frac{1}{2}\|x\|^2$ subject to $Gx \geq h$ require that the gradient vector $\hat{x}$ must be representable as a nonnegative linear combination of the rows of G that are associated with equality conditions in Eq. (23.34), i.e., the rows of G indexed in set $\mathcal{S}$.

From Steps 2 and 5 and Eq. (23.32) we have

$$\hat{x} = \begin{bmatrix} r_1 \\ \cdot \\ \cdot \\ \cdot \\ r_n \end{bmatrix}(-r_{n+1})^{-1} = G^T \hat{u}(-r_{n+1})^{-1}$$

$$= G^T \hat{u}\|r\|^{-2}$$

Noting the sign conditions on $\hat{u}$ given in Eq. (23.30) completes the proof that $\hat{x}$ is a solution of Problem LDP.

It is clearly the unique solution vector since, if $\tilde{x}$ is a different solution vector, then $\|\tilde{x}\| = \|\hat{x}\|$ and the vector $\bar{x} = \frac{1}{2}(\tilde{x} + \hat{x})$ would be a feasible vector having a strictly smaller norm than $\hat{x}$, which contradicts the fact that $\hat{x}$ is a feasible vector of minimum norm.

Now consider the case of $\|r\| = 0$ at Step 3. We must show that the inequalities $Gx \geq h$ are inconsistent. Assume the contrary, i.e., that there exists a vector $\tilde{x}$ satisfying $G\tilde{x} \geq h$. Define

$$q = G\tilde{x} - h = [G:h]\begin{bmatrix} \tilde{x} \\ -1 \end{bmatrix} \geq 0$$

Then

$$0 = [\tilde{x}^T : -1]r = [\tilde{x}^T : -1]\left\{ \begin{bmatrix} G^T \\ h^T \end{bmatrix} \hat{u} - f \right\}$$

$$= q^T\hat{u} + 1$$

This last expression cannot be zero, however, because $q \geq 0$ and $\hat{u} \geq 0$. From this contradiction we conclude that the condition $\|r\| = 0$ implies the inconsistency of the system $Gx \geq h$. This completes the mathematical verification of Algorithm LDP.

Section 5. CONVERTING PROBLEM LSI
TO PROBLEM LDP

Let k denote the rank of the $m_2 \times n$ matrix E of Problem LSI (23.1). In various ways as described in Chapters 2 to 4 one can obtain an orthogonal decomposition of the matrix E:

$$(23.35) \qquad E = Q\begin{bmatrix} R & 0 \\ 0 & 0 \end{bmatrix}K^T \equiv [Q_1 : \underbrace{Q_2}_{m_2 - k}]\begin{matrix} \\ k \end{matrix}\begin{bmatrix} R & 0 \\ 0 & 0 \end{bmatrix}\begin{bmatrix} K_1^T \\ K_2^T \end{bmatrix}\begin{matrix} \}k \\ \}n-k \end{matrix}$$

where Q is $m_2 \times m_2$ orthogonal, K is $n \times n$ orthogonal, and R is $k \times k$ nonsingular. Furthermore, the matrix R can be obtained in triangular or diagonal form.

Introduce the orthogonal change of variables

$$(23.36) \qquad\qquad\qquad x = K_1 y$$

The objective function to be minimized in Problem LSI can then be written as

$$(23.37) \qquad \varphi(x) = \| f - Ex \|^2 = \left\| \begin{bmatrix} Q_1^T f \\ Q_2^T f \end{bmatrix} - \begin{bmatrix} Ry \\ 0 \end{bmatrix} \right\|^2$$

$$= \| \tilde{f}_1 - Ry \|^2 + \| \tilde{f}_2 \|^2$$

where

$$(23.38) \qquad \tilde{f}_i = Q_i^T f \qquad i = 1, 2$$

With a further change of variables,

$$(23.39) \qquad z = Ry - \tilde{f}_1$$

we may write

$$(23.40) \qquad \varphi(x) = \| z \|^2 + \| \tilde{f}_2 \|^2$$

The original problem LSI of minimizing $\| f - Ex \|$ subject to $Gx \geq h$ is thus equivalent, except for the additive constant $\| \tilde{f}_2 \|^2$ in the objective function, to the following Problem LDP:

$$(23.41) \qquad \text{Minimize } \| z \|$$

$$\text{subject to } GK_1 R^{-1} z \geq h - GK_1 R^{-1} \tilde{f}_1$$

If a vector $\hat{z}$ is computed as a solution of this Problem LDP, then a solution vector $\hat{x}$ for the original Problem LSI can be computed from Eq. (23.39) and (23.36). The squared residual vector norm for the original problem can be computed from Eq. (23.40).

Section 6. PROBLEM LSI WITH EQUALITY CONSTRAINTS

Consider Problem LSI (23.1) with the addition of a system of equality constraints, say $C_{m_1 \times n} x = d$, with $\text{Rank}(C) = m_1 < n$. These constraint equations can be eliminated initially with a corresponding reduction in the number of independent variables. Either the method of Chapter 20 or that of Chapter 21 is suitable for this purpose.

Using the method of Chapter 20, introduce the orthogonal change of variables

$$(23.42) \qquad x = K \begin{bmatrix} y_1 \\ y_2 \end{bmatrix} \begin{matrix} \}m_1 \\ \}n - m_1 \end{matrix}$$

where K triangularizes C from the right:

$$(23.43) \qquad \begin{bmatrix} C \\ E \\ G \end{bmatrix} K = \begin{bmatrix} \tilde{C}_1 & 0 \\ \tilde{E}_1 & \tilde{E}_2 \\ \tilde{G}_1 & \tilde{G}_2 \end{bmatrix}$$

$$\underbrace{\phantom{\tilde{G}_1}}_{m_1} \quad \underbrace{\phantom{\tilde{G}_2}}_{n - m_1}$$

Then $\hat{y}_1$ is determined as the solution of the lower triangular system $\tilde{C}_1 y_1 = d$, and $\hat{y}_2$ is the solution of the following Problem LSI:

$$(23.44) \qquad \text{Minimize} \, \| \tilde{E}_2 y_2 - (f - \tilde{E}_1 \hat{y}_1) \|$$
$$\text{subject to } \tilde{G}_2 y_2 \geq h - \tilde{G}_1 y_1$$

After solving problem (23.44) for $\hat{y}_2$ the solution $\hat{x}$ can be computed using Eq. (23.42).

If the method of Chapter 21 is used, one would compute $Q_1, \tilde{C}_1, \tilde{C}_2, \tilde{d}, \tilde{E}_1,$ $\tilde{E}_2$, and $\tilde{f}$ using Eq. (21.11) to (21.14) and additionally solve for the matrix $\tilde{G}_1$ in the upper triangular system

$$(23.45) \qquad \tilde{G}_1 \tilde{C}_1 = G_1$$

Then $\hat{x}_2$ is the solution of the following Problem LSI:

$$(23.46) \qquad \text{Minimize} \, \| \tilde{E}_2 x_2 - \tilde{f} \|$$
$$\text{subject to } (G_2 - \tilde{G}_1 \tilde{C}_2) x_2 \geq h - \tilde{G}_1 \tilde{d}$$

and $\hat{x}_1$ would be computed by solving the upper triangular system

$$(23.47) \qquad \tilde{C}_1 x_1 = \tilde{d} - \tilde{C}_2 \hat{x}_2$$

Section 7. AN EXAMPLE OF CONSTRAINED CURVE FITTING

As an example illustrating a number of the techniques that have been described in this chapter we consider a problem of fitting a straight line to a set of data points where the line must satisfy certain constraints. **PROG6**, a Fortran main program that performs the computation for this example, is given in Appendix C.

Let the data be given as follows:

t	w
0.25	0.5
0.50	0.6
0.50	0.7
0.80	1.2

We wish to find a line of the form

(23.48) $$f(t) = x_1 t + x_2$$

which fits these data in a least squares sense subject to the constraints

(23.49) $$f'(t) \geq 0$$
(23.50) $$f(0) \geq 0$$
(23.51) $$f(1) \leq 1$$

This problem can be written as Problem LSI:

(23.52) $$\text{Minimize } \| Ex - f \|$$
$$\text{subject to } Gx \geq h$$

where

$$E = \begin{bmatrix} 0.25 & 1 \\ 0.50 & 1 \\ 0.50 & 1 \\ 0.80 & 1 \end{bmatrix}, \quad f = \begin{bmatrix} 0.5 \\ 0.6 \\ 0.7 \\ 1.2 \end{bmatrix}$$

$$G = \begin{bmatrix} 1 & 0 \\ 0 & 1 \\ -1 & -1 \end{bmatrix}, \quad h = \begin{bmatrix} 0 \\ 0 \\ -1 \end{bmatrix}$$

We compute an orthogonal decomposition of the matrix E in order to convert the Problem LSI to a Problem LDP as described in Section 5 of this chapter. Either a QR or a singular value decomposition of E could be used. We shall illustrate the use of a singular value decomposition.

$$E = U_{4 \times 4} \begin{bmatrix} S_{2 \times 2} \\ 0_{2 \times 2} \end{bmatrix} V_{2 \times 2}^T$$

$$S = \begin{bmatrix} 2.255 & 0.0 \\ 0.0 & 0.346 \end{bmatrix}$$

$$V = \begin{bmatrix} -0.467 & 0.884 \\ -0.884 & -0.467 \end{bmatrix}$$

$$\begin{matrix} 2\{ \\ 2\{ \end{matrix} \begin{bmatrix} \tilde{f}_1 \\ \tilde{f}_2 \end{bmatrix} = U^T f = \begin{bmatrix} -1.536 \\ 0.384 \\ -0.054 \\ 0.174 \end{bmatrix}$$

Introduce the change of variables

$$(23.53) \qquad\qquad z = SV^T x - \tilde{f}_1$$

We then wish to solve the following Problem LDP:

$$(23.54) \qquad\qquad \text{Minimize } \|z\|$$
$$\text{subject to } \tilde{G}z \geq \tilde{h}$$

where

$$\tilde{G} = GVS^{-1} = \begin{bmatrix} -0.207 & 2.558 \\ -0.392 & -1.351 \\ 0.599 & -1.206 \end{bmatrix}$$

and

$$\tilde{h} = h - \tilde{G}\tilde{f}_1 = \begin{bmatrix} -1.300 \\ -0.084 \\ 0.384 \end{bmatrix}$$

A graphical interpretation of this Problem LDP is given in Fig. 23.1. Each row of the augmented matrix $[\tilde{G} : \tilde{h}]$ defines one boundary line of the feasible region. The solution point $\hat{z}$ is the point of minimum euclidean norm

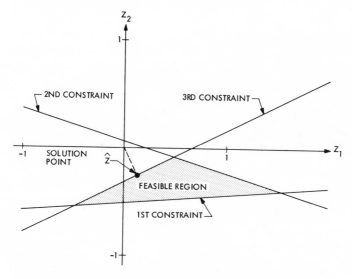

Fig. 23.1 Graphical interpretation of the sample Problem LDP (23.54).

within the feasible region. This point, as computed by subroutine **LDP**, is

$$\hat{z} = \begin{bmatrix} 0.127 \\ -0.255 \end{bmatrix}$$

Then using Eq. (23.53) we finally compute

$$\hat{x} = VS^{-1}(\hat{z} + \tilde{f}_1) = \begin{bmatrix} 0.621 \\ 0.379 \end{bmatrix}$$

The residual vector for the solution vector $\hat{x}$ is

$$\hat{r} = f - E\hat{x} = \begin{bmatrix} -0.034 \\ -0.089 \\ 0.011 \\ 0.324 \end{bmatrix}$$

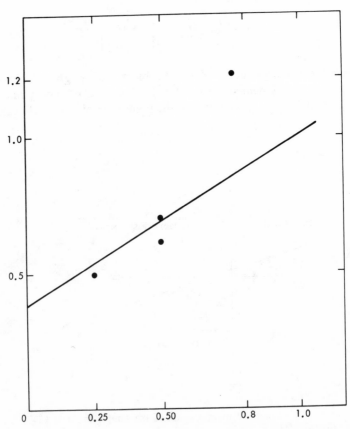

Fig. 23.2   Graph of solution line for the sample problem (23.48)–(23.51).

and the residual norm is

$$\|\hat{r}\| = 0.338$$

The given data points (t_i, w_i), $i = 1, \ldots, 4$, and the fitted line, $f(t) = 0.621t + 0.379$, are shown in Fig. 23.2. Note that the third constraint, $f(1) \leq 1$, is active in limiting how well the fitted line approximates the data points.

The numerical values shown in describing this example were computed using a UNIVAC 1108 computer. Executing the same Fortran code on an IBM 360/67 resulted in opposite signs in the intermediate quantities V, $\tilde{f}_1$, $\tilde{G}$, and $\hat{z}$. This is a consequence of the fact that the signs of columns of the matrix V in a singular value decomposition are not uniquely determined. The difference in the number of iterations required to compute the singular value decomposition of the matrix E on the two computers having different word lengths resulted in a different assignment of signs in the matrices U and V.

EXERCISES

(23.55) Prove that if a Problem LSE has a unique solution *without* inequality constraints, then it has a unique solution *with* inequality constraints.

(23.56) Show that the problem of minimizing a quadratic function $f(x) = \frac{1}{2}x^T Bx + a^T x$, for positive definite B can be transformed to the problem of minimizing $\frac{1}{2}\|w\|^2$ by letting $w = Fx - g$ for an appropriate choice of the nonsingular matrix F and vector g.

(23.57) If the function f of Ex. (23.56) is to be minimized subject to the constraints $Cx = d$ and $Gx \geq h$, what are the corresponding constraints for the problem of minimizing $\frac{1}{2}\|w\|^2$?

24 MODIFYING A QR DECOMPOSITION TO ADD OR REMOVE COLUMN VECTORS

The key to a successful algorithm for the constrained Problem LSI of Chapter 23 is the ability to compute solutions to a sequence of least squares problems in an efficient manner. In Algorithm NNLS (23.10) it can be seen that the coefficient matrix has the property that one linearly independent column vector is added or removed at each step.

Thus we discuss the following problem: Let $n > 0$. Let $a_1, \ldots, a_n$ be a set of m-vectors. Consider the matrix

$$(24.1) \qquad A_k = [a_1, \ldots, a_k]_{m \times k}$$

Introduce a QR decomposition for A_k:

$$(24.2) \qquad Q A_k = \begin{bmatrix} R_k \\ 0 \end{bmatrix}$$

Here R_k is $k \times k$ upper triangular and nonsingular, while Q is orthogonal. Once Q and R_k have been computed, the least squares solution of the system

$$(24.3) \qquad A_k x \cong b$$

is given by

$$(24.4) \qquad x = [R_k^{-1} : 0] Q b \equiv A_k^+ b$$

which requires very little additional work.

We shall consider the problem of computing a QR decomposition for a matrix obtained by deleting a column vector from A_k or adjoining a new column vector to A_k. We wish to take advantage of the prior availability of a QR decomposition of A_k.

We shall discuss three useful methods for updating the QR decomposition. In view of Eq. (24.4) the updating of Q and R effectively provides an updating of A_k^+.

METHOD 1

We have Q stored as an explicit $m \times m$ orthogonal matrix and R stored as a $k \times k$ upper triangle.

Adjoining a Vector

Let a_{k+1} be a vector linearly independent of the columns of A_k. Form the augmented matrix

$$(24.5) \qquad A_{k+1} = [A_k : a_{k+1}]$$

Compute the product

$$b_{k+1} = Q a_{k+1}$$

using Q of Eq. (24.2). Introduce a Householder matrix Q_{k+1} so that the vector

$$r_{k+1} = Q_{k+1} b_{k+1}$$

has zero components in positions $k + 2, \ldots, m$. Then a QR decomposition of A_{k+1} is given by

$$\tilde{Q} A_{k+1} = \begin{bmatrix} \tilde{R} \\ 0 \end{bmatrix}$$

where

$$\tilde{Q} = Q_{k+1} Q$$

and

$$\begin{bmatrix} \tilde{R} \\ 0 \end{bmatrix} = \begin{bmatrix} \begin{bmatrix} R \\ 0 \end{bmatrix} : r_{k+1} \end{bmatrix}$$

Removing a Vector

If the matrix A_k of Eq. (24.1) is modified by the removal of the column vector a_j, forming the matrix

(24.6) $$A_{k-1} = [a_1, \ldots, a_{j-1}, a_{j+1}, \ldots, a_k]$$

we see that

(24.7) $$QA_{k-1} = [r_1, \ldots, r_{j-1}, r_{j+1}, \ldots, r_k]$$

where each r_i is zero in entries $i + 1, \ldots, m$. [See Golub and Saunders (1969) and Stoer (1971).] Define

(24.8) $$\hat{Q} = G_{k-1} \cdots G_{j+1} G_j Q$$

where the G_i are Givens rotation matrices chosen so that the matrix

(24.9) $$\hat{Q} A_{k-1} = \begin{bmatrix} \hat{R} \\ 0 \end{bmatrix}$$

is upper triangular. The matrix G_i operates on rows i and $i + 1$ and produces a zero in position $(i + 1, i)$ of the product matrix $(G_i \cdots G_j Q A_{k-1})$.

Method 1 requires m^2 computer storage locations for Q and $l(l + 1)/2$ locations for R where $l \leq m$ is the largest number of column vectors to be used simultaneously.

METHOD 2

Instead of storing Q explicitly as in Method 1 we now assume that Q is given as a product of k Householder transformations.

(24.10) $$Q = Q_k \cdots Q_1$$

where each

(24.11) $$Q_i = I_m + b_i^{-1} u_i u_i^T$$

is stored by retaining only the nonzero entries of the vectors u_i plus one additional scalar associated with each vector. The matrix R is again stored as a $k \times k$ upper triangle.

Adjoining a Vector

If the set is enlarged to the linearly independent set $[a_1, \ldots, a_k, a_{k+1}]$ compute the product $b_{k+1} = Q_k \cdots Q_1 a_{k+1}$. Compute a Householder transformation Q_{k+1} so that $Q_{k+1} b_{k+1} = r_{k+1}$ is zero in entries $k + 2, \ldots, m$. It follows that

(24.12)
$$Q_{k+1}(Q_k \cdots Q_1)[A_k : a_{k+1}] = \left[\begin{bmatrix} R \\ 0 \end{bmatrix} : r_{k+1}\right]$$

is a QR decomposition for the enlarged set.

Removing a Vector

Again define the matrix A_{k-1} by Eq. (24.6). The data defining the QR decomposition of A_k is of the form

(24.13)
$$Q_k \cdots Q_1 A_k = \begin{bmatrix} R \\ 0 \end{bmatrix} = \begin{bmatrix} \overbrace{R_{11}}^{j-1} & \overbrace{R'_{12}}^{k-j+1} \\ 0 & R'_{22} \\ 0 & 0 \end{bmatrix} \begin{matrix} \}j-1 \\ \}k-j+1 \\ \}m-k \end{matrix}$$

Denote the QR decomposition of A_{k-1} by $QA_{k-1} = \begin{bmatrix} \hat{R} \\ 0 \end{bmatrix}$. The matrix $\hat{R}$ will contain the same submatrix R_{11} appearing in Eq. (24.13) as its leading $(j-1) \times (j-1)$ submatrix. Thus only the last $k-j$ columns of $\hat{R}$ need to be determined.

One approach is to replace

$$\begin{bmatrix} R'_{12} \\ R'_{22} \\ 0 \end{bmatrix}$$

in storage by $[a_{j+1}, \ldots, a_k]$. Then compute

(24.14)
$$Q_{j-1} \cdots Q_1[a_{j+1}, \ldots, a_k] = S_{m \times (k-j)}$$

Next premultiply S by new Householder transformations, Q'_i, so that

$$Q'_{k-1} \cdots Q'_j S = \begin{bmatrix} \overbrace{\hat{R}_{12}}^{k-j} \\ \hat{R}_{22} \\ 0 \end{bmatrix} \begin{matrix} \}j-1 \\ \}k-j \\ \}m-k+1 \end{matrix}$$

with $\hat{R}_{22}$ upper triangular.

The QR decomposition of A_{k-1} is then represented as

$$Q'_{k-1} \cdots Q'_j Q_{j-1} \cdots Q_1 A_{k-1} = \begin{bmatrix} R_{11} & \hat{R}_{12} \\ 0 & \hat{R}_{22} \\ 0 & 0 \end{bmatrix}$$

Alternatively, the matrix S of Eq. (24.14) could be produced by the formula

$$(24.15) \qquad Q_j Q_{j+1} \cdots Q_{k-1} Q_k \begin{bmatrix} R'_{12} \\ R'_{22} \\ 0 \end{bmatrix} = [a_j : S]$$

where R'_{12} and R'_{22} are defined in Eq. (24.13).

Method 2 requires $(m + 1)l$ storage locations to retain the data defining Q and R. Here $l \le m$ is the maximum number of columns being used simultaneously. The version of Method 2 using Eq. (24.14) requires an additional $m \times n$ storage array to retain $[a_1, \ldots, a_n]$ so that any of these vectors will be available when needed in Eq. (24.14).

The second version of Method 2, using Eq. (24.15), obviates the need to retain copies of the vectors $a_j, a_{j+1}, \ldots, a_k$ in storage. Thus all the data representing the QR decomposition of a matrix A_k plus all the vectors a_i not presently being regarded as columns of A_k can be stored in one $m \times n$ storage array plus about $3m$ additional locations for bookkeeping purposes.

METHOD 3

In this method the entire contents of a storage array, $W_{m \times (n+1)}$, which initially contains the data $[A_{m \times n} : b_{m \times 1}]$, is modified each time a new column vector is adjoined to or deleted from the basis. The column vectors of the upper triangular matrix R_k of Eq. (24.2) will occupy some set of k columns of the array W. No record will be kept of the matrix Q of Eq. (24.2).

Let $\mathcal{P}_k = \{p_1, p_2, \ldots, p_k\}$ be a subset of $\{1, 2, \ldots, n\}$ identifying the columns of A that have been introduced into the basis. The order of the integers p_i in the set $\mathcal{P}_k$ is significant. Let A_k denote the $m \times k$ matrix consisting of those column vectors of A indexed in $\mathcal{P}$ and ordered in the sequence $p_1, p_2, \ldots, p_k$. Let Q denote the orthogonal matrix satisfying Eq. (24.2). Then W will contain the $m \times (n + 1)$ matrix $Q[A : b]$. This implies that the jth column vector of R_k will be stored in column number p_j of W. In this method it may be found convenient to store explicit zeros for elements that are zeroed by Householder or Givens transformations.

Adjoining a Vector

Let p_{k+1} denote the index of a column of A to be adjoined to the basis. Form a new index set $\mathcal{P}_{k+1}$ consisting of $p_1, \ldots, p_k$ from $\mathcal{P}_k$ plus p_{k+1}. Form the Householder transformation H that transforms column number p_{k+1} of W to have zeros below row $k + 1$. Premultiply H times the entire array W.

Removing a Vector

Assume again that $\mathcal{P}_k$ identifies the current basis. Let $1 \leq j \leq k$ and assume column number p_j is to be removed from the basis. For $i = j + 1, j + 2, \ldots, k$ form the Givens rotation matrix G_i which operates on rows $i - 1$ and i and zeros the (i, p_{i+1}) element of W. Premultiply G_i times the entire array W.

Form the new index set $\mathcal{P}_{k-1}$ by setting $p_i := p_i$ for $i = 1, 2, \ldots, j - 1$ and setting $p_i := p_{i+1}$ for $i = j, j + 1, \ldots, k - 1$. This method requires only the $m \times (n + 1)$ storage array W which initially contains the data $[A : b]$. If a copy of the initial data will be needed at later stages of the computation, then it must be saved in additional storage.

25 PRACTICAL ANALYSIS OF LEAST SQUARES PROBLEMS

Section 1. GENERAL CONSIDERATIONS

In this chapter we discuss the strategy of planning and interpreting the practical solution of a least squares problem. We consider the case of $m \geq n$, as this is the central problem under consideration in this book.

The person who originates a least squares problem must acquire data and develop or select a mathematical model and frequently a statistical model also. At some point in his work he reaches the stage at which he has a matrix A, and a vector b, and he desires to find a vector x that minimizes the euclidean norm of the residual vector

$$(25.1) \qquad r = b - Ax$$

The problem originator should also have information about the uncertainty of the data constituting A and b. Frequently he may have a priori information about the solution of his matrix problem. This may involve some knowledge as to what would constitute reasonable values for some or all of the solution components. It might also include information such as a requirement that some or all solution components must be nonnegative.

In very general terms there are two major considerations in the design of a computational approach to the problem and it is important to keep these points separate.

1. Computational error can be kept down to the point where it is negligible compared with uncertainty in the solution caused by uncertainty in the initial data.

2. The combined effect of the a priori uncertainty in A and b and the condition number of A may produce a situation in which there are many significantly different vectors x that reduce the norm of r to an acceptably small value. We shall discuss techniques for detecting this situation and some methods for controlling the selection of a particular solution from this set of candidate solutions.

Expanding on point 1, we assume that the information regarding the uncertainty in A and b can be expressed by the statement that there are known numbers φ and ψ such that any matrix of the form $A + E$ with

$$(25.2) \qquad\qquad \|E\| \leq \varphi$$

and any vector of the form $b + db$ with

$$(25.3) \qquad\qquad \|db\| \leq \psi$$

would be acceptable to the problem originator as substitutes for the specific data A and b.

For comparison with these inherent error parameters φ and ψ we may use Eq. (16.3) and (16.4) to define the following two functions of η:

$$(25.4) \qquad \varphi'(\eta) = (6m - 3n + 41)n^{3/2} \|A\| \eta$$

$$(25.5) \qquad \psi'(\eta) = (6m - 3n + 40)n \|b\| \eta$$

Here we have replaced $\|A\|_F$ of Eq. (16.3) by its upper bound $n^{1/2} \|A\|$.

Ignoring the terms $O(\eta^2)$ in Eq. (16.3) and (16.4), we infer from Theorem (16.1) that if η is chosen small enough so that

$$(25.6) \qquad\qquad \varphi'(\eta) < \varphi$$

and

$$(25.7) \qquad\qquad \psi'(\eta) < \psi$$

then the solution computed using arithmetic of uniform precision η is the exact solution of some problem whose initial data differs from the given data by amounts satisfying Eq. (25.2) and (25.3).

Similarly to select an η and $\omega \leq \eta^2$ for mixed precision computation one can use Eq. (17.13) and (17.14) to define

$$(25.8) \qquad \varphi''(\eta) = 30n^{3/2} \|A\| \eta$$

and

$$(25.9) \qquad\qquad \psi''(\eta) = 29n \, \| b \| \, \eta$$

and choose η so that

$$(25.10) \qquad\qquad \varphi''(\eta) < \varphi$$

and

$$(25.11) \qquad\qquad \psi''(\eta) < \psi$$

With this choice of η and $\omega \leq \eta^2$ we infer from Theorem (17.11) that the computed solution is the exact solution of a perturbed problem with perturbations smaller than the a priori uncertainty as described by Eq. (25.2) and (25.3).

Recall that the multipliers of $\| A \| \, \eta$ and $\| b \| \, \eta$ in Eq. (25.4), (25.5), (25.8), and (25.9) were derived by considering worst-case propagation of computational errors. In practice, replacing these multipliers by their square roots will generally give more realistic estimates of the norms of the virtual perturbations due to the computational errors.

We now turn to point 2. It will be convenient to formalize the notion of an "acceptably small" residual vector. Suppose we are willing to accept residual vectors with norms as large as some number ρ. Then we may define the acceptable solution set as

$$(25.12) \quad X = \{ x : \| (A + E)x - (b + db) \| \leq \rho, \, \| E \| \leq \varphi, \, \| db \| \leq \psi \}$$

It is important to note that the definition of the set X depends upon the three tolerance parameters, φ, ψ, and ρ, which should be chosen on the basis of actual knowledge about the problem and its data.

Our purpose in writing Eq. (25.12) is not so much to establish this particular definition of an acceptable solution set as it is to give some degree of concreteness to the general idea of an acceptable solution set and to identify some of the quantities that determine the "size" of this set.

To assess the "size" of the set X we first observe that if Rank $(A) < n$, or if φ and κ (κ denotes the condition number of A) are so large that $\kappa\varphi/\| A \| \geq 1$, then $A + E$ will be singular for some $\| E \| \leq \varphi$ and the set X is unbounded.

On the other hand, if Rank $(A) = n$ and $\kappa\varphi/\| A \| < 1$, then the perturbation bound (9.13) or (9.14) may be used to obtain useful information regarding the diameter of the set X. The presence of the parameter ρ in the definition of the set X leads to some further possible increase in its size.

If this set X is "large" in the sense that it contains vectors which are significantly different from each other, then some selection must be made of a particular solution vector from X. This selection process can be viewed very broadly as including any steps that the problem originator might take to change the problem to one having a smaller solution set, generally contained in X.

The criterion that one uses to reduce the size of the set X depends upon the application. A situation that occurs very commonly is one in which the problem $Ax \cong b$ arises as a local linearization of a nonlinear least squares problem. This was mentioned in Chapter 1. In this case one is likely to prefer the use of the $\hat{x} \epsilon X$ having least norm in order to reduce the likelihood of leaving the region in which $b - Ax$ is a good approximation to the nonlinear problem.

Although there are many different motivations (statistical, mathematical, numerical, heuristic, etc.) that have been proposed for different specific procedures for changing a given least squares problem, most of these procedures consist of performing one or more of the following four operations, not necessarily in the order listed.

1. Left-multiply A and b by an $m \times m$ matrix G.
2. Right-multiply A by an $n \times n$ matrix H with the corresponding change of variables $x = H\tilde{x}$ or $x = H\tilde{x} + \xi$.
3. Append additional rows to A and additional elements to b.
4. Assign fixed values (often zero) to some components of the solution vector. This may be done either with respect to the original set of variables or a transformed set of variables.

We shall expand on each of these four items in Sections 2, 3, 4, and 5, respectively.

Finally in Section 6 we describe singular value analysis. By singular value analysis we mean the computation of a number of quantities derived from the singular value decomposition of the matrix A and the interpretation of these quantities as an aid in understanding the indeterminacies of the problem $Ax \cong b$ and in selecting a useful solution vector. The problem $Ax \cong b$ to which singular value analysis is applied may of course be a problem resulting from preliminary application of various of the operations described in Sections 2 to 5.

Section 2. LEFT MULTIPLICATION OF A AND b
BY A MATRIX G

This operation changes the norm by which the size of the residual vector is assessed. Thus instead of seeking x to minimize $(b - Ax)^T(b - Ax)$, one changes the problem to that of minimizing $(Gb - GAx)^T(Gb - GAx)$, which also can be written as $(b - Ax)^T(G^TG)(b - Ax)$, or as $(b - Ax)^T W(b - Ax)$, where $W = G^TG$.

A commonly used special case is that in which the matrix G is diagonal. Then the matrix W is diagonal. In this case left-multiplication of the augmented matrix $[A : b]$ by G can be interpreted as being a row-scaling operation in which row i of $[A : b]$ is multiplied by the number g_{ii}.

Define

$$(25.13) \qquad\qquad r = b - Ax$$

and

$$(25.14) \qquad\qquad \tilde{r} = Gr = Gb - GAx$$

Then if G is diagonal the quantity to be minimized is

$$(25.15) \qquad\qquad \|\tilde{r}\|^2 = \sum_{i=1}^{m} g_{ii}^2 r_i^2 = \sum_{i=1}^{m} w_{ii} r_i^2$$

Loosely speaking, assigning a relatively large weight, $|g_{ii}|$ (or equivalently w_{ii}) to the ith equation will tend to cause the resulting residual component $|r_i|$ to be smaller. Thus if some components of the data vector b_i are known with more absolute accuracy than others, one may wish to introduce relatively larger weights, $|g_{ii}|$, or w_{ii}, in the ith row of $[A : b]$.

This procedure, in which G is diagonal, is generally referred to as weighted least squares. As a systematic scheme for assigning the weights suppose that one can associate with each component, b_i, of the data vector b a positive number, σ_i, indicating the approximate size of the uncertainty in b_i. If one has the appropriate statistical information about b it would be common to take σ_i to be the standard deviation of the uncertainty in b_i. Then it would be usual to define weights by

$$g_{ii} = \frac{1}{\sigma_i}$$

or equivalently

$$w_{ii} = \frac{1}{\sigma_i^2}$$

Note that with this scaling all components of the modified vector $\tilde{b}$ defined by

$$\tilde{b}_i = g_{ii} b_i = \frac{b_i}{\sigma_i}$$

have unit standard deviation.

More generally, if one has sufficient statistical information about the uncertainty in b to place numerical values on the correlation between the errors in different components of b, one may express this information as an $m \times m$ positive definite symmetric covariance matrix C. [See Plackett (1960) for a detailed discussion of these statistical concepts.] In this case one can compute the Cholesky factorization of C as described in Eq. (19.5) and (19.12) to (19.14) obtaining a lower triangular matrix F that satisfies

$$C = FF^T$$

Then the weighting matrix G can be defined by

$$G = F^{-1}$$

Operationally it is not necessary to compute the explicit inverse of F. Rather one would compute the weighted matrix $[\tilde{A} : \tilde{b}]$ by solving the equations

$$F[\tilde{A} : \tilde{b}] = [A : b]$$

This process is straightforward because F is triangular.

If G is derived from an a priori covariance matrix for the errors in b as described above, then the covariance matrix for the errors in the transformed vector $\tilde{b}$ is the identity matrix.

Whether on the basis of statistical arguments or otherwise, it is desirable to use a matrix G such that the uncertainty is of about the same size in all components of the transformed vector $\tilde{b} = Gb$. This has the effect of making the euclidean norm a reasonable measure of the "size" of the error vector db, as for instance in Eq. (25.3).

Section 3. RIGHT MULTIPLICATION OF A BY A MATRIX H AND CHANGE OF VARIABLES $x = H\tilde{x} + \xi$

Here one replaces the problem

(25.16) $Ax \cong b$

by the problem

(25.17) $\tilde{A}\tilde{x} \cong \tilde{b}$

where

(25.18) $\tilde{A} = AH$

(25.19) $\tilde{b} = b - A\xi$

(25.20) $x = H\tilde{x} + \xi$

The matrix H is $n \times l$, with $l \leq n$. The vector $\tilde{x}$ is l-dimensional and the matrix $\tilde{A}$ is $m \times l$. If H is $n \times n$ diagonal, this transformation may be interpreted as a column scaling operation applied to A.

If H is $n \times n$ nonsingular, this transformation does not change the problem mathematically. Thus the set of vectors satisfying $x = H\tilde{x} + \xi$ where $\tilde{x}$ minimizes $\|\tilde{b} - \tilde{A}\tilde{x}\|$ is the same as the set of vectors x minimizing $\|b - Ax\|$.

However, unless H is orthogonal, the condition number of $\tilde{A}$ will generally be different from that of A. Therefore, if one is using an algorithm, such as HFTI (14.9) or singular value analysis [Eq. (18.36) to (18.45)], that makes a determination of the pseudorank of A, the algorithm may make a different determination using $\tilde{A}$.

Furthermore, if the pseudorank is determined to have a value $k < n$ and one proceeds to compute the minimal length solution of the rank k problem, then the use of the transformation matrix H alters the norm by which the "size" of the solution vector is measured. This will, in general, lead to a different vector being selected as the "minimum length" solution. Thus in problem (25.16) the minimal length solution is a solution that minimizes $\|x\|$, whereas using the transformed problem in Eq. (25.17) the minimal length solution is a solution $\tilde{x}$ of Eq. (25.17) that minimizes $\|\tilde{x}\|$. This amounts to minimizing $\|H^{-1}(x - \xi)\|$ rather than minimizing $\|x\|$.

As to criteria for selecting H we first note that the use of the spectral norm of the perturbation matrix, E, in Eq. (25.2) is a realistic mathematical model only if the absolute uncertainties in different components of A are all of about the same magnitude. Thus, if one has estimates of the uncertainty of individual elements of A, the matrix H can be selected as a column scaling matrix to balance the size of the uncertainties in the different columns of $\tilde{A}$.

A somewhat similar idea can be based on a priori knowledge of the solution vector x. Suppose it is known that the solution vector should be close to a known vector ξ (the a priori expected value of x). Suppose further that one has an a priori estimate, σ_i, of the uncertainty of ξ_i as an estimate of x_i. One can take H to be the $n \times n$ diagonal matrix with diagonal components

$$(25.21) \qquad h_{ii} = \sigma_i$$

Then the transformed variables

$$(25.22) \qquad \tilde{x}_i = \frac{x_i - \xi_i}{\sigma_i}$$

have unit a priori uncertainty and zero a priori expected values.

More generally, if one has sufficient a priori statistical information to define an $n \times n$ positive-definite a priori covariance matrix K describing the uncertainty of ξ, then H can be computed as the upper triangular Cholesky factor of C [see Eq. (19.16) to (19.18)]:

$$(25.23) \qquad\qquad C = HH^T$$

Then the a priori covariance of the transformed variable vector $\tilde{x}$ of Eq. (25.20) is the $n \times n$ identity matrix, and the a priori expected value of $\tilde{x}$ is the zero vector.

The situation in which the separate components of $\tilde{x}$ have approximately equal uncertainty is a desirable scaling of the variables if the problem is judged to have pseudorank $k < n$ and a minimal length solution is to be computed.

Since the condition number of $\tilde{A}$ of Eq. (25.18) will, in general, be different from that of A, one may decide to choose H with the objective of making the condition number of $\tilde{A}$ small. If A is nonsingular, there exists a matrix H such that Cond $(AH) = 1$. Such a matrix is given by $H = R^{-1}$, where R is the triangular matrix resulting from Householder triangularization. One is not likely to have this matrix, R, available a priori. It is of interest to note, however, that if the data that lead to the determination of the a priori covariance matrix, C, of Eq. (25.23) were approximately the same as the present data $[A : b]$, then the matrix H of Eq. (25.23) will be approximately equal to R^{-1}. Then Cond (AH) is likely to be fairly small.

If no estimate of C is available, there remains the possibility of using a diagonal matrix H as a column scaling matrix to balance the euclidean norms (or any other selected norms) of the columns of $\tilde{A}$. This is accomplished by setting

$$(25.24) \qquad\qquad h_{jj} = \begin{cases} \|a_j\|^{-1} & \text{if } \|a_j\| \neq 0 \\ 1 & \text{if } \|a_j\| = 0 \end{cases}$$

where a_j denotes the jth column vector of A. It has been proved by van der Sluis (1969) that with H defined by Eq. (25.24), using euclidean column norms, Cond (AH) does not exceed the minimal condition number obtainable by column scaling by more than a factor of $n^{1/2}$.

Improving the condition number has the advantage that perturbation bounds such as Eq. (9.10) will give less pessimistic results. It also may lead to a determination of pseudorank $k = n$, whereas without the improvement of the condition number by scaling a pseudorank of $k < n$ might have been determined leading to unnecessary and inappropriate complications in obtaining a solution.

A transformation matrix H can be selected so that $\tilde{A}$ or a submatrix of $\tilde{A}$ has an especially convenient form, such as triangular or diagonal. Thus if one has computed the singular value decomposition, $A = USV^T$, then left-multiplying $[A : b]$ by U^T and right-multiplying A by $H = V$ transforms A to the diagonal matrix S. Operationally one would not usually do this pre- and postmultiplication explicitly as it would be done along with the computation of the singular value decomposition as described in Chapter 18. Systematic use of the singular value decomposition will be discussed in Section 6 of this chapter.

Section 4. APPEND ADDITIONAL ROWS TO [A:b]

Here we discuss replacing the problem $Ax \cong b$ by the problem

$$(25.25) \qquad \begin{bmatrix} A \\ F \end{bmatrix} x \cong \begin{bmatrix} b \\ d \end{bmatrix}$$

where F is an $l \times n$ matrix and d is an l-dimensional vector.

Suppose one prefers that the solution vector x should be close to a known vector ξ. Setting $F = I_n$ and $d = \xi$ in (25.25) expresses this preference. In particular if $d = \xi = 0$ this expresses a preference for $\|x\|$ to be small.

The intensity of this preference can be indicated by a scaling parameter, say σ, incorporated into the definition of F and d as

$$F = \sigma^{-1} I_n \quad \text{and} \quad d = \sigma^{-1} \xi$$

The number σ can be regarded as an estimate of the size of the uncertainty in ξ. Thus setting σ small causes the solution to be closer to ξ.

A further refinement of this idea is to assign separate numbers, σ_i, $i = 1, \ldots, n$, where σ_i is an estimate of the size of the uncertainty of the ith component of ξ. Thus one would define $d_i = \xi_i / \sigma_i$ for $i = 1, \ldots, n$ and let F be a diagonal matrix with diagonal elements $f_{ii} = \sigma_i^{-1}$.

Finally, if one has sufficient a priori statistical information about the expected value, ξ, of the solution, x, to define a symmetric positive definite $n \times n$ covariance matrix, K, for $(x - \xi)$, then it is reasonable to set

$$(25.26) \qquad F = L^{-1}$$

where L is the lower triangular Cholesky factor of K [see Eq. (19.12) to (19.14)]

$$(25.27) \qquad K = LL^T$$

and set

$$(25.28) \qquad d = F\xi$$

Note that the case of $d \neq 0$ can be reduced to the case of $d = 0$ by a simple translation if the system $Fw = d$ is consistent. To see this let w be a solution of $Fw = d$ and make the change of variables $x = w + \tilde{x}$ in Eq. (25.25), which yields the transformed problem

$$(25.29) \qquad \begin{bmatrix} A \\ F \end{bmatrix} \tilde{x} \cong \begin{bmatrix} b - Aw \\ 0 \end{bmatrix}$$

It is important to make this change of variables if one intends to apply a minimal length solution method since the preference for $\|\tilde{x}\|$ to be small is consistent with the conditions $F\tilde{x} \cong 0$.

We consider now the question of the relative weighting of the two sets of conditions $Ax \cong b$ and $Fx \cong d$ in Eq. (25.25). From a formal statistical point of view one might say that the appropriate relative weighting is established by setting up the problem in the form

$$(25.30) \qquad \begin{bmatrix} G & 0 \\ 0 & F \end{bmatrix} \cdot \begin{bmatrix} A \\ I \end{bmatrix} x \cong \begin{bmatrix} G & 0 \\ 0 & F \end{bmatrix} \cdot \begin{bmatrix} b \\ \xi \end{bmatrix}$$

where $(G^T G)^{-1}$ is the a priori covariance matrix of the uncertainty in the data vector b and $(F^T F)^{-1}$ is the a priori covariance matrix of the uncertainty in the a priori expected value ξ of the solution vector x. In practice, however, these a priori covariance matrices, particularly $(F^T F)^{-1}$, may not be known with much certainty and one may wish to explore the changes that different relative weighting produces in the solution vector and in the residual vector.

For this purpose introduce a nonnegative scalar weighting parameter λ into problem (25.30) and consider the new problem

$$(25.31) \qquad \begin{bmatrix} \tilde{A} \\ \lambda F \end{bmatrix} x \cong \begin{bmatrix} \tilde{b} \\ \lambda d \end{bmatrix}$$

where

$$(25.32) \qquad \tilde{A} = GA$$
$$(25.33) \qquad \tilde{b} = Gb$$

and

$$(25.34) \qquad d = F\xi$$

For the convenience of readers who may be more accustomed to other ways of motivating and stating the least squares problem involving a priori covariance matrices we note that problem (25.31) is the problem of finding a vector x to minimize the quadratic form

$$\|\tilde{A}x - \tilde{b}\|^2 + \lambda^2 \|Fx - d\|^2$$

or equivalently

$$(Ax - b)^T(G^TG)(Ax - b) + \lambda^2(x - \xi)^T(F^TF)(x - \xi)$$

The idea of using a relative weighting parameter λ in this context was discussed by Levenberg (1944). Further elaboration and applications of the technique are given by Hoerl (1959, 1962, and 1964), Morrison (1960), Marquardt (1963 and 1970), and Hoerl and Kennard (1970a and 1970b). As a result of the 1963 paper and a computer program contributed to the computer information exchange organization SHARE by Marquardt, the use of this idea in problems arising from the linearization of nonlinear least squares is often called *Marquardt's method*. The study of problem (25.31) as a function of λ has also been called *ridge regression* or *damped least squares*.

To analyze the dependence of the solution vector and the residual vector in problem (25.31) upon the parameter λ we first introduce the change of variables

(25.35) $x = \xi + F^{-1}y$

and obtain the transformed problem

(25.36) $$\begin{bmatrix} \hat{A} \\ \lambda I \end{bmatrix} y \cong \begin{bmatrix} \hat{b} \\ 0 \end{bmatrix}$$

where

$$\hat{A} = \tilde{A}F^{-1}$$

and

$$\hat{b} = \tilde{b} - \tilde{A}\xi$$

The translation by ξ and scaling by F^{-1} used in Eq. (25.35) have been discussed in Section 3 of this chapter. Their purpose is to produce the new variable y that is better scaled for meaningful assessment of the "size" of the solution vector.

Write a singular value decomposition of $\hat{A}$:

(25.37) $$\hat{A}_{m \times n} = U_{m \times m}\begin{bmatrix} S_{n \times n} \\ 0_{(m-n) \times n} \end{bmatrix} V^T_{n \times n}$$

Recall that $S = \text{Diag}\{s_1, \ldots, s_n\}$. If Rank $(A) = k < n$, then $s_i = 0$ for $i > k$. Introduce the orthogonal change of variables

(25.38) $y = Vp$

and left-multiply Eq. (25.36) by the orthogonal matrix

$$\begin{bmatrix} U^T & 0 \\ 0 & V^T \end{bmatrix}$$

obtaining the new least squares problem

(25.39)
$$
\begin{bmatrix} S_{n \times n} \\ 0_{(m-n) \times n} \\ \lambda I_n \end{bmatrix} p \cong \begin{bmatrix} g \\ 0 \end{bmatrix} \begin{matrix} \}m \\ \}n \end{matrix}
$$

where

(25.40)
$$
g = U^T \hat{b}
$$

Finally, for $\lambda > 0$, the submatrix, λI_n, in problem (25.39) can be eliminated by means of left-multiplication by appropriate Givens rotation matrices. Thus to eliminate the ith diagonal element of λI_n, left-multiply Eq. (25.39) by a matrix that differs from an $(m + n)$th-order identity matrix only in the following four positions:

$$
\begin{array}{ccc}
 & \text{Col. } i & \text{Col. } m + i \\
 & \vdots & \vdots \\
\text{Row } i \cdots & \dfrac{s_i}{(s_i^2 + \lambda^2)^{1/2}} \cdots & \dfrac{\lambda}{(s_i^2 + \lambda^2)^{1/2}} \cdots \\
 & \vdots & \vdots \\
\text{Row } m + i \cdots & -\dfrac{\lambda}{(s_i^2 + \lambda^2)^{1/2}} \cdots & \dfrac{s_i}{(s_i^2 + \lambda^2)^{1/2}} \cdots \\
 & \vdots & \vdots
\end{array}
$$

After this elimination operation for $i = 1, \ldots, n$ the resulting equivalent problem is

(25.41)
$$
\begin{bmatrix} S_{n \times n}^{(\lambda)} \\ 0_{(m-n) \times n} \\ 0_{n \times n} \end{bmatrix} p \cong \begin{bmatrix} g^{(\lambda)} \\ h^{(\lambda)} \end{bmatrix} \begin{matrix} \}m \\ \}n \end{matrix}
$$

where

(25.42)
$$
g_i^{(\lambda)} = \begin{cases} \dfrac{g_i s_i}{(s_i^2 + \lambda^2)^{1/2}} & i = 1, \ldots, n \\ g_i & i = n + 1, \ldots, m \end{cases}
$$

(25.43)
$$
h_i^{(\lambda)} = -\dfrac{g_i \lambda}{(s_i^2 + \lambda^2)^{1/2}} \qquad i = 1, \ldots, n
$$

and

$$
S^{(\lambda)} = \text{Diag} \{s_1^{(\lambda)}, \ldots, s_n^{(\lambda)}\}
$$

with

(25.44) $$s_i^{(\lambda)} = (s_i^2 + \lambda^2)^{1/2} \qquad i = 1, \dots, n$$

For $\lambda = 0$ problem (25.39) has a diagonal matrix of rank k and the solution vector $p^{(0)}$ is given by

(25.45) $$p_i^{(0)} = \begin{cases} \dfrac{g_i}{s_i} & i = 1, \dots, k \\ 0 & i = k+1, \dots, n \end{cases}$$

For $\lambda > 0$ we use Eq. (25.41) to (25.44) and obtain the solution components

(25.46) $$p_i^{(\lambda)} = \begin{cases} \dfrac{g_i^{(\lambda)}}{s_i^{(\lambda)}} = \dfrac{g_i s_i}{s_i^2 + \lambda^2} = p_i^{(0)}\left[\dfrac{s_i^2}{s_i^2 + \lambda^2}\right] & i = 1, \dots, k \\ 0 & i = k+1, \dots, n \end{cases}$$

Furthermore we have

(25.47) $$\|p^{(\lambda)}\|^2 = \sum_{i=1}^{k}\left[\frac{g_i s_i}{(s_i^2 + \lambda^2)}\right]^2$$
$$= \sum_{i=1}^{k}[p_i^{(0)}]^2\left[\frac{s_i^2}{(s_i^2 + \lambda^2)}\right]^2$$

Note that Eq. (25.46) and (25.47) remain valid for $\lambda = 0$.

The vector $p^{(\lambda)}$, which is a solution for problem (25.39), can be transformed by the linear equations (25.38) and (25.35) to obtain vectors $y^{(\lambda)}$ and $x^{(\lambda)}$, which are solutions for problems (25.36) and (25.31), respectively.

We are primarily interested in the problem $\tilde{A}x \cong \tilde{b}$ and have introduced problem (25.31) only as a technical device for generating solutions to a particular family of related problems. Thus formulas will also be derived for the quantity $\omega_\lambda = \|\tilde{b} - \tilde{A}x^{(\lambda)}\|$. With $\lambda \geq 0$ and $p^{(\lambda)}$ defined by Eq. (25.46) we obtain

(25.48) $$\omega_\lambda^2 = \|\tilde{b} - \tilde{A}x^{(\lambda)}\|^2 = \|\hat{b} - \hat{A}y^{(\lambda)}\|^2$$
$$= \left\|g - \begin{bmatrix} S \\ 0 \end{bmatrix}p^{(\lambda)}\right\|^2$$
$$= \sum_{i=1}^{k}[g_i - s_i p_i^{(\lambda)}]^2 + \sum_{i=k+1}^{m} g_i^2$$
$$= \sum_{i=1}^{k} g_i^2\left[\frac{\lambda^2}{s_i^2 + \lambda^2}\right]^2 + \sum_{i=k+1}^{m} g_i^2$$

Note that as λ increases, $\|p^{(\lambda)}\|$ decreases and ω_λ increases. Thus the problem originator has the opportunity of selecting λ to obtain some acceptable compromise between the size of the solution vector $p^{(\lambda)}$ and the size of the residual norm ω_λ.

The set of solutions of problem (25.31) given by this technique has an important optimality property expressed by the following theorem.

(25.49) THEOREM [*Morrison (1960); Marquardt (1963)*]

> *For a fixed nonnegative value of λ, say, $\bar{\lambda}$, let $\bar{y}$ be the solution vector for problem (25.36), and let $\bar{\omega} = \|\hat{b} - \hat{A}\bar{y}\|$. Then $\bar{\omega}$ is the minimum value of $\|\hat{b} - \hat{A}y\|$ for all vectors y satisfying $\|y\| \leq \|\bar{y}\|$.*

Proof: Assume the contrary. Then there exists a vector $\tilde{y}$ with $\|\tilde{y}\| \leq \|\bar{y}\|$ satisfying $\|\hat{b} - \hat{A}\tilde{y}\| < \|\hat{b} - \hat{A}\bar{y}\|$. Therefore, $\|\hat{b} - \hat{A}\tilde{y}\|^2 + \bar{\lambda}^2\|\tilde{y}\|^2 < \|\hat{b} - \hat{A}\bar{y}\|^2 + \bar{\lambda}^2\|\bar{y}\|$, which contradicts the assumption that $\bar{y}$ is the least squares solution of problem (25.36) and thus minimizes $\|\hat{b} - \hat{A}y\|^2 + \bar{\lambda}^2\|y\|^2$.

A simple tabular or graphic presentation of the quantities $\|p^{(\lambda)}\|$ and ω_λ given by Eq. (25.47) and (25.48) can be very useful in studying a particular least squares problem. Further detail can be obtained by tabular or graphic presentation of the individual solution components [Eq. (25.46), (25.38), and (25.35)] as functions of λ. An example of this type of analysis is given in Chapter 26.

One may wish to solve Eq. (25.47) or (25.48) to find a value of λ which produces a prespecified solution norm or residual norm. If Newton's method is used for this purpose the following expressions will be useful:

$$t = \lambda^2$$

$$\varphi_i(t) = \frac{g_i s_i}{s_i^2 + t}$$

$$\|p^{(\lambda)}\|^2 = \sum_{i=1}^{k} \varphi_i^2(t)$$

$$\frac{d(\|p^{(\lambda)}\|^2)}{dt} = -2\sum_{i=1}^{k} \frac{\varphi_i^2(t)}{s_i^2 + t}$$

and

$$u = \lambda^{-2}$$

$$\psi_i(u) = \frac{g_i}{u s_i^2 + 1}$$

$$\omega_\lambda^2 = \sum_{i=1}^{k} \psi_i^2(u) + \sum_{t=k+1}^{m} g_i^2$$

$$\frac{d(\omega_\lambda^2)}{du} = -2\sum_{i=1}^{k} \frac{\psi_i^2(u)s_i^2}{u s_i^2 + 1}$$

If one solves problem (25.36) directly, using any algorithm that determines the norm, say ρ_λ, of the residual vector, then it should be noted that ρ_λ satisfies

(25.50) $$\rho_\lambda^2 = \omega_\lambda^2 + \lambda^2 \| y^{(\lambda)} \|^2$$

Thus if one wishes to obtain the quantity ω_λ, this can be done by computing $\| y^{(\lambda)} \|^2$ and then solving Eq. (25.50) for ω_λ.

Section 5. DELETING VARIABLES

Deleting a variable from a problem is equivalent to fixing the value of that variable at zero. If one variable, say x_n, is deleted, then the problem

(25.51) $$Ax \cong b$$

is changed to

(25.52) $$\tilde{A}\tilde{x} \cong b$$

where $\tilde{A}$ is the $m \times (n-1)$ matrix that consists of the first $n-1$ column vectors of A and $\tilde{x}$ is an $(n-1)$-dimensional vector.

Assuming $m \geq n$, the separation theorem (5.12) implies that the singular values of $\tilde{A}$ interlace with those of A, from which it follows that Cond $(\tilde{A}) \leq$ Cond (A). By repeated application of this process it follows that removing any proper subset of the variables leads to a matrix whose condition number is no larger than that of the original matrix.

Clearly the minimum residual obtainable in problem (25.52) is not smaller than the minimum residual attainable in problem (25.51).

Thus like some of the techniques discussed previously, removing variables is a device that reduces, or at least does not increase, the condition number of a matrix at the expense of increasing, or at least not decreasing, the norm of the residual vector.

Although we feel it is useful for the purpose of comparing variable deletion or variable augmentation with other stabilization methods to note the properties stated above, it is nevertheless true that these properties are not usually taken as the motivating concepts for this type of procedure. More commonly one uses variable deletion or augmentation because one wishes to find the smallest number of parameters that can be used in the solution and still give an acceptably small residual norm.

In some cases there is a natural ordering of the variables, for instance, when the variables are coefficients of a polynomial in one variable. Then it is a straightforward matter to obtain a sequence of solutions, first using only the first variable, next the first and second variable, and so forth. In fact, in the special case of fitting data by a polynomial or by truncated Fourier series there are quite specialized algorithms available [e.g., Forsythe (1957)].

If there is not a natural ordering of the variables, then the following problem of *subset selection* is sometimes considered. For each value of $k = 1, 2, \ldots, n$, find the set J_k consisting of k indices such that the residual norm ρ_k obtained by solving for only the k variables x_i, $i \in J_k$, is as small as can be attained by any set of k variables. Usually one would also introduce a *linear independence tolerance* τ and limit consideration to sets of k variables such that the associated matrix satisfies some conditioning test (such as size of pivot element) involving τ.

The sequence $\{\rho_k\}$ is obviously nonincreasing with increasing k and one might introduce a number $\bar{\rho}$ with the meaning that the computational procedure is to stop if a value of k is reached such that $\rho_k < \bar{\rho}$. Alternatively, one might stop when $\rho_k - \rho_{k+1}$ is smaller than some tolerance, possibly depending upon k. This latter type of termination test arises from statistical considerations involving the "F" distribution [e.g., see Plackett (1960)].

This problem, as stated, appears to require for each value of k an exhaustive procedure for solving for all $\binom{n}{k}$ combinations of k variables. This exhaustive approach becomes prohibitively expensive if $\binom{n}{k}$ is large.

Methods have been devised that exploit the partial ordering (of set inclusion) among the subsets to determine the optimal subsets J_k without explicitly testing every subset. See LaMotte and Hocking (1970) and Garside (1971) and references cited there for an exposition of these ideas. Suggestions for basing subset selection on ridge regression (see Section 4, this chapter) are given in Hoerl and Kennard (1970b).

An alternative course of action is to compromise by solving the following more restricted problem instead. Solve the problem as stated above for $k = 1$. Let $\tilde{J}_1$ be the solution set J_1. After the set $\tilde{J}_k$ has been determined, consider only candidate sets of the form $\tilde{J}_k \cup \{j\}$ for $j \notin \tilde{J}_k$ in searching for a preferred set of $k + 1$ variables that will constitute the set $\tilde{J}_{k+1}$. Let us denote the sequence of sets formed in this way by $\tilde{J}_k$, $k = 1, \ldots, n$, and the associated residual norms by $\tilde{\rho}_k$. Note that $\tilde{J}_k = J_k$ and $\tilde{\rho}_k = \rho_k$ for $k = 1$ and $k = n$, but for $1 < k < n$, $\tilde{J}_k$ is generally not the same as J_k, and $\tilde{\rho}_k \geq \rho_k$.

This type of algorithm is referred to as *stepwise regression* [Efroymson, pp. 191–203 in Ralston and Wilf (1960)]. The computation can be organized so that the selection of the new variable to be introduced at stage k is only slightly more involved than the pivot selection algorithm normally used in solving a system of linear equations.

This idea may be elaborated further to consider at each stage the possibility of removing variables whose contribution to making the residual small has become insignificant due to the effect of variables subsequently introduced into the solution (Efroymson, *loc. cit.*).

The primary mathematical complication of stepwise regression is the fact that the amount that one coefficient, say x_j, contributes to reducing the residual norm depends in general upon what other set of coefficients are being included in the solution at the same time. This drawback can be circumvented by performing a linear change of variables to obtain a new set of variables whose individual effects upon the residual vector are mutually independent. In statistical terminology the new set of variables is uncorrelated. In algebraic terms this amounts to replacing the A matrix by a new matrix $\tilde{A} = AC$ such that the columns of $\tilde{A}$ are orthogonal. There are a variety of distinct matrices, C, that would accomplish such a transformation and in general different sets of uncorrelated variables would be produced by different such matrices. One such transformation matrix that has some additional desirable properties is $C = V$ where V is the matrix arising in a singular value decomposition, $A = USV^T$.

In particular, if the transformation matrix C is required to be orthogonal, then for AC to have orthogonal columns it follows that C must be the matrix V of some singular value decomposition of A.

Furthermore the numbers t_i that occur as the square roots of the diagonal elements of the diagonal matrix $[(AC)^T(AC)]^{-1}$ have a statistical interpretation as being proportional to the standard deviations of the new coefficients. The choice of $C = V$ causes these numbers t_i to be reciprocals of the singular values of A. This choice minimizes the smallest of the standard deviations, and also minimizes each of the other standard deviations subject to having selected the preceeding variables.

The use of the singular value decomposition of A will be discussed further in the next section.

Section 6. SINGULAR VALUE ANALYSIS

Suppose a singular value decomposition (see Chapters 4 and 18) is computed for the matrix A.

$$(25.53) \qquad A = U\begin{bmatrix} S \\ 0 \end{bmatrix} V^T$$

One can then compute

$$(25.54) \qquad g = U^T b$$

and consider the least squares problem

$$(25.55) \qquad \begin{bmatrix} S \\ 0 \end{bmatrix} p \cong g$$

where p is related to x by the orthogonal linear transformation

(25.56)
$$x = Vp$$

Problem (25.55) is equivalent to the problem $Ax \cong b$ in the sense discussed in Chapter 2 for general orthogonal transformations of least squares problems.

Since S is diagonal $(S = \mathrm{Diag}\ \{s_1, \ldots, s_n\})$ the effect of each component of p upon the residual norm is immediately obvious. Introducing a component p_j with the value

(25.57)
$$\hat{p}_j = \frac{g_j}{s_j}$$

reduces the sum of squares of residuals by the amount g_j^2.

Assume the singular values are ordered so that $s_k \geq s_{k+1}$, $k = 1, \ldots, n-1$. It is then natural to consider "candidate" solutions for problem (25.55) of the form

(25.58)
$$p^{(k)} = \begin{bmatrix} \hat{p}_1 \\ \cdot \\ \cdot \\ \cdot \\ \hat{p}_k \\ 0 \\ \cdot \\ \cdot \\ \cdot \\ 0 \end{bmatrix} \qquad k = 0, 1, \ldots, n$$

where $\hat{p}_j$ is given by Eq. (25.57). The candidate solution vector $p^{(k)}$ is the pseudoinverse solution (i.e., the minimal length solution) of problem (25.55) under the assumption that the singular values s_j for $j > k$ are regarded as being zero.

From the candidate solution vectors $p^{(k)}$ one obtains candidate solution vectors $x^{(k)}$ for the problem $Ax \cong b$ as

(25.59)
$$x^{(k)} = Vp^{(k)} = \sum_{j=1}^{k} \hat{p}_j v^{(j)} \qquad k = 0, \ldots, n$$

where $v^{(j)}$ denotes the jth column vector of V. Note that

(25.60)
$$\|x^{(k)}\|^2 = \|p^{(k)}\|^2 = \sum_{j=1}^{k} \hat{p}_j^2 = \sum_{j=1}^{k} \left(\frac{g_j}{s_j}\right)^2$$

hence $\|x^{(k)}\|$ is a nondecreasing function of k. The squared residual norm associated with $x^{(k)}$ is given by

(25.61) $$\rho_k^2 = ||b - Ax^{(k)}||^2 = \sum_{j=k+1}^{m} g_j^2$$

Inspection of the columns of the matrix V associated with small singular values is a very effective technique for identifying the sets of columns of A that are nearly linearly dependent [see Eq. (12.23) and (12.24)].

It has been our experience that computing and displaying the matrix V, the quantities s_k, s_k^{-1}, $\hat{p}_k$, $x^{(k)}$, and $||x^{(k)}||$ for $k = 1, \ldots, n$ and g_k, g_k^2, ρ_k^2, and ρ_k for $k = 0, 1, \ldots, n$ is extremely helpful in analyzing difficult practical least squares problems. For certain statistical interpretations [see Eq. (12.2)] the quantities

(25.62) $$\sigma_k = \left[\frac{\rho_k^2}{(m-k)}\right]^{1/2} \qquad k = 0, 1, \ldots, n$$

are also of interest.

Suppose the matrix A is ill-conditioned; then some of the later singular values are significantly smaller than the earlier ones. In such a case some of the later $\hat{p}_j$ values may be undesirably large. Typically one hopes to locate an index k such that all coefficients p_j for $j \leq k$ are acceptably small, all singular values s_j for $j \leq k$ are acceptably large, and the residual norm ρ_k is acceptably small. If such an index k exists, then one can take the candidate vector $x^{(k)}$ as an acceptable solution vector.

This technique has been used successfully in a variety of applications. For example, see Hanson (1971) for applications to the numerical solution of Fredholm integral equations of the first kind.

Once the singular values and the vector g have been computed it is also a simple matter to compute the numbers $||p^{(\lambda)}||$ and ω_λ of the Levenberg–Marquardt stabilization method [see Eq. (25.47) and (25.48)] for a range of values of λ. These quantities are of interest due to the optimality property stated in Theorem (25.49).

An example in which all the quantities mentioned here are computed and interpreted for a particular set of data is given in the following chapter.

26

EXAMPLES OF SOME METHODS OF ANALYZING A LEAST SQUARES PROBLEM

Consider the least squares problem $Ax \cong b$ where the data matrix $[A_{15\times5} : b_{15\times1}]$ is given by Table 26.1. (These data are given also as the data element, **DATA4**, in Appendix C.) We shall assume there is uncertainty of the order of 0.5×10^{-8} in the elements of A and 0.5×10^{-4} in the components of b.

First consider a singular value analysis of this problem as described in Section 6 of Chapter 25. The computation and printing of the quantities used in singular value analysis is accomplished by the Fortran subroutine **SVA** (Appendix C). The Fortran main program **PROG4** (Appendix C) applies the subroutine **SVA** to the particular data of the present example. Executing **PROG4** on a UNIVAC 1108 computer resulted in the output reproduced in Fig. 26.1.

The subroutine **SVA** in general performs a change of variables $x = Dy$ and treats the problem $(AD)y \cong b$. In this example we have set $D = I$ so the label y in Fig. 26.1 is synonymous with x for this example.

Recall that the vector g (output column headed "G COEF") is computed as $g = U^T b$, where U is orthogonal. Since the fourth and fifth components of g are smaller than the assumed uncertainty in b, we are willing to treat these two components of g as zero. This leads us to regard the third candidate solution, $x^{(3)}$ (output column headed "SOLN 3") as the most satisfactory solution.

Alternatively one could compare the numbers σ_k [see Eq. (25.62)] in the column headed "N.S.R.C.S.S" (meaning Normalized Square Root of Cumulative Sum of Squares) with the assumed uncertainty (0.5×10^{-4}) in b. Note that $\sigma_2 (= 1.1107 \times 10^{-2})$ is significantly larger than this assumed uncertainty while $\sigma_3 (= 4.0548 \times 10^{-5})$ is slightly smaller than this uncertainty. This

Table 26.1   THE DATA MATRIX $[A:b]$

−.13405547	−.20162827	−.16930778	−.18971990	−.17387234	−.4361
−.10379475	−.15766336	−.13346256	−.14848550	−.13597690	−.3437
−.08779597	−.12883867	−.10683007	−.12011796	−.10932972	−.2657
.02058554	.00335331	−.01641270	.00078606	.00271659	−.0392
−.03248093	−.01876799	.00410639	−.01405894	−.01384391	.0193
.05967662	.06667714	.04352153	.05740438	.05024962	.0747
.06712457	.07352437	.04489770	.06471862	.05876455	.0935
.08687186	.09368296	.05672327	.08141043	.07302320	.1079
.02149662	.06222662	.07213486	.06200069	.05570931	.1930
.06687407	.10344506	.09153849	.09508223	.08393667	.2058
.15879069	.18088339	.11540692	.16160727	.14796479	.2606
.17642887	.20361830	.13057860	.18385729	.17005549	.3142
.11414080	.17259611	.14816471	.16007466	.14374096	.3529
.07846038	.14669563	.14365800	.14003842	.12571177	.3615
.10803175	.16994623	.14971519	.15885312	.14301547	.3647

could be taken as a reason to select $x^{(3)}$ as the preferred candidate solution.

If the assumed uncertainty in b is given the statistical interpretation of being the standard deviation of the errors in b, then multiplication of this quantity (0.5×10^{-4}) times each reciprocal singular value (column headed "RECIP. S.V.") gives the standard deviation of the corresponding component of the vector p (column headed "P COEF"). Thus the first three components of p exceed their respective standard deviations in magnitude while the last two components are smaller in magnitude than their standard deviations. This could be taken as a reason to prefer the candidate solution $x^{(3)}$.

As still another method for choice of an $x^{(k)}$, one would probably reject the candidate solutions $x^{(1)}$ and $x^{(2)}$ because their associated residual norms are too large (column headed "RNORM"). One would probably reject $x^{(5)}$ because $\|x^{(5)}\|$ is too large (column headed "YNORM"). The choice between $x^{(3)}$ and $x^{(4)}$ is less decisive. The vector $x^{(3)}$ would probably be preferred to $x^{(4)}$ since $\|x^{(3)}\| < \|x^{(4)}\|$ and the residual norm associated with $x^{(4)}$ is only slightly smaller than that associated with $x^{(3)}$.

While these four methods of choosing a preferred $x^{(k)}$ have led us to the same selection, $x^{(3)}$, this would not always be the case for other sets of data. The user must decide which of these criteria (possibly others included) are most meaningful for his application.

It is of interest to compare some other methods of obtaining stabilized solutions to this same problem. The Levenberg–Marquardt analysis (Chapter 25, Section 4) provides a continuum of candidate solutions. Using the Levenberg–Marquardt data shown in Fig. 26.1 one can produce the graph, Fig. 26.2, of RNORM versus YNORM. Following Theorem (25.49) this curve constitutes a boundary line in the (YNORM–RNORM)-plane such

SINGULAR VALUE ANALYSIS OF THE LEAST SQUARES PROBLEM, A*X=B, SCALED AS (A*D)*Y=B .

M = 15, N = 5, MDATA = 15

SCALING OPTION NO. 1. D IS THE IDENTITY MATRIX.

V-MATRIX OF THE SINGULAR VALUE DECOMPOSITION OF A*D.
(ELEMENTS OF V SCALED UP BY A FACTOR OF 10**4)

	COL 1	COL 2	COL 3	COL 4	COL 5
1	3742.	-7526.	3382.	-1981.	-3741.
2	5196.	-636.	2301.	6349.	5195.
3	4123.	6510.	4741.	-1067.	-4123.
4	4796.	689.	-2493.	-6877.	4797.
5	4359.	302.	-7388.	2707.	-4359.

INDEX	SING. VALUE	P COEF	RECIP. S.V.	G COEF	G**2	C.S.S.	N.S.R.C.S.S.
0						1.0412+00	2.6347-01
1	1.0000	9.9981-01	1.0000+00	9.9981-01	9.9963-01	4.1617-02	5.4522-02
2	.1000	2.0003+00	1.0000+01	2.0003-01	4.0013-02	1.6038-03	1.1117-02
3	.0100	-4.0047+00	1.0000+02	-4.0047-02	1.6038-03	1.9730-08	4.0548-05
4	.9997-05	1.7755+00	1.0003+05	1.7750-05	3.1506-10	1.9415-08	4.2012-05
5	.9904-07	1.6761+02	1.0097+07	1.6599-05	2.7554-10	1.9139-08	4.3748-05

INDEX	YNORM	RNORM	LOG10(YNORM)	LOG10(RNORM)
0	.0u000	.10204+01	-1000.00000	.00878
1	.99981+00	.20400+00	-.00008	-.69036
2	.22363+01	.40047-01	.34953	-1.39743
3	.45868+01	.14046-03	.66151	-3.85244
4	.49184+01	.13934-03	.69183	-3.85593
5	.16768+03	.13834-03	2.22449	-3.85904

NORMS OF SOLUTION AND RESIDUAL VECTORS FOR A RANGE OF VALUES OF THE LEVENBERG-MARQUARDT PARAMETER, LAMBDA.

LAMBDA	YNORM	RNORM	LOG10(LAMBDA)	LOG10(YNORM)	LOG10(RNORM)
.10000+02	.99012+00	.10107+01	1.00000	-2.00431	.00463
.35464+01	.73657-01	.94834+00	.54979	-1.13279	-.02303
.12577+01	.36746+00	.64525+00	.09958	-.41178	-.19027
.44603+00	.83939+00	.25574+00	-.35063	-.07604	-.59221
.15818+00	.11304+01	.15037+00	-.80084	.05325	-.82283
.56098-01	.16231+01	.61718-01	-1.25105	.26080	-1.20959
.19895-01	.23138+01	.32867-01	-1.70126	.36433	-1.48324
.70555-02	.34800+01	.13348-01	-2.15147	.54158	-1.87460
.25022-02	.43817+01	.23671-02	-2.60168	.64165	-2.62579
.88737-03	.45594+01	.34333-03	-3.05189	.65891	-3.46429
.31470-03	.45833+01	.14596-03	-3.50210	.66114	-3.83578
.11161-03	.45864+01	.14053-03	-3.95231	.66147	-3.85222
.39580-04	.45880+01	.14033-03	-4.40252	.66162	-3.85264
.14037-04	.46256+01	.13983-03	-4.85273	.66516	-3.85439
.49780-05	.48028+01	.13938-03	-5.30295	.68150	-3.85580
.17654-05	.49274+01	.13933-03	-5.75316	.69262	-3.85595
.62609-06	.63958+01	.13929-03	-6.20337	.80590	-3.85608
.22204-06	.28244+02	.13904-03	-6.65358	1.45092	-3.85687
.78743-07	.10281+03	.13849-03	-7.10379	2.01203	-3.85857
.27926-07	.15534+03	.13835-03	-7.55400	2.19129	-3.85902
.99036-08	.16602+03	.13834-03	-8.00421	2.22017	-3.85904

SEQUENCE OF CANDIDATE SOLUTIONS, X

	SOLN 1	SOLN 2	SOLN 3	SOLN 4	SOLN 5
1	.37409643+00	-.11313896+00	-.24857328+01	-.28373911+01	-.65543824+02
2	.51951898+00	.39226783+00	-.52913252+00	.59819891+00	.87667385+02
3	.41223419+00	.17145393+01	-.11914114+00	-.37354642+00	-.69476752+02
4	.47949432+00	.61736754+00	.16156794+01	.39464945+00	.80802341+02
5	.43580915+00	.49625358+00	.34547871+01	.39353651+01	-.69133711+02

Fig. 26.1 Output from the subroutine SVA for the sample problem of Chapter 26.

that for any vector y the coordinate pair $(\|y\|, \|b - Ay\|)$ lies on or above the curve.

For more detailed information Eq. (25.46), (25.38), and (25.35) can be used to compute and plot values of the individual solution components as functions of λ. Figures 26.3 and 26.4 show this information for the present example. Graphs of this type are extensively discussed in Hoerl and Kennard (1970b).

Figure 26.5 provides a comparison of solution norms and residual norms of the five candidate solutions obtained by singular value analysis with the corresponding data for the continuum of Levenberg–Marquardt solutions.

We have also applied the subroutine, **HFTI** (see Appendix C) to this example. The magnitudes of the diagonal elements of the triangular matrix

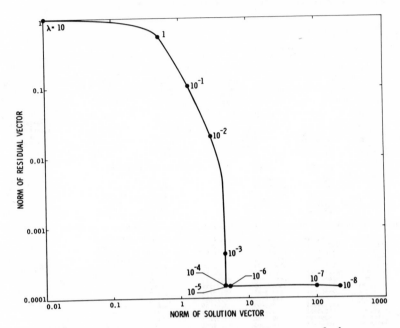

Fig. 26.2   Residual norm versus solution norm for a range of values of the Levenberg-Marquardt stabilization parameter λ.

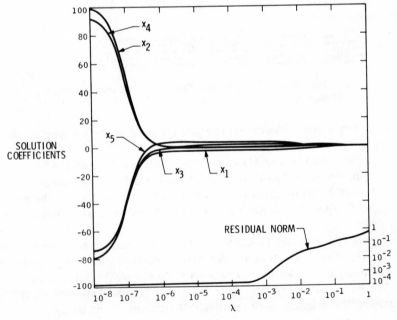

Fig. 26.3   Solution coefficients and residual norm as a function of λ.

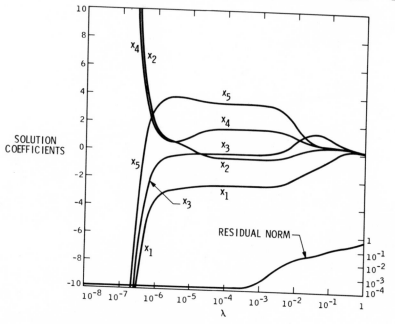

Fig. 26.4 Same data as Fig. 26.3 with expanded vertical scale.

R to which the matrix A is transformed are 0.52, 0.71×10^{-1}, 0.91×10^{-2}, 0.14×10^{-4}, and 0.20×10^{-6}. It is interesting to note that these numbers agree with the respective singular values of A to within a factor of two. From Theorem (6.31) it is known that the singular values, s_i, and the diagonal elements r_{ii} must satisfy

$$1.00 \leq \frac{s_1}{r_{11}} \leq 2.24$$

$$0.58 \leq \frac{s_2}{r_{22}} \leq 2.00$$

$$0.33 \leq \frac{s_3}{r_{33}} \leq 1.73$$

$$0.18 \leq \frac{s_4}{r_{44}} \leq 1.41$$

$$0.09 \leq \frac{s_5}{r_{55}} \leq 1.00$$

By setting the pseudorank tolerance parameter successively to the values $\tau = 0.29$, 0.040, 0.0046, 0.0000073, and 0.0, the subroutine **HFTI** was used to produce five different solutions, say $z^{(k)}$, associated with pseudoranks of $k = 1, 2, 3, 4$, and 5. The solution norms and residual norms for these five vectors are shown in Table 26.2.

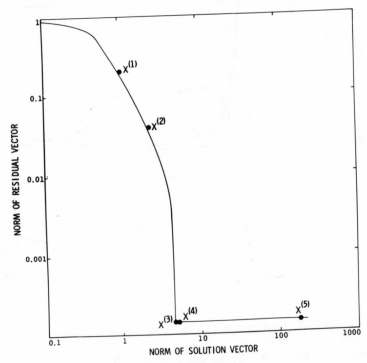

Fig. 26.5 Candidate solutions $x^{(i)}$ from singular value analysis compared with the continuum of Levenberg-Marquardt solutions.

Table 26.2 SOLUTION AND RESIDUAL NORMS USING **HFTI** ON THE SAMPLE PROBLEM

k	$\|z^{(k)}\|$	$\|b - Az^{(k)}\|$
1	0.99719	0.216865
2	2.24495	0.039281
3	4.58680	0.000139
4	4.92951	0.000139
5	220.89008	0.000138

Note that the data of Table 26.2 are quite similar to the corresponding data (columns headed "YNORM" and "RNORM") given in Fig. 26.1 for the candidate solutions obtained by singular value analysis.

As still another way of analyzing this problem, solutions were computed using each of the 31 possible nonnull subsets of the five columns of the matrix A. Solution norms and residual norms for each of these 31 solutions are listed in Table 26.3 and are plotted in Fig. 26.6.

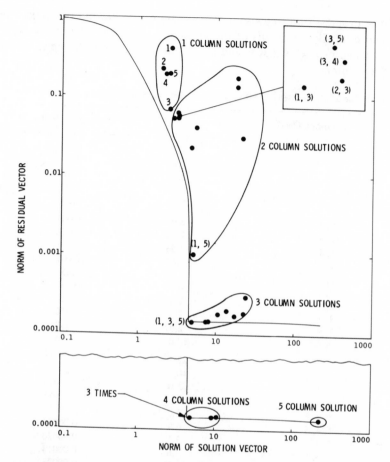

Fig. 26.6 Solutions using subsets of the columns of A.

We shall use the notation $w^{(i, j, \dots)}$ to denote the solution obtained using columns $i, j, \dots$.

From Fig. 26.6 we see that the simplest form of stepwise regression would select solution $w^{(3)}$ as the preferred solution among those involving only one column.

It would next consider solutions $w^{(1,3)}$, $w^{(2,3)}$, $w^{(3,4)}$, and $w^{(3,5)}$. It would select $w^{(1,3)}$ as the one of these four solutions providing the smallest residual norm. Note that this solution has a residual norm larger by a factor of 85 than that corresponding to $w^{(1,5)}$. This latter solution clearly gives the smallest norm obtainable using any set of two columns.

At the next stage simple stepwise regression would select $w^{(1,3,5)}$, which is quite similar in solution norm and residual norm to the previously mentioned $x^{(3)}$ and $z^{(3)}$.

The progress of a stepwise regression algorithm beyond this stage would depend critically on details of the particular algorithm due to the very ill-conditioned problems that would be encountered in considering sets of four or five columns.

Table 26.3 SOLUTION AND RESIDUAL NORMS USING SUBSETS OF THE COLUMNS

Columns Used	$\|w\|$	$\|b - Aw\|$
1	2.46	0.40
2	1.92	0.22
3	2.42	0.07
4	2.09	0.19
5	2.30	0.19
1, 2	5.09	0.039
1, 3	2.72	0.052
1, 4	4.53	0.023
1, 5	5.07	0.001
2, 3	3.03	0.053
2, 4	20.27	0.030
2, 5	17.06	0.128
3, 4	3.07	0.056
3, 5	2.97	0.058
4, 5	17.05	0.175
1, 2, 3	22.1	0.00018
1, 2, 4	10.8	0.00018
1, 2, 5	5.0	0.00014
1, 3, 4	8.1	0.00015
1, 3, 5	5.0	0.00014
1, 4, 5	4.9	0.00014
2, 3, 4	13.5	0.00020
2, 3, 5	7.6	0.00014
2, 4, 5	24.0	0.00028
3, 4, 5	17.3	0.00017
1, 2, 3, 4	10.3	0.00014
1, 2, 3, 5	5.0	0.00014
1, 2, 4, 5	5.0	0.00014
1, 3, 4, 5	5.0	0.00014
2, 3, 4, 5	9.0	0.00014
1, 2, 3, 4, 5	220.9	0.00014

MODIFYING A QR DECOMPOSITION TO ADD OR REMOVE ROW VECTORS WITH APPLICATION TO SEQUENTIAL PROCESSING OF PROBLEMS HAVING A LARGE OR BANDED COEFFICIENT MATRIX

In this chapter we present adaptations of orthogonal transformations to the sequential processing of data for Problem LS. Such methods provide a means of conserving computer storage for certain types of problems having voluminous data.

These methods also have important application to problems in which one wishes to obtain a sequence of solutions for a data set to which data are being sequentially added or deleted. A requirement for this type of computation arises in the class of problems called *sequential estimation, filtering,* or *process identification.* The augmentation methods presented in this chapter have excellent properties of numerical stability. This is in contrast to many published approaches to this problem that are based on the notion of updating the inverse of a matrix.

The problem of deleting data can be an inherently unstable operation depending upon the relationship of the rows to be deleted to the whole data set. Some methods will be presented that attempt to avoid introducing any additional unnecessary numerical instability.

Write the least squares problem in the usual way:

$$(27.1) \qquad Ax \cong b, \ A_{m \times n}, \ b_{m \times 1}$$

In Section 1 we treat the problem in which the number of rows of A is large compared with the number of columns. In Section 2 attention is devoted to the case in which A has a "banded" structure. The methods presented in both Sections 1 and 2 are applicable to the sequential estimation problem since the solution vector or its associated covariance matrix can readily be computed at each step of the sequential matrix accumulation if required. In Section 3 an application of these ideas is presented using the example of curve fitting using *line splines*. In Section 4 a procedure is described for least squares data fitting by *cubic splines* having equally spaced breakpoints. This provides a further example of the applicability of the banded sequential accumulation method of Section 2.

Methods of deleting data will be described in Section 5.

Section 1. SEQUENTIAL ACCUMULATION

In this section we shall describe an algorithm for transforming the matrix $[A : b]$ to upper triangular form without requiring that the entire matrix $[A : b]$ be in computer storage at one time. A formal description of this procedure will be given as Algorithm SEQHT (27.10).

We first establish the notation to be used and give an informal description of the process. Write the matrix A and the vector b in partitioned form:

$$(27.2) \qquad A = \begin{bmatrix} A_1 \\ A_2 \\ \cdot \\ \cdot \\ \cdot \\ A_q \end{bmatrix}$$

$$(27.3) \qquad b = \begin{bmatrix} b_1 \\ b_2 \\ \cdot \\ \cdot \\ \cdot \\ b_q \end{bmatrix}$$

where each A_i is $m_i \times n$ and each b_i is a vector of length m_i. We have $m = m_1 + \cdots + m_q$, of course. The integers m_i can be as small as 1, which permits the greatest economy of storage. Remarks on the dependence of the total number of arithmetic operations on the size of m_i are given on page 211.

The algorithm will construct a sequence of triangular matrices $[R_i : d_i]$, $i = 1, \ldots, q$, with the property that the least squares problem

$$R_i x \cong d_i$$

has the same solution set and the same residual norm as the problem

$$\begin{bmatrix} A_1 \\ \cdot \\ \cdot \\ \cdot \\ A_t \end{bmatrix} x \cong \begin{bmatrix} b_1 \\ \cdot \\ \cdot \\ \cdot \\ b_t \end{bmatrix}$$

The significant point permitting a saving of storage is the fact that for each i, the matrix $[R_i : d_i]$ can be constructed and stored in the storage space previously occupied by the augmented matrix

(27.4) $$[\bar{A}_i : \bar{b}_i] \equiv \begin{bmatrix} R_{i-1} : d_{i-1} \\ A_i \quad : b_i \end{bmatrix}$$

Thus the maximum number of rows of storage required is $\max_j \{m_j + \min [n + 1, \sum_{i=1}^{j-1} m_i]\}$.

For notational convenience let $[R_0 : d_0]$ denote a (null) matrix having no rows. At the beginning of the ith stage of the algorithm one has the $\hat{m}_{i-1} \times (n + 1)$ matrix $[R_{i-1} : d_{i-1}]$ from the $(i - 1)$st stage and the new $m_i \times (n + 1)$ data matrix $[A_i : b_i]$. Let $\bar{m}_i = \hat{m}_{i-1} + m_i$. Form the $\bar{m}_i \times (n + 1)$ augmented matrix of Eq. (27.4) and reduce this matrix to triangular form by Householder triangularization as follows:

(27.5) $$Q_i[\bar{A}_i : \bar{b}_i] = \begin{bmatrix} R_i : d_i \\ 0 : 0 \end{bmatrix} \begin{matrix} \}\hat{m}_i \\ \}\bar{m}_i - \hat{m}_i \end{matrix}$$

where $\hat{m}_i = \min \{n + 1, \bar{m}_i\}$.

This completes the ith stage of the algorithm. For future reference, let the result at the qth stage of the algorithm be denoted by

(27.6) $$\begin{matrix} n & 1 \\ \overbrace{\begin{bmatrix} R & d \\ 0 & e \end{bmatrix}} \end{matrix} \begin{matrix} \}n \\ \}1 \end{matrix} \equiv [R_q, d_q]$$

It is easily verified that there exists an orthogonal matrix Q such that

$$Q[A : b] = \begin{bmatrix} R & d \\ 0 & e \\ 0 & 0 \end{bmatrix}$$

Therefore, the least squares problem

(27.7) $$\begin{bmatrix} R \\ 0 \end{bmatrix} x \cong \begin{bmatrix} d \\ e \end{bmatrix}$$

is equivalent to the original problem in Eq. (27.1) in that it has the same set of solution vectors and the same minimal residual norm. Furthermore the matrix R has the same set of singular values as the matrix A of Eq. (27.1).

We shall now describe a computing algorithm that formalizes this procedure. To this end set

$$(27.8) \qquad v_j = \begin{cases} 0 & \text{if } j = 0 \\ \sum_{i=1}^{j} m_i & \text{if } j > 0 \end{cases}$$

and

$$(27.9) \qquad \mu = \max_{1 \le j \le q} \{m_j + \min [n + 1, v_{j-1}]\}$$

The processing will take place in a computer storage array W having at least μ rows and $n + 1$ columns. We shall use the notation $W(i_1 : i_2, j_1 : j_2)$ to denote the subarray of W consisting of rows i_1 through i_2 of columns j_1 through j_2. We use the notation $W(i, j)$ to denote the (i, j) element of the array W. The symbol ρ identifies a single storage location.

(27.10) ALGORITHM SEQHT (*Sequential Householder Triangularization*)

Step	Description
1	Set $l := 0$.
2	For $t := 1, \ldots, q$, do Steps 3–6.
3	Set $p := l + m_t$.
4	Set $W(l + 1 : p, 1 : n + 1) := [A_t : b_t]$ [see Eq. (27.2) and (27.3)].
5	For $i := 1, \ldots, \min (n + 1, p - 1)$, execute Algorithm H1 $(i, \max (i + 1, l + 1), p, W(1, i), \rho, W(1, i + 1), n - i + 1)$.
6	Set $l := \min (n + 1, p)$.
7	*Remark:* The matrix $[A : b]$ has been reduced to upper triangular form as given in Eq. (27.6).

A sequence of Fortran statements implementing Step 5 of this algorithm is given as an example in the user's guide for the subroutine **H12** in Appendix C.

Notice, in Step 4, that only the submatrix $[A_t : b_t]$ needs to be introduced into computer stores at one time. All these data can be processed in the $\mu \times (n + 1)$ working array W. By more complicated programming one could further reduce the storage required by exploiting the fact that each matrix $[R_j : d_j]$ is upper triangular.

For the purpose of discussing operation counts let α denote an addition or subtraction and μ denote a multiplication or division. Define

$$v = \frac{mn^2 - n^3/3}{2}$$

It was noted in Table 19.1 that if quadratic and lower-order terms in m and n are neglected, the number of operations required to triangularize an $m \times n$ matrix $(m > n)$ using Householder transformations is approximately $v(2\alpha + 2\mu)$. If the Householder processing is done sequentially involving q stages as in Algorithm SEQHT (27.10), then the operation count is increased to approximately $v(2\alpha + 2\mu)(m + q)/m$. If the entering blocks of data each contain k rows $(kq = m)$, then the operation count can be written as $v(2\alpha + 2\mu)(k + 1)/k$.

Thus sequential Householder accumulation increases in cost as the block size is decreased. In the extreme case of a block size of $k = 1$, the operation count is approximately doubled relative to the count for nonsequential Householder processing.

For small block sizes it may be more economical to replace the Householder method used at Step 5 of Algorithm SEQHT (27.10) with one of the methods based on 2×2 rotations or reflections. The operation counts for these methods are independent of the block size. The operation count for triangularization using the Givens method [Algorithms G1 (10.25) and G2 (10.26)] is $v(2\alpha + 4\mu)$. The 2×2 Householder transformation expressed as in Eq. (10.27) has an operation count of $v(3\alpha + 3\mu)$.

Gentleman's modification (see Chapter 10) of the Givens method reduces the count to $v(2\alpha + 2\mu)$. Thus this method is competitive with the standard Householder method for nonsequential processing and requires fewer arithmetic operations than the Householder method for sequential processing.

Actual comparative performance of computer programs based on any of the methods will depend strongly on coding details.

Following an application of Algorithm SEQHT (27.10), if the upper triangular matrix R of Eq. (27.6) is nonsingular, we may compute the solution, $\hat{x}$, by solving

(27.11)
$$Rx = d$$

The number e in Eq. (27.6) will satisfy

(27.12)
$$|e| = \|A\hat{x} - b\|$$

In many applications one needs the unscaled covariance matrix [see Eq. (12.1)]

(27.13)
$$C = (A^T A)^{-1} = R^{-1}(R^{-1})^T$$

Computation of R^{-1} can replace R in storage and then computation of (the upper triangular part of) $R^{-1}(R^{-1})^T$ can replace R^{-1} in storage. Thus the matrix C of Eq. (27.13) can be computed with essentially no additional storage. See Chapter 12 for a more detailed discussion of this computation. Furthermore, the quantity σ^2 of Eq. (12.2) can be computed as

$$\sigma^2 = \frac{e^2}{m - n}$$

where e is defined by Eq. (27.6).

For definiteness, Algorithm SEQHT was presented as a loop for $t = 1,$ $\ldots, q$. In an actual application using this sequential processing it is more likely that Steps 3–6 of the algorithm would be implemented as a subroutine. The number q and the total data set represented by the matrix $[A : b]$ need not be known initially. The calling program can use the current triangular matrix at any stage at which R is nonsingular to solve for a solution based on the data that have been accumulated up to that point. By using an additional set of $n(n + 1)/2$ storage locations, the calling program can also compute the upper triangular part of the unscaled covariance matrix $C_t = R_t^{-1}(R_t^{-1})^T$ at any stage at which R_t is nonsingular. Thus Steps 3–6 of Algorithm SEQHT provide the algorithmic core around which programs for sequential estimation can be constructed.

Section 2. SEQUENTIAL ACCUMULATION
OF BANDED MATRICES

In some problems, possibly after preliminary interchange of rows or columns, the data matrix $[A : b]$ of Eq. (27.2) and (27.3) has a banded structure in the following sense. There exists an integer $n_b \leq n$ and a nondecreasing set of integers $j_1, \ldots, j_q$ such that all nonzero elements in the submatrix A_t occur in columns j_t through $j_t + n_b - 1$.

Thus A_t is of the form

$$(27.14) \qquad A_t = [\ \overbrace{0}^{j_t - 1} : \overbrace{C_t}^{n_b} : \overbrace{0}^{n - n_b - j_t + 1}\ ]\}m_t$$

We shall refer to n_b as the *bandwidth* of A.

It can be easily verified that all nonzero elements in the ith row of the upper triangular matrix R of Eq. (27.6) will occur in column i through $i + n_b - 1$. Furthermore, rows 1 through $j_t - 1$ of R will not be affected by the processing of the submatrices $A_t, \ldots, A_q$ in Algorithm (27.10).

These observations allow us to modify Algorithm (27.10) by managing the storage in a working array G that needs only $n_b + 1$ columns. Specifically, let G denote a $\mu \times (n_b + 1)$ array of storage with μ satisfying Eq. (27.9).

The working array G will be partitioned by the algorithm into three sub-arrays G_1, G_2, and G_3 as follows:

(27.15) $G_1 = $ rows 1 through $i_p - 1$ of G

(27.16) $G_2 = $ rows i_p through $i_r - 1$ of G

(27.17) $G_3 = $ rows i_r through $i_r + m_t - 1$ of G

The integers i_p and i_r are determined by the algorithm and their values change in the course of the processing. These integers are limited by the inequalities $1 \leq i_p \leq i_r \leq n + 2$.

At the various stages of the algorithm the $(n_b + 1)$st column of the G array contains either the vector $\bar{b}_t$ [e.g., see the left side of Eq. (27.4)] or the processed vector d_t [e.g., see the right side of Eq. (27.5)].

For $1 \leq j \leq n$ the identification of matrix elements with storage locations is as follows:

(27.18)   In G_1: storage location (i, j) contains matrix element
            $(i, i + j - 1)$

(27.19)   In G_2: storage location (i, j) contains matrix element
            $(i, i_p + j - 1)$

(27.20)   In G_3: storage location (i, j) contains matrix element
            $(i, j_t + j - 1)$

To indicate the primary idea of a compact storage algorithm for the banded least squares problem consider Fig. 27.1 and 27.2. If Algorithm (27.10) were applied to the block diagonal problem with $n_b = 4$, then at the step in which the data block $[C_t, b_t]$ is introduced the working array might appear as in Fig. 27.1. Taking advantage of the limited bandwidth these data can be packed in storage as in Fig. 27.2.

In this illustration the inequalities $i_p < j_t \leq i_r$ are satisfied. For further diagrammatic illustration of this algorithm the reader may wish to look ahead to Section 3 and Fig. 27.4 and 27.5.

A detailed statement of the algorithm follows. In describing the algorithm we shall use the notation $G(i, j)$ to denote the (i, j) element of the array G and the notation $G(i_1 : i_2, j_1 : j_2)$ to denote the subarray of G consisting of all elements $G(i, j)$ with $i_1 \leq i \leq i_2$ and $j_1 \leq j \leq j_2$.

(27.21) ALGORITHM BSEQHT (*Banded Sequential Householder Triangularization*)

Step	Description
1	Set $i_r := 1$ and $i_p := 1$.
2	For $t := 1, \ldots, q$, do Steps 3–24.

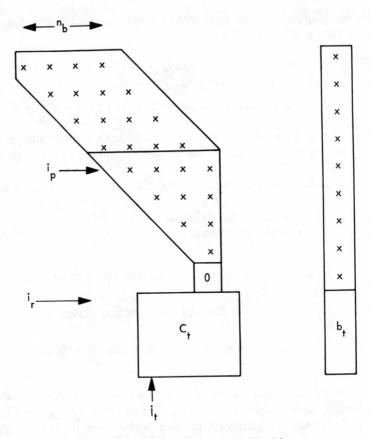

Fig. 27.1 Introducing the data C_t and b_t.

Step	Description
3	*Remarks:* At this point the data $[C_t : b_t]$ and the integers m_t and j_t must be made available to the algorithm. It is assumed that $m_t > 0$ for all t and $j_q \geq j_{q-1} \geq \cdots \geq j_1 \geq 1$.
4	Set $G(i_r : i_r + m_t - 1, 1 : n_b + 1) := [C_t : b_t]$ [See Eq. (27.14) for a definition of C_t. Note that the portion of the G array into which data are being inserted is the subarray G_3 of Eq. (27.17).]
5	If $j_t = i_p$, go to Step 18.
6	*Remark:* Here the monotonicity assumption on j_t assures that $j_t > i_p$.
7	If $j_t \leq i_r$, go to Step 13.

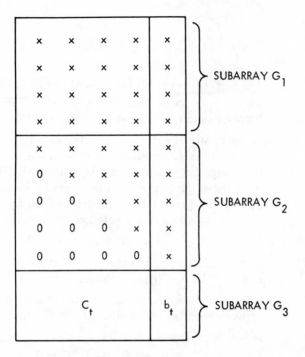

Fig. 27.2 The same data as in Fig. 27.1 but packed into the storage array, G.

Step	Description
8	*Remark:* Here j_t exceeds i_r. This indicates rank deficiency since the diagonal element in position (i_r, i_r) of the triangular matrix R [see Eq. (27.6)] will be zero. Nevertheless, the triangularization can be completed. Some methods for solving this rank-deficient least squares problem will be discussed in the text following this algorithm.
9	Move the contents of $G(i_r : i_r + m_t - 1, 1 : n_b + 1)$ into $G(j_t : j_t + m_t - 1, 1 : n_b + 1)$.
10	Zero the subarray $G(i_r : j_t - 1, 1 : n_b + 1)$.
11	Set $i_r := j_t$.
12	Set $\mu := \min(n_b - 1, i_r - i_p - 1)$; if $\mu = 0$, go to Step 17.
13	For $l := 1, \ldots, \mu$, do Steps 14–16.
14	Set $k := \min(l, j_t - i_p)$.

Step	*Description*

15   For $i := l + 1, \ldots, n_b$, set $G(i_p + l, i - k) := G(i_p + l, i)$.

16   For $i := 1, \ldots, k$, set $G(i_p + l, n_b + 1 - i) := 0$.

17   Set $i_p := j_t$. [Note that this redefines the partition line between subarrays G_1 and G_2 of Eq. (27.15) and (27.16).]

18   *Remark:* Steps 19 and 20 apply Householder triangularization to rows i_p through $i_r + m_t - 1$. As was discussed following the statement of Algorithm SEQHT (27.10), some reduction of execution time may be attainable by replacing the general Householder method by one of the methods based on 2×2 rotations or reflections.

19   Set $\hat{m} := i_r + m_t - i_p$. Set $\hat{k} := \min (n_b + 1, \hat{m})$.

20   For $i := 1, \ldots, \hat{k}$, execute Algorithm H1 $(i, \max (i + 1, i_r - i_p + 1), \hat{m}, G(i_p, i), p, G(i_p, i + 1), n_b + 1 - i)$.

21   Set $i_r := i_p + \hat{k}$. [Note that this redefines the partition line between subarrays G_2 and G_3 of Eq. (27.16) and (27.17).]

22   If $\hat{k} < n_b + 1$, go to Step 24.

23   For $j := 1, \ldots, n_b$, set $G(i_r - 1, j) := 0$.

24   Continue.

25   *Remark:* The main loop is finished. The triangular matrix of Eq. (27.6) is stored in the subarrays G_1 and G_2 [see Eq. (27.15) and (27.16)] according to the storage mapping of Eq. (27.18) and (27.19).

            If the main diagonal elements of the matrix R of Eq. (27.6) are all nonzero, the solution of problem (27.7) can be computed by back substitution using Steps 26–31. Note that these diagonal elements of the matrix R are used as divisors at Step 31. Discussion of some techniques for handling the alternative (singular) case are given in the text following this algorithm.

26   For $i := 1, \ldots, n$, set $X(i) := G(i, n_b + 1)$.

27   For $i := n, n - 1, \ldots, 1$ do Steps 28–31.

28   Set $s := 0$ and $l := \max (0, i - i_p)$.

29   If $i = n$, go to Step 31.

30   For $j := 2, \ldots, \min (n + 1 - i, n_b)$, set $s := s + G(i, j + l) \times X(i - 1 + j)$.

31   Set $X(i) := [X(i) - s]/G(i, l + 1)$.

Step	Description
32	*Remark:* The solution vector x is now stored in the array X. If, as would usually be the case, the full data matrix $[A : b]$ [see Eq. (27.2) and (27.3)] has more than n rows, the scalar quantity, e, of Eq. (27.6) will be stored in location $G(n + 1, n_b + 1)$. The magnitude of e is the norm of the residual associated with the solution vector, x.

The banded structure of A does not generally imply that the unscaled covariance matrix C of Eq. (12.1) will have any band-limited structure. However, its columns, $c_j, j = 1, \ldots, n$, can be computed, one at a time, without requiring any additional storage. Specifically the vector c_j can be computed by solving the two banded triangular systems

(27.22) $$R^T w_j = e_j$$

(27.23) $$R c_j = w_j$$

where e_j is the jth column of the $n \times n$ identity matrix.

The computational steps necessary to solve Eq. (27.22) or more generally the problem $R^T y = z$ for arbitrary z are given in Algorithm (27.24). Here it is assumed that the matrix R occupies the array G in the storage arrangement resulting from the execution of Algorithm (27.21). The right-side vector, for example e_j for Eq. (27.22), must be placed in the array X. Execution of Algorithm (27.24) replaces the contents of the array X by the solution vector, for example, w_j of Eq. (27.22).

(27.24) ALGORITHM *Solving $R^T y = z$*

Step	Description
1	For $j := 1, \ldots, n$, do Steps 2–6.
2	Set $s := 0$.
3	If $j = 1$, go to Step 6.
4	Set $i_1 := \max(1, j - n_b + 1)$. Set $i_2 := j - 1$.
5	For $i := i_1, \ldots, i_2$, set $s := s + X(i) \times G(i, j - i + 1 + \max(0, i - i_p))$.
6	Set $X(j) := (X(j) - s)/G(j, 1 + \max(0, j - i_p))$.

To solve Eq. (27.23) for c_j after solving Eq. (27.22) for w_j one can simply execute Steps 27–31 of Algorithm (27.21).

Furthermore, the quantity σ^2 of Eq. (12.2) may be computed as

$$\sigma^2 = \frac{e^2}{m - n}$$

where e is defined in Eq. (27.6). The quantity e is computed by the Algorithm BSEQHT as noted at Step 32.

Fortran subroutines **BNDACC** and **BNDSOL** are included in Appendix C as an implementation of Algorithm BSEQHT (27.21) and Algorithm (27.24). Use of these subroutines is illustrated by the Fortran program **PROG5** of Appendix C, which fits a cubic spline curve to discrete data and computes the covariance matrix of the coefficients using the technique of Eq. (27.22) and (27.23).

The remarks at the end of Section 1 regarding possible implementations of Algorithm SEQHT (27.10) for sequential estimation apply with some obvious modification to Algorithm BSEQHT also. The significant difference is the fact that in Algorithm BSEQHT the number of columns having nonzero entries is typically increasing as additional blocks of data are introduced. Thus the number of components of the solution vector that can be determined is typically less at earlier stages of the process than it is later.

At Steps 8 and 25 mention was made of the possibility that the problem might be rank-deficient. We have found that the Levenberg idea (see Chapter 25, Section 4) provides a very satisfactory approach to this situation without increasing the bandwidth of the problem. We refer to the notation of Eq. (25.30) to (25.34).

Assume the matrix $\tilde{A}$ of Eq. (25.31) is banded with bandwidth n_b. For our present consideration assume that F is also banded with bandwidth not exceeding n_b. It is common in practice for the F matrix to be diagonal. By appropriate row interchanges the matrix

$$\begin{bmatrix} \tilde{A} & \tilde{b} \\ \lambda F & \lambda d \end{bmatrix}$$

of Eq. (25.31) can be converted to a new matrix

$$[\hat{A} : \hat{b}]$$

in which the matrix $\hat{A}$ has bandwidth n_b. The matrix $[\hat{A} : \hat{b}]$ can then be processed by Algorithm BSEQHT. If the matrix F is nonsingular, then $\hat{A}$ is of full rank. If the matrix F is nonsingular and λ is sufficiently large, then $\hat{A}$ will be of full pseudorank so that Algorithm BSEQHT can be completed including the computation of a solution.

If one wishes to investigate the effect of different values of λ in problem (25.31), one could first process only the data $[\tilde{A} : \tilde{b}]$ of problem (25.31), reducing this matrix to a banded triangular array

$$T = \begin{bmatrix} R : \bar{d} \\ 0 : e \end{bmatrix}$$

Assume that this matrix T is saved, possibly in the computer's secondary storage if necessary.

Then to solve problem (25.31) for a particular value of λ, note that an equivalent problem is

$$
\begin{bmatrix} R \\ 0 \\ \lambda F \end{bmatrix} x \cong \begin{bmatrix} \bar{d} \\ e \\ \lambda d \end{bmatrix}
$$

The latter problem can be subjected to row interchanges so that the coefficient matrix has bandwidth n_b, and this transformed problem is then in the form such that its solution can be computed using Algorithm BSEQHT.

This technique of interleaving rows of the matrix $[\lambda F : \lambda d]$ with the rows of the previously computed triangular matrix $[R : \bar{d}]$, to preserve a limited bandwidth, can also be used as a method for introducing new data equations into a previously solved problem. This avoids the need for triangularization using the entire original data set for the expanded problem.

In the case where $F = I_n$ and $d = 0$, the choice of parameter λ can be facilitated by computing the singular values and transformed right side of the least squares problem of Eq. (27.7). Thus with a singular value decomposition

$$
R = USV^T
$$

the vector

(27.25)
$$
g = U^T \bar{d}
$$

and the matrix

$$
S = \mathrm{Diag}\{s_1, \ldots, s_n\}
$$

can be computed without additional arrays of storage other than the array that contains the banded matrix R. The essential two ideas are: post- and premultiply R by a finite sequence of Givens rotation matrices J_i and T_i so that the matrix $B = T_v \cdots T_1 R J_1 \cdots J_v$ is bidiagonal. The product $T_v \cdots T_1 \bar{d}$ replaces $\bar{d}$ in storage. Compute the SVD of the matrix B using Algorithm QRBD (18.31). The premultiplying rotations are applied to $\bar{d}$ in storage, ultimately producing the vector g of Eq. (27.25). A value of λ can then be determined to satisfy a prespecified residual norm or solution vector norm by using Eq. (25.48) or (25.47), respectively.

Section 3. AN EXAMPLE: LINE SPLINES

To fix some of the ideas that were presented in Sections 1 and 2, consider the following data-fitting (or data-compression) problem. Suppose that we have m data pairs $\{(t_i, y_i)\}$ whose abscissas occur in an interval of t such that $a \leq t_i \leq b$, $i = 1, \ldots, m$. It is desired that these data be fit by a function $f(t)$ whose representation reduces the storage needed to represent the data. Probably the simplest continuous function (of some generality) that can be (least squares) fit to these data is the piecewise linear continuous function defined as follows:

(27.26)   Partition the interval $[a, b]$ into $n - 1$ subintervals with breakpoints $t^{(l)}$ satisfying $a = t^{(1)} < t^{(2)} < \cdots < t^{(n)} = b$.

For

(27.27) $t^{(l)} \leq t \leq t^{(l+1)}$

define

(27.28) $w_i(t) = \dfrac{t - t^{(l)}}{t^{(l+1)} - t^{(l)}}$

(27.29) $z_i(t) = 1 - w_i(t) = \dfrac{t^{(l+1)} - t}{t^{(l+1)} - t^{(l)}}$

and

(27.30) $f(t) = x_i z_i(t) + x_{i+1} w_i(t)$ $i = 1, \ldots, n - 1$

The n parameters x_i, $i = 1, \ldots, n$, of Eq. (27.29) are to be determined as variables of a linear least squares problem.

As an example, let us take $m = 10$ and $n = 4$ with breakpoints evenly spaced. A schematic graph for such a problem is given in Fig. 27.3.

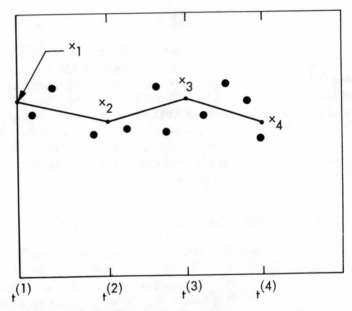

Fig. 27.3   A schematic example of a line spline with $m = 10$, $n = 4$, and data points (x_i, y_i) indicated by dots.

The least squares problem for the x_i then has the form shown in Fig. 27.4. Note that the coefficient matrix is banded with bandwidth $n_b = 2$.

$$
\begin{bmatrix}
z_1(t_1) & w_1(t_1) & 0 & 0 \\
z_1(t_2) & w_1(t_2) & 0 & 0 \\
z_1(t_3) & w_1(t_3) & 0 & 0 \\
0 & z_2(t_4) & w_2(t_4) & 0 \\
0 & z_2(t_5) & w_2(t_5) & 0 \\
0 & z_2(t_6) & w_2(t_6) & 0 \\
0 & 0 & z_3(t_7) & w_3(t_7) \\
0 & 0 & z_3(t_8) & w_3(t_8) \\
0 & 0 & z_3(t_9) & w_3(t_9) \\
0 & 0 & z_3(t_{10}) & w_3(t_{10})
\end{bmatrix}
\cdot
\begin{bmatrix}
x_1 \\ x_2 \\ x_3 \\ x_4
\end{bmatrix}
\cong
\begin{bmatrix}
y_1 \\ y_2 \\ y_3 \\ y_4 \\ y_5 \\ y_6 \\ y_7 \\ y_8 \\ y_9 \\ y_{10}
\end{bmatrix}
$$

Fig. 27.4

The progress of the band-limited algorithm BSEQHT (27.21) for this example is diagramed in Fig. 27.5.

$$
(1)\begin{bmatrix} a & a & a \\ a & a & a \\ a & a & a \end{bmatrix} \longrightarrow
(2)\begin{bmatrix} b & b & b \\ 0 & b & b \\ 0 & 0 & b \end{bmatrix}
$$

$$
(3)\begin{bmatrix} b & b & b \\ 0 & b & b \\ 0 & 0 & b \\ c & c & c \\ c & c & c \\ c & c & c \end{bmatrix} \longrightarrow
(4)\begin{bmatrix} b & b & b \\ b & 0 & b \\ 0 & 0 & b \\ c & c & c \\ c & c & c \\ c & c & c \end{bmatrix} \longrightarrow
(5)\begin{bmatrix} b & b & b \\ d & d & d \\ 0 & d & d \\ 0 & 0 & d \\ 0 & 0 & 0 \\ 0 & 0 & 0 \end{bmatrix}
$$

$$
(6)\begin{bmatrix} b & b & b \\ d & d & d \\ 0 & d & d \\ 0 & 0 & d \\ e & e & e \\ e & e & e \\ e & e & e \\ e & e & e \end{bmatrix} \longrightarrow
(7)\begin{bmatrix} b & b & b \\ d & d & d \\ d & 0 & d \\ 0 & 0 & d \\ e & e & e \\ e & e & e \\ e & e & e \\ e & e & e \end{bmatrix} \longrightarrow
(8)\begin{bmatrix} b & b & b \\ d & d & d \\ f & f & f \\ 0 & f & f \\ 0 & 0 & f \\ 0 & 0 & 0 \\ 0 & 0 & 0 \\ 0 & 0 & 0 \end{bmatrix}
$$

$$
(9)\begin{bmatrix} b & b & 0 & 0 \\ 0 & d & d & 0 \\ 0 & 0 & f & f \\ 0 & 0 & 0 & f \\ 0 & 0 & 0 & 0 \end{bmatrix}
\cdot
\begin{bmatrix} x_1 \\ x_2 \\ x_3 \\ x_4 \end{bmatrix}
\cong
\begin{bmatrix} b \\ d \\ f \\ f \\ f \end{bmatrix}
$$

Fig. 27.5

Stage (1)   Initially $i_p = i_r = 1$. The first block of data, consisting of the nontrivial data in the first three rows of Fig. 27.4, is introduced. Set $j_1 = 1$, $m_1 = 3$.

Stage (2)   This block is triangularized using Householder transformations. Set $i_r = 4$.

Stage (3)   Introduce second block of data from Fig. 27.4. Set $j_2 = 2$, $m_2 = 3$.

Stage (4)   Left shift of second row exclusive of the last column, which represents the right-side vector. Set $i_p = j_2 (= 2)$.

Stage (5)   Triangularization of rows 2–6. Set $i_r = 5$.

Stage (6)   Introduce third block of data from Fig. 27.4. Set $j_3 = 3$, $m_3 = 4$.

Stage (7)   Left shift of row 3 exclusive of last column. Set $i_p = j_3 (= 3)$.

Stage (8)   Triangularization of rows 3–8. Set $i_r = 6$.

The data appearing at stage (8) represent the least squares problem shown in diagram (9). This problem can now be solved by back substitution.

As an illustration of the extent of storage that can be saved with the use of this band-matrix processing of Section 2, consider an example of line-spline fitting with $m = 1000$ data points using 100 intervals. Thus the least squares problem will have $n = 101$ unknowns. Let us suppose further that the row dimension of each block we shall process does not exceed 10. Then the maximum size of the working array does not have to exceed $[n + 1 + \max(m_t)] \times 3 = (101 + 1 + 10) \times 3 = 336$. The use of a less specialized sequential accumulation algorithm such as that of Section 1 of this chapter would require a working array of dimension at least $[n + 1 + \max(m_t)] \times (n + 1) = (101 + 1 + 10) \times 102 = 11,424$. If all rows were brought in at once, the working array would have to have dimensions $m \times (n + 1) = 1000 \times 101 = 101,000$.

Section 4.   DATA FITTING USING CUBIC SPLINE FUNCTIONS

As a further example of the use of sequential processing for banded least squares problems we shall discuss the problem of data fitting using cubic spline functions with uniformly spaced breakpoints. The Fortran program **PROG5** given in Appendix C implements the approach described in this section.

Let numbers $b_1 < b_2 < \cdots < b_n$ be given. Let S denote the set of all cubic spline functions defined on the interval $[b_1, b_n]$ and having internal breakpoints (often called knots) $b_2, \ldots, b_{n-1}$. A function f is a member of S if and only if f is a cubic polynomial on each subinterval $[b_k, b_{k+1}]$ and is continuous together with its first and second derivatives throughout the interval $[b_1, b_n]$.

It can be shown that the set S constitutes an $(n + 2)$-dimensional linear space of functions. Thus any set of $n + 2$ linearly independent members of S, say $\{q_j : j = 1, \ldots, n + 2\}$, is a basis for S. This implies that each $f \in S$ has a unique representation in the form

$$f(x) = \sum_{j=1}^{n+2} c_j q_j(x)$$

Using such a basis, the problem of finding a member of S that best fits a set of data $\{(x_i, y_i) : x_i \in [b_1, b_n]; i = 1, \ldots, m\}$ in a least squares sense takes the form

$$Ac \cong y$$

where

$$a_{ij} = q_j(x_i) \qquad i = 1, \ldots, m$$
$$j = 1, \ldots, n + 2$$

Here c is the $(n + 2)$-vector with components c_j and y is the m-vector with components y_i.

There exist bases for S with the property that if the data are ordered so that $x_1 \leq x_2 \leq \cdots \leq x_m$, the matrix A will be band-limited with a bandwidth of four. Methods for computing such basis functions for unequally spaced breakpoint sets are described in Carasso and Laurent (1968), de Boor (1971 and 1972), and Cox (1971). Here we restrict the discussion to uniformly spaced breakpoint sets in order to avoid complexities not essential to the illustration of banded sequential processing.

To define such a basis for a uniformly spaced breakpoint set let

$$h = b_{k+1} - b_k \qquad k = 1, \ldots, n - 1$$

Define the two cubic polynomials

$$p_1(t) = 0.25t^3$$

and

$$p_2(t) = 1 - 0.75(1 + t)(1 - t)^2$$

Let I_1 denote the closed interval $[b_1, b_2]$ and let I_k denote the half-open interval (b_k, b_{k+1}) for $k = 2, \ldots, n - 1$.

In the interval I_k only four of the functions q_j have nonzero values. These four functions are defined for $x \in I_k$ by

$$t = \frac{x - b_k}{h}$$
$$q_k(x) = p_1(1 - t)$$
$$q_{k+1}(x) = p_2(1 - t)$$
$$q_{k+2}(x) = p_2(t)$$
$$q_{k+3}(x) = p_1(t)$$

The Fortran program **PROG5** in Appendix C solves a sample data-fitting problem using this set of basis functions and sequential Householder accumulation of the resulting band-limited matrix. The data for this curve fitting example are given in Table 27.1.

Table 27.1 DATA FOR CUBIC SPLINE-FITTING EXAMPLE

x	y	x	y
2	2.2	14	3.8
4	4.0	16	5.1
6	5.0	18	6.1
8	4.6	20	6.3
10	2.8	22	5.0
12	2.7	24	2.0

A parameter NBP is set to the values 5, 6, 7, 8, 9, and 10, respectively, to specify the number of breakpoints for each of the six cases. In terms of the parameter NBP the breakpoints for a case are defined as

$$b_i = 2 + \frac{22(i - 1)}{\text{NBP} - 1} \qquad i = 1, \ldots, \text{NBP}$$

The number of coefficients to be determined is NC = NBP + 2. Note that in the sixth case the number of coefficients is equal to the number of data points. In this sixth case the fitted curve interpolates the data.

Define

$$\text{RMS} = \left(\frac{\sum_{i=1}^{12} r_i^2}{12}\right)^{1/2}$$

where r_i is the residual at the ith data point. The value of RMS for each case is given in Table 27.2. Note that RMS is not a monotone function of NBP.

Table 27.2 RMS AS A FUNCTION OF NBP

NBP	5	6	7	8	9	10
RMS	0.254	0.085	0.134	0.091	0.007	0.0

Plots of each of the six curve fits are given in Fig. 27.6.

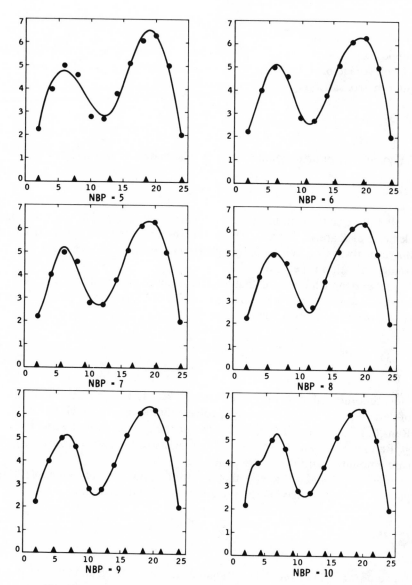

Fig. 27.6 Spline curves computed by **PROG5** as fits to the data of Table 27.1. Triangles along the baseline of each graph indicate breakpoint abcissas.

Section 5. REMOVING ROWS OF DATA

Three methods will be described for removing a row of data from a least squares problem, $Ax \cong b$. For notational convenience let

$$(27.31) \qquad\qquad C = [A : b]$$

Suppose C has m rows and n columns and is of rank k. Suppose further that a Cholesky factor R for C has been computed. Here R is a $k \times n$ upper triangular matrix satisfying

$$(27.32) \qquad\qquad QC = \begin{bmatrix} R \\ 0 \end{bmatrix}$$

for some $m \times m$ orthogonal matrix Q. The matrix R also satisfies

$$(27.33) \qquad\qquad R^T R = C^T C$$

It will be convenient to make the further assumption that the k diagonal elements of R are all nonzero. This will necessarily be true if $k = n$. If $k < n$, this can always be achieved by appropriate column interchanges and re-triangularization if necessary.

Let v^T denote a row of C that is to be removed. Without loss of generality we shall assume that v^T is the last row of C. Let $\tilde{C}$ denote the submatrix consisting of the remaining rows of C.

$$(27.34) \qquad\qquad C = \begin{bmatrix} \tilde{C} \\ v^T \end{bmatrix}$$

Note that the rank of $\tilde{C}$ may be either k or $k - 1$. The practical applications of data removal most commonly involve only the case in which Rank $(C) =$ Rank $(\tilde{C}) = n$; however, we shall consider the general case of Rank $(C) - 1 \leq$ Rank $(\tilde{C}) \leq$ Rank $(C) = k \leq n$ as long as it does not require a significant amount of additional discussion.

We wish to find a Cholesky factor $\tilde{R}$ for $\tilde{C}$. Thus $\tilde{R}$ will be an upper triangular $k \times n$ matrix having the same rank as $\tilde{C}$ and satisfying

$$(27.35) \qquad\qquad \tilde{Q}\tilde{C} = \begin{bmatrix} \tilde{R} \\ 0 \end{bmatrix}$$

for some $(m - 1) \times (m - 1)$ orthogonal matrix $\tilde{Q}$. Such a matrix $\tilde{R}$ will also satisfy

$$(27.36) \qquad \begin{aligned} \tilde{R}^T \tilde{R} &= \tilde{C}^T \tilde{C} = C^T C - v v^T \\ &= R^T R - v v^T \end{aligned}$$

The first two methods to be described appear in Gill, Golub, Murray, and Saunders (1972).

Row Removal Method 1

In this method it is assumed that the matrix Q of Eq. (27.32) is available as well as the matrix R. Partition Q and rewrite Eq. (27.32) as

$$(27.37) \qquad \begin{array}{c} k\{ \\ 1\{ \\ m-k-1\{ \end{array} \begin{bmatrix} Q_1 & p \\ u^T & \alpha \\ Q_2 & q \end{bmatrix} \cdot \begin{bmatrix} \tilde{C} \\ v^T \end{bmatrix} = \begin{bmatrix} R \\ 0 \\ 0 \end{bmatrix}$$
$$\underbrace{}_{m-1} \,\, 1$$

Left-multiply both sides of Eq. (27.37) by an orthogonal matrix that transforms the vector q to zero, modifies α, and leaves the first k rows of Q unchanged. This can be accomplished, for instance, by a single Householder transformation or by $(m - k - 1)$ Givens transformations. Equation (27.37) is thereby transformed to

$$(27.38) \qquad \begin{bmatrix} Q_1 & p \\ \hat{u}^T & \hat{\alpha} \\ \hat{Q}_2 & 0 \end{bmatrix} \cdot \begin{bmatrix} \tilde{C} \\ v^T \end{bmatrix} = \begin{bmatrix} R \\ 0 \\ 0 \end{bmatrix}$$

Next left-multiply both members of Eq. (27.38) by a sequence of k Givens transformations that successively transform each component of p to zero while modifying $\hat{\alpha}$. Let G_{ij} denote a Givens transformation matrix operating on rows i and j. First left-multiply by $G_{k,k+1}$, next by $G_{k-1,k+1}$, and so forth with the final transformation matrix being $G_{1,k+1}$. This sequence of operations assures that R is transformed to an upper triangular matrix, say, $\bar{R}$.

By these transformations Eq. (27.38) is transformed to

$$(27.39) \qquad \begin{bmatrix} \bar{Q}_1 & 0 \\ \bar{u}^T & \bar{\alpha} \\ \hat{Q}_2 & 0 \end{bmatrix} \cdot \begin{bmatrix} \tilde{C} \\ v^T \end{bmatrix} = \begin{bmatrix} \bar{R} \\ \bar{w}^T \\ 0 \end{bmatrix}$$

Since $\bar{\alpha}$ is the only nonzero element in its column and the column is of unit euclidean length it follows that $\bar{\alpha} = \pm 1$. Furthermore since the rows also have unit euclidean length $\bar{u}$ must be zero. This in turn implies that $\bar{w}^T = \bar{\alpha}v^T = \pm v^T$. Thus Eq. (27.39) can be written as

$$(27.40) \qquad \begin{bmatrix} \bar{Q}_1 & 0 \\ 0 & \pm 1 \\ \hat{Q}_2 & 0 \end{bmatrix} \cdot \begin{bmatrix} \tilde{C} \\ v^T \end{bmatrix} = \begin{bmatrix} \bar{R} \\ \pm v^T \\ 0 \end{bmatrix}$$

Defining

$$(27.41) \qquad \bar{Q} = \begin{bmatrix} \bar{Q}_1 \\ \hat{Q}_2 \end{bmatrix}$$

it follows that $\bar{Q}$ is an $(m-1) \times (m-1)$ orthogonal matrix, $\bar{R}$ is a $k \times n$ upper triangular matrix and

(27.42)
$$\bar{Q}\tilde{C} = \begin{bmatrix} \bar{R} \\ 0 \end{bmatrix}$$

Thus $\bar{Q}$ and $\bar{R}$ satisfy the conditions required of $\tilde{Q}$ and $\tilde{R}$ in Eq. (27.35) and (27.36).

The question of whether the rank of $\bar{R}$ is k or $k-1$ depends upon whether $\hat{\alpha}$ in Eq. (27.38) is nonzero or zero. By assumption the k diagonal elements of R are each nonzero. If $\hat{\alpha}$ is nonzero, it is easily verified by noting the structure and effect of the Givens transformation matrices used in passing from Eq. (27.38) to Eq. (27.40) that the k diagonal elements of $\bar{R}$ will also be nonzero.

Suppose on the other hand that $\hat{\alpha}$ is zero. Then $\|p\| = 1$, which assures that some components of p are nonzero. Let p_l denote the last nonzero component of p, i.e., $p_l \neq 0$, and if $l \neq k$, then $p_i = 0$ for $l < i \leq k$. In the order in which the matrices G_{ij} are applied in transforming Eq. (27.38) to Eq. (27.40), the matrix $G_{l,k+1}$ will be the first transformation matrix that is not simply an identity matrix. This matrix $G_{l,k+1}$ will be a (signed) permutation matrix that interchanges rows l and $k+1$, possibly changing the sign of one of these rows. In particular, its effect on R will be to replace row l with a row of zero elements. Subsequent transformations will not alter this row. Therefore, row l of $\bar{R}$ in Eq. (27.40) will be zero.

Thus Rank $(\bar{R})$ will be less than k. The rank of $\bar{R}$ cannot be less than $k-1$ since Rank $(\bar{R}) = $ Rank $(\tilde{C}) \geq $ Rank $(C) - 1 = k - 1$.

Row Removal Method 2

In a problem in which m is very large or in which data are being accumulated sequentially one would probably not retain the matrix Q of Eq. (27.32). In this case Method 1 cannot be used directly. Note, however, that the only information from the matrix Q that contributes to the determination of $\bar{R}$ in Method 1 is the vector p of Eq. (27.37) and (27.38) and the number $\hat{\alpha}$ of Eq. (27.38). We shall see that these quantities p and $\hat{\alpha}$ can be computed from R and v.

Rewrite Eq. (27.32) as

(27.43)
$$C = Q^T \begin{bmatrix} R \\ 0 \end{bmatrix}$$

Then using the partitions defined in Eq. (27.37) it follows that

(27.44)
$$v^T = p^T R$$

or equivalently

(27.45) $$R^T p = v$$

If R is of rank n, Eq. (27.45) constitutes a nonsingular triangular system that can be solved for p.

If the rank k of R is less than n, the system in Eq. (27.45) is still consistent and has a unique k-dimensional solution vector p. By assumption the first k rows of R^T constitute a $k \times k$ triangular submatrix with nonzero diagonal elements. This submatrix along with the first k components of v defines a system of equations that can be directly solved for p.

Having determined p the number $\hat{\alpha}$ can be computed as

(27.46) $$\hat{\alpha} = (1 - \|p\|^2)^{1/2}$$

as a consequence of the unit euclidean length of the column vector $(p^T, \hat{\alpha}, 0)^T$ in Eq. (27.38). The sign of $\hat{\alpha}$ is arbitrary and so can be taken to be nonnegative, as is implied by Eq. (27.46).

Having computed p and $\hat{\alpha}$ one can compute $\bar{R}$ by using k Givens transformations as described following Eq. (27.38) in Method 1. This part of the computation is expressed by the equation

(27.47) $$G_{1,k+1} G_{2,k+1} \cdots G_{k,k+1} \begin{bmatrix} p & R \\ \hat{\alpha} & 0 \end{bmatrix} = \begin{bmatrix} 0 & R \\ \pm 1 & \pm v^T \end{bmatrix}$$

It is to be expected that Method 2 will be a somewhat less accurate computational procedure than Method 1 because of the necessity of computing p and $\hat{\alpha}$ from Eq. (27.45) and (27.46). The accuracy with which p is determined will be affected by the condition number of R, which, of course, is the same as the condition number of C.

This accuracy limitation appears to be inherent, however, in any method of updating the Cholesky factor R to reflect data removal that does not make use of additional stored matrices, such as Q or C. For a given precision of computer arithmetic, updating methods that operate directly on the matrix $(A^T A)^{-1}$ may be expected to be significantly less reliable than Method 2.

Row Removal Method 3

The third method to be described is of interest because it requires fewer operations than Method 2 and because it is algorithmically very similar to one of the methods for adding data. This permits the design of fairly compact code that càn handle either addition or removal of data. The comparative numerical reliability of Methods 2 and 3 has not been studied.

This method has been discussed by Golub (1969), Chambers (1971), and Gentleman (1972a and 1972b). Following the practice of these authors we shall restrict our discussion to the case in which Rank $(\tilde{C}) = $ Rank $(C) \equiv k = n$. Properties of the more general case will be developed in the Exercises.

Letting i denote the imaginary unit ($i^2 = -1$), Eq. (27.36) can be written as

$$(27.48) \qquad \tilde{R}^T\tilde{R} = \begin{bmatrix} R \\ iv^T \end{bmatrix}^T \begin{bmatrix} R \\ iv^T \end{bmatrix}$$

By formal use of the Givens equations (3.5), (3.8), and (3.9), matrices $F^{(l)}$ can be defined that will triangularize $\begin{bmatrix} R \\ iv^T \end{bmatrix}$ as follows:

$$(27.49) \qquad \begin{bmatrix} R^{(0)} \\ iv^{(0)T} \end{bmatrix} = \begin{bmatrix} R \\ iv^T \end{bmatrix}$$

$$(27.50) \qquad \begin{bmatrix} R^{(l)} \\ iv^{(l)T} \end{bmatrix} = F^{(l)} \begin{bmatrix} R^{(l-1)} \\ iv^{(l-1)T} \end{bmatrix} \qquad l = 1, \ldots, k$$

$$(27.51) \qquad \begin{bmatrix} \bar{R} \\ 0 \end{bmatrix} = \begin{bmatrix} R^{(k)} \\ iv^{(k)T} \end{bmatrix}$$

Here the matrix $F^{(l)}$ differs from the identity matrix only at the intersections of rows l and $k + 1$ with columns l and $k + 1$. In these four positions $F^{(l)}$ contains the submatrix

$$(27.52) \qquad \begin{bmatrix} \sigma^{(l)} & i\tau^{(l)} \\ -i\tau^{(l)} & \sigma^{(l)} \end{bmatrix}$$

where

$$(27.53) \qquad \delta^{(l)} = [r_{ii}^{(l-1)}]^2 - [v_i^{(l-1)}]^2$$

$$(27.54) \qquad \rho^{(l)} = [\delta^{(l)}]^{1/2}$$

$$(27.55) \qquad \sigma^{(l)} = \frac{r_{ii}^{(l-1)}}{\rho^{(l)}}$$

$$(27.56) \qquad \tau^{(l)} = \frac{v_i^{(l-1)}}{\rho^{(l)}}$$

It can be shown that our assumption that both C and $\tilde{C}$ are of full rank n implies that the numbers $\delta^{(l)}$ defined by Eq. (27.53) are positive.

The multiplication by $F^{(l)}$ indicated in Eq. (27.50) can be expressed entirely in real arithmetic as follows:

(27.57) $\left. \begin{array}{l} r_{ij}^{(l)} = \sigma^{(l)} r_{ij}^{(l-1)} - \tau^{(l)} v_j^{(l-1)} \\[6pt] v_j^{(l)} = -\tau^{(l)} r_{ij}^{(l-1)} + \sigma^{(l)} v_j^{(l-1)} \end{array} \right\}$ $j = l, \dots, n$

(27.58)

The method of defining $F^{(l)}$ assures that $r_{ll}^{(l)} = \rho^{(l)}$ and $v_l^{(l)} = 0$. Furthermore, it can be easily verified that $v^{(k)} = 0$, that each matrix $R^{(l)}$ is upper triangular, that $F^{(l)T} F^{(l)} = I$, and that

(27.59)
$$\bar{R}^T \bar{R} = \begin{bmatrix} R \\ iv \end{bmatrix}^T \begin{bmatrix} R \\ iv \end{bmatrix}$$

Thus $\bar{R}$ satisfies the conditions required of $\tilde{R}$; i.e., $\bar{R}$ is an upper triangular Cholesky factor for the matrix $\tilde{C}$ of Eq. (27.34).

In practice the inherent potential instability of the data-removal problem will be reflected in loss of digits in the subtraction by which $\delta^{(l)}$ is computed and in the fact that the multipliers $\sigma^{(l)}$ and $\tau^{(l)}$ in Eq. (27.57) and (27.58) can be large.

Gentleman (1972a and 1972b) points out that modifications of the Givens method that maintain a separate array of squared row scale factors [see Eq. (10.32)] are particularly adaptable to Row Removal Method 3. Simply by permitting a squared row scale factor to be negative one can represent a row such as iv^T of Eq. (27.48) that has an imaginary scale factor. Thus by permitting negative as well as positive values of d_2 in Eq. (10.31) and $\tilde{d}_2$ in Eq. (10.33) the method given in Eq. (10.31) to (10.44) will handle data removal as well as data augmentation. Note that when d_2 is negative the quantity t of Eq. (10.35) is no longer bounded in the interval $[0, 1]$; hence the problem of avoiding overflow or underflow of $\tilde{D}$ and $\tilde{B}$ becomes more involved.

EXERCISES

(27.60)   [Chambers (1971)]
  (a) Show that the numbers $\sigma^{(l)}$ and $\tau^{(l)}$ of Eq. (27.55) and (27.56) can be interpreted as being the secant and tangent, respecitvely, of an angle $\theta^{(l)}$.
  (b) Show that these angles $\theta^{(l)}$ are the same angles that appear in the Givens transformations $G^{(l)}$ ($c^{(l)} = \cos\theta^{(l)}$, $s^{(l)} = \sin\theta^{(l)}$) which would occur in the process of adjoining the row v^T to $\bar{R}$ to produce R, i.e., in the operations represented by

$$G^{(k)} \cdots G^{(1)} \begin{bmatrix} \bar{R} \\ v^T \end{bmatrix} = \begin{bmatrix} R \\ 0 \end{bmatrix}$$

(c) With $\theta^{(l)}$, $c^{(l)}$, and $s^{(l)}$ defined as above, show that as an alternative to Eq. (27.58) $v_j^{(l)}$ could be computed as

$$v_j^{(l)} = -s^{(l)}r_{lj}^{(l)} + c^{(l)}v_j^{(l-1)}$$

where $c^{(l)} = 1/\sigma^{(l)}$ and $s^{(l)} = \tau^{(l)}/\sigma^{(l)}$.

(27.61) Show that Row Removal Method 3 can theoretically be extended to the general case of Rank $(C) - 1 \leq$ Rank $(\tilde{C}) \leq$ Rank $(C) \equiv k \leq n$ using the following observations. (Continue to assume, as at the beginning of Section 5, that the k diagonal elements of R are non-zero.)

(a) If $\delta^{(l)}$ is nonpositive for some $l = 1, \ldots, k$, let h be the index of the first such value of $\delta^{(l)}$, i.e., $\delta^{(h)} \leq 0$ and if $h \neq 1$, $\delta^{(l)} > 0$ for $1 \leq l < h$. Show that $\delta^{(h)} = 0$.

(b) With h as above it follows that $v_h^{(h-1)} = \epsilon r_{hh}^{(h-1)}$ where $\epsilon = +1$ or $\epsilon = -1$. In this case show that it must also be true that $v_j^{(h-1)} = \epsilon r_{hj}^{(h-1)}$ for $j = h + 1, \ldots, n$.

(c) With h as above, show that it is valid to terminate the algorithm by defining the final matrix $\bar{R}$ to be the matrix $R^{(h-1)}$ with row h set to zero.

BASIC LINEAR ALGEBRA
INCLUDING PROJECTIONS

In this appendix we list the essential facts of linear algebra that are used in this book. No effort is made to present a logically complete set of concepts. Our intention is to introduce briefly just those concepts that are directly connected with the development of the book's material.

For a real number x define

$$\operatorname{sgn}(x) = \begin{cases} 1 & \text{if } x \geq 0 \\ -1 & \text{if } x < 0 \end{cases}$$

An n-*vector* x is an ordered n-tuple of (real) numbers, $x_1, \ldots, x_n$.

An $m \times n$ *matrix* A is a rectangular array of (real) numbers having m rows and n columns. The element at the intersection of row i and column j is identified as a_{ij}. We shall often denote this $m \times n$ matrix by $A_{m \times n}$.

The *transpose* of an $m \times n$ matrix, A, denoted A^T, is the $n \times m$ matrix whose component at the intersection of row i and column j is a_{ji}.

The *matrix product* of an $m \times n$ matrix A and an $l \times k$ matrix B, written AB, is defined only if $l = n$, in which case $C = AB$ is the $m \times k$ matrix with components

$$c_{ij} = a_{i1}b_{1j} + a_{i2}b_{2j} + \cdots + a_{in}b_{nj} = \sum_{p=1}^{n} a_{ip}b_{pj}$$

Frequently it is convenient to regard an n-vector as an $n \times 1$ matrix. In particular, the product of a matrix and a vector can be defined in this way.

It is also often convenient to regard an $m \times n$ matrix as consisting of m n-vectors, as its rows, or n m-vectors as its columns.

The *inner product* (also called scalar product or dot product) of two n-

dimensional vectors u and v is defined by

$$s = u^T v = \sum_{i=1}^{n} u_i v_i$$

Two vectors are *orthogonal* to each other if their inner product is zero.

The *euclidean length* or *euclidean norm* or l_2 *norm* of a vector v, denoted by $\| v \|$, is defined as

$$\| v \| = (v^T v)^{1/2} = \left(\sum_{i=1}^{n} v_i^2 \right)^{1/2}$$

This norm satisfies the *triangle inequality*,

$$\| u + v \| \leq \| u \| + \| v \|$$

positive homogeneity,

$$\| \alpha u \| = | \alpha | \cdot \| u \|$$

where α is a number and u is a vector, and *positive definiteness*,

$$\| u \| > 0 \quad \text{if } u \neq 0$$
$$\| u \| = 0 \quad \text{if } u = 0$$

These three properties characterize the abstract concept of a *norm*.

The *spectral norm* of a matrix A, denoted by $\| A \|$, is defined as

(A.1) $$\| A \| = \max \{ \| Av \| : \| v \| = 1 \}$$

We shall also have occasion to use the *Frobenious norm* (also called the Schur or euclidean matrix norm) of a matrix A, denoted by $\| A \|_F$ and defined as

(A.2) $$\| A \|_F = \left(\sum_{i=1}^{m} \sum_{j=1}^{n} a_{ij}^2 \right)^{1/2}$$

The spectral and Frobenious norms satisfy

(A.3) $$\max_{ij} | a_{ij} | \leq \| A \| \leq \| A \|_F \leq k^{1/2} \| A \|$$

where A is an $m \times n$ matrix and $k = \min(m, n)$.

Both of these matrix norms satisfy the three properties of an abstract norm plus the *multiplicative inequalities*

$$\| AB \| \leq \| A \| \cdot \| B \|$$
$$\| AB \|_F \leq \| A \|_F \cdot \| B \|_F$$

We shall let the numeral 0 denote the *zero vector* or the *zero matrix* with the distinction and dimension to be determined by the context.

A set of vectors $v_1, \ldots, v_k$ is *linearly dependent* if there exist scalars $\alpha_1, \ldots, \alpha_k$, not all zero, such that

$$(A.4) \qquad \sum_{i=1}^{k} \alpha_i v_i = 0$$

Conversely if condition (A.4) holds only with $\alpha_1 = \cdots = \alpha_k = 0$, the vectors are *linearly independent*.

The set of all n-dimensional vectors is an n-*dimensional vector space*. Note that if u and v are members of this vector space, then so is $u + v$ and αv where α is any scalar. These two conditions of closure under vector addition and scalar-vector multiplication characterize the abstract definition of a vector space; however, we shall be concerned exclusively with the particular finite-dimensional vector space whose members are n-tuples of real numbers.

If a subset T of a vector space S is closed under vector addition and scalar-vector multiplication, then T is called a *subspace*. There is a maximal number of vectors in a subspace T that can be linearly independent. This number, m, is the *dimension* of the subspace T. A maximal set of linearly independent vectors in a subspace T is a *basis* for T. Every subspace T of dimension $m \geq 1$ has a basis. In fact, given any set of $k < m$ linearly independent vectors in an m-dimensional subspace T, there exist $m - k$ additional vectors in T such that these m vectors together constitute a basis for T. If the vectors $u_1, \ldots, u_m$ constitute a basis for T and $v \in T$, then there exists a unique set of scalars α_i such that $v = \sum_{i=1}^{m} \alpha_i u_i$.

The *span* of a set of vectors $v_1, \ldots, v_k$ is the set of all linear combinations of these vectors, i.e., the set of all vectors of the form $u = \sum_{i=1}^{k} \alpha_i v_i$ for arbitrary scalars α_i. The span of a set of k vectors is a subspace of dimension $m \leq k$.

There are certain subspaces that arise naturally in connection with matrices. Thus with an $m \times n$ matrix A we associate the *range* or *column space*, which is the span of its column vectors; the *null space* or *kernel*, which is the set $\{x : Ax = 0\}$; and the *row space*, which is the span of the row vectors. Note that the row space of A is the range of A^T.

It is often convenient to note that the range of a product matrix, say, $A = UVW$, is a subspace of the range of the leftmost factor in the product, here U. Similarly the row space of A is a subspace of the row space of the rightmost factor, here W.

The row and column spaces of a matrix A have the same dimension. This number is called the *rank* of the matrix, denoted by Rank (A). A matrix $A_{m \times n}$ is *rank deficient* if Rank $(A) < \min(m, n)$ and of *full rank* if Rank $(A) = \min(m, n)$. A square matrix $A_{n \times n}$ is *nonsingular* if Rank $(A) = n$ and *singular* if Rank $(A) < n$.

A *vector v is orthogonal to a subspace T* if v is orthogonal to every vector in T. It suffices for v to be orthogonal to every vector in a basis of T. A *subspace T is orthogonal to a subspace U* if t is orthogonal to u for every $t \in T$ and $u \in U$. If two subspaces T and U are orthogonal, then the *direct sum* of T and U, written $V = T \oplus U$, is the subspace consisting of all vectors $\{v: v = t + u, t \in T, u \in U\}$. The dimension of V is the sum of the dimensions of T and U.

If T and U are mutually orthogonal subspaces of an n-dimensional vector space S and $S = T \oplus U$, then T and U are called *orthogonal complements* of each other. This is written as $T = U^{\perp}$ and $U = T^{\perp}$. For any subspace T, $T^{\perp}$ exists. If T is a subspace of S and $s \in S$, then there exist unique vectors $t \in T$ and $u \in T^{\perp}$ such that $s = t + u$. For such vectors one has the *Pythagorean condition* $\|s\|^2 = \|t\|^2 + \|u\|^2$.

A *linear flat* is a translated subspace; i.e., if T is a subspace of S and $s \in S$, then the set $L = \{v: v = s + t, t \in T\}$ is a linear flat. The *dimension* of a linear flat is the dimension of the unique associated subspace T. A *hyperplane* H in an n-dimensional vector space S is an $(n-1)$-dimensional linear flat.

If H is a hyperplane and $h_0 \in H$, then $T = \{t: t = h - h_0, h \in H\}$ is an $(n-1)$-dimensional subspace and $T^{\perp}$ is one-dimensional. If $u \in T^{\perp}$, then $u^T h$ takes the same value, say, d, for all $h \in H$. Thus given the vector u and the scalar d, the hyperplane H can be characterized as the set $\{x: u^T x = d\}$. It is from this type of characterization that hyperplanes arise in practical computation.

A *halfspace* is the set of vectors lying on one side of a hyperplane, i.e., a set of the form $\{x: u^T x \geq d\}$. A *polyhedron* is the intersection of a finite number of halfspaces, i.e., a set of the form $\{x: u_i^T x \geq d_i, i = 1, \ldots, m\}$ or equivalently $\{x: Ux \geq d\}$ where U is an $m \times n$ matrix with rows u_i^T, d is an m vector with components d_i, and the inequality is interpreted termwise.

The elements a_{ii} of a matrix $A_{m \times n}$ are called the *diagonal elements* of A. This set of elements, $\{a_{ii}\}$, is called the *main diagonal* of A. If all other elements of A are zero, A is a *diagonal matrix*. If $a_{ij} = 0$ for $|i - j| > 1$, A is *tridiagonal*.

If $a_{ij} = 0$ for $j < i$ and $j > i + 1$, A is *upper bidiagonal*. If $a_{ij} = 0$ for $j < i$, A is *upper triangular*. If $a_{ij} = 0$ for $j < i - 1$, A is *upper Hessenberg*. If A^T is upper bidiagonal, triangular, or Hessenberg, then A is *lower* bidiagonal, triangular, or Hessenberg, respectively.

A square $n \times n$ diagonal matrix with all diagonal elements equal to unity is an *identity* matrix and is denoted by I_n or by I.

If $BA = I$, then B is a *left inverse* of A. A matrix $A_{m \times n}$ has a left inverse if and only if Rank $(A) = n \leq m$. The left inverse is unique if and only if Rank $(A) = n = m$. By transposition one similarly defines a *right inverse*.

If A is square and nonsingular, there exists a unique matrix, denoted A^{-1}, that is the unique left inverse and the unique right inverse of A and is called the *inverse* of A.

A generalization of the concept of an inverse matrix is the *pseudoinverse*, denoted A^+, that is uniquely defined for any matrix $A_{m \times n}$ and agrees with the inverse when A is square nonsingular. The row and column spaces of A^+ are the same as those of A^T. The pseudoinverse is intimately related to the least squares problem and is defined and discussed in Chapter 7.

A square matrix Q is *orthogonal* if $Q^T Q = I$. It follows from the uniqueness of the inverse matrix that $QQ^T = I$ also.

A set of vectors is *orthonormal* if the vectors are mutually orthogonal and of unit euclidean length. Clearly the set of column vectors of an orthogonal matrix is orthonormal and so is the set of row vectors.

If $Q_{m \times n}$, $m \geq n$, has orthonormal columns, then $\| Q \| = 1$, $\| Q \|_F = n^{1/2}$, $\| QA \| = \| A \|$, and $\| QA \|_F = \| A \|_F$. If $Q_{m \times n}$, $m \leq n$, has orthonormal rows, then $\| Q \| = 1$, $\| Q \|_F = m^{1/2}$, $\| AQ \| = \| A \|$, and $\| AQ \|_F = \| A \|_F$. Note that if $Q_{n \times n}$ is orthogonal, it satisfies both these sets of conditions.

Some particular orthogonal matrices that are computationally useful are the *Givens rotation matrix*,

$$\begin{bmatrix} \cos \theta & \sin \theta \\ -\sin \theta & \cos \theta \end{bmatrix}$$

the *Givens reflection matrix*,

$$\begin{bmatrix} \cos \theta & \sin \theta \\ \sin \theta & -\cos \theta \end{bmatrix}$$

and the *Householder reflection matrix*,

$$H = I - 2 \frac{uu^T}{\| u \|^2}$$

for an arbitrary nonzero vector u.

A *permutation matrix* is a square matrix whose columns are some permutation of those of the identity matrix. A permutation matrix is orthogonal.

A square matrix A is *symmetric* if $A^T = A$. A symmetric matrix has an *eigenvalue–eigenvector decomposition* of the form

$$A = QEQ^T$$

where Q is orthogonal and E is (real) diagonal. The diagonal elements of E are the *eigenvalues* of A and the column vectors of Q are the *eigenvectors* of A. The jth column vector of Q, say, q_j, is associated with the jth eigenvalue, e_{jj}, and satisfies the equation $Aq_j = e_{jj}q_j$. The matrix $A - e_{jj}I$ is singular. The eigenvalues of a symmetric $n \times n$ matrix A are unique. If a number λ occurs m times among the n eigenvalues of A, then the m-dimensional subspace spanned by the m eigenvectors (columns of Q) associated with λ is uniquely determined and is called the *eigenspace* of A associated with the eigenvalue λ.

A symmetric matrix is *positive definite* if all its eigenvalues are positive. A positive definite matrix P is also characterized by the fact that $x^T P x > 0$ for all $x \neq 0$.

A symmetric matrix S is *nonnegative definite* if its eigenvalues are non-negative. Such a matrix satisfies $x^T S x \geq 0$ for all $x \neq 0$.

For any $m \times n$ matrix A the matrix $S = A^T A$ is symmetric and nonnegative definite. It is positive definite if A is of rank n.

An *invariant space* of a square matrix A is a subspace T such that $x \in T$ implies $Ax \in T$. If S is a symmetric matrix, then every invariant space of S is the span of some set of eigenvectors of S, and conversely the span of any set of eigenvectors of S is an invariant space of S.

A symmetric matrix P is a *projection matrix* if all its eigenvalues are either unity or zero. A matrix P is *idempotent* if $P^2 = P$ [equivalently $P(I - P) = 0$]. A matrix P is a projection matrix if and only if P is symmetric and idempotent.

Let $P_{n \times n}$ be a projection matrix with k unit eigenvalues. Let T denote the k-dimensional eigenspace of P associated with the unit eigenvalues. The subspace T is the (unique) *subspace associated with the projection matrix P* and P is the (unique) *projection matrix associated with the subspace T*. The subspace T is both the row space and column space of its associated projection matrix P and is characterized by

$$T = \{x : Px = x\}$$

If P is a projection matrix with associated subspace T and $\|Px\| = \|x\|$, then $Px = x$ and $x \in T$. Furthermore $\|Px\| < \|x\|$ for $x \notin T$ and thus $\|P\| = 1$ unless $P = 0$.

For any matrix $A_{m \times n}$ the matrix $A^+ A$ is the $n \times n$ projection matrix associated with the row space of A, $(I - A^+ A)$ is the projection matrix associated with the null space of A, and AA^+ is the $m \times m$ projection matrix associated with the range (column space) of A.

Given any n-vector x the unique representation of x as $x = t + u$ where $t \in T$ and $u \in T^\perp$ is given by $t = Px$ and $u = (I - P)x$. The vector t is the nearest vector to x in T in the sense that $\|x - t\| = \min \{\|x - v\| : v \in T\}$. The vector t is called the *projection* of x onto T.

For a discussion of projection operators including proofs of most of the assertions above see pages 43–44 of Halmos (1957).

An arbitrary matrix $A_{m \times n}$ has a *singular value decomposition* $A = U_{m \times m} S_{m \times n} V_{n \times n}^T$ where U and V are orthogonal and S is diagonal with nonnegative diagonal elements. The diagonal elements of S ($s_{ii}, i = 1, \ldots, k$, where $k = \min [m, n]$) are the *singular values* of A. This set of numbers is uniquely determined by A. The number of nonzero singular values is the rank of A. Furthermore

$$\|A\| = \max_i \{s_{ii}\} \quad \text{and} \quad \|A\|_F = \left(\sum_{i=1}^n s_{ii}^2\right)^{1/2}$$

which provides a simple proof of the two rightmost inequalities (A.3).

Let $A = USV^T$ be a singular value decomposition of $A_{m \times n}$. Then eigenvalue decompositions of A^TA and AA^T are given by $A^TA = V(S^TS)V^T$ and $AA^T = U(SS^T)U^T$.

The singular value decomposition provides information useful in the practical analysis of linear algebraic problems and is discussed further in Chapters 4, 5, 18, 25, and 26.

B

PROOF OF GLOBAL QUADRATIC CONVERGENCE OF THE QR ALGORITHM

The purpose of this appendix is to give a proof of Theorem (18.5). The proof given is a more detailed version of that given in Wilkinson (1968a and 1968b). Conclusion (a) of Theorem (18.5) is left as Exercise (18.46).

The content of Lemma (B.1) will be found, with proof, in Wilkinson (1965a).

(B.1) LEMMA

> Let A be an $n \times n$ symmetric tridiagonal matrix with diagonal terms $a_1, \ldots, a_n$ and super- and subdiagonal terms $b_2, \ldots, b_n$ where each b_i is nonzero. Then the eigenvalues of A are distinct.

We also need the following simple lemma whose proof is left as Exercise (B.52). Use will be made of notation introduced in the text and equations of Chapter 18 from the beginning through Eq. (18.4).

(B.2) LEMMA

> The elements of the matrices A_k and the shift parameters σ_k are bounded in magnitude by $\|A\|$, and the elements of the matrices $(A_k - \sigma_k I_n)$ and R_k are bounded in magnitude by $2\|A\|$, for all $k = 1, 2, \ldots$.

We must examine the basic operations of the QR method with shifts to establish properties of certain intermediate quantities and ultimately to estimate the magnitude of certain of the off-diagonal terms of the matrices A_k. Denote the diagonal terms of the shifted matrix $(A_k - \sigma_k I_n)$ by

$$\bar{a}_i^{(k)} = a_i^{(k)} - \sigma_k \qquad i = 1, \ldots, n$$

By the choice rule for σ_k [see Eq. (18.4)] it follows that the eigenvalue of the lower right 2×2 submatrix of $(A_k - \sigma_k I_n)$ which is closest to $\bar{a}_n^{(k)}$ is zero. Thus

(B.3)
$$\bar{a}_n^{(k)} \bar{a}_{n-1}^{(k)} = (b_n^{(k)})^2$$

and

(B.4)
$$|\bar{a}_n^{(k)}| \le |b_n^{(k)}| \le |\bar{a}_{n-1}^{(k)}|$$

The orthogonal matrix Q_k is determined in the form

(B.5)
$$Q_k = J_{n-1,n}^{(k)} \cdots J_{2,3}^{(k)} J_{1,2}^{(k)}$$

where each $J_{i-1,i}^{(k)}$ is a rotation in the plane determined by the $(i-1)$st and ith coordinate axes. The scalars $c_i^{(k)}$ and $s_i^{(k)}$ of the rotation matrices, as given in Lemma (3.4), define the rotation by means of the identity

(B.6)
$$J_{i-1,i}^{(k)} = \begin{bmatrix} I_{i-2} & 0 & 0 \\ 0 & P_i^{(k)} & 0 \\ 0 & 0 & I_{n-i} \end{bmatrix} \qquad i = 2, \ldots, n$$

Here each $P_i^{(k)}$ is defined as in Eq. (3.5) with c and s appropriately super- and subscripted.

Following premultiplication of $(A_k - \sigma_k I_n)$ by the first $i - 2$ of the rotation matrices we shall have

(B.7)
$$J_{i-2,i-1}^{(k)} \cdots J_{1,2}^{(k)}(A_k - \sigma_k I_n)$$

$$= \begin{bmatrix} p_1^{(k)} & q_1^{(k)} & r_1^{(k)} & & & & & \\ 0 & p_2^{(k)} & q_2^{(k)} & r_2^{(k)} & & & & \\ & & \cdot & & & & & \\ & & & \cdot & & & & \\ & & & & p_{i-2}^{(k)} & q_{i-2}^{(k)} & r_{i-2}^{(k)} & \\ & & & & 0 & x_{i-1}^{(k)} & y_{i-1}^{(k)} & \\ & & & & & b_i^{(k)} & \bar{a}_i^{(k)} & b_{i+1}^{(k)} \\ & & & & & & \cdot & \\ & & & & & & b_{n-1}^{(k)} & \bar{a}_{n-1}^{(k)} & b_n^{(k)} \\ & & & & & & & 0 & b_n^{(k)} & \bar{a}_n^{(k)} \end{bmatrix}$$

Premultiplying both sides of Eq. (B.7) by $J_{i-1,i}^{(k)}$ shows, by means of Eq. (3.7) to (3.9), that the following recurrence equations hold.

(B.8) $p_{i-1}^{(k)} = [(x_{i-1}^{(k)})^2 + (b_i^{(k)})^2]^{1/2}$ $i = 2, \ldots, n$

(B.9) $p_n^{(k)} = x_n^{(k)}$

(B.10) $c_i^{(k)} = \dfrac{x_{i-1}^{(k)}}{p_{i-1}^{(k)}}$ $i = 2, \ldots, n$

(B.11) $s_i^{(k)} = \dfrac{b_i^{(k)}}{p_{i-1}^{(k)}}$ $i = 2, \ldots, n$

(B.12) $0 = -s_i^{(k)} x_{i-1}^{(k)} + c_i^{(k)} b_i^{(k)}$ $i = 2, \ldots, n$

(B.13) $x_i^{(k)} = -s_i^{(k)} y_{i-1}^{(k)} + c_i^{(k)} \bar{a}_i^{(k)}$ $i = 2, \ldots, n$

(B.14) $y_i^{(k)} = c_i^{(k)} b_{i+1}^{(k)}$ $i = 2, \ldots, n-1$

The similarity transformation is completed by forming

(B.15)
$$A_{k+1} = R_k Q_k^T + \sigma_k I_n$$
$$= R_k (J_{1,2}^{(k)})^T \cdots (J_{n-1,n}^{(k)})^T + \sigma_k I_n$$

In this process the new off-diagonal elements are computed as

(B.16) $b_i^{(k+1)} = s_i^{(k)} p_i^{(k)}$ $i = 2, \ldots, n$

We wish to analyze the convergence of $b_n^{(k+1)}$ as $k \longrightarrow \infty$.

In the Eq. (B.17) to (B.19) we suppress the superscript (k) but write the superscript $(k + 1)$. Beginning with Eq. (B.16) with $i = n$ we have

(B.17) $b_n^{(k+1)} = s_n p_n$

$= s_n x_n$ using (B.9)

$= s_n[-s_n y_{n-1} + c_n \bar{a}_n]$ using (B.13)

$= s_n[-s_n c_{n-1} b_n + c_n \bar{a}_n]$ using (B.14)

$= s_n\left[-s_n c_{n-1} b_n + s_n\left(\dfrac{\bar{a}_n}{b_n}\right) x_{n-1} \right]$ using (B.12)

$= s_n\left[-s_n c_{n-1} b_n + s_n\left(\dfrac{b_n}{\bar{a}_{n-1}}\right)(-s_{n-1} c_{n-2} b_{n-1} + c_{n-1} \bar{a}_{n-1}) \right]$

using (B.3), (B.13), and (B.14)

$= -\dfrac{s_n^2 s_{n-1} c_{n-2} b_{n-1} b_n}{\bar{a}_{n-1}}$

Eliminating p_{n-1} between Eq. (B.11) and (B.16) gives

(B.18) $b_{n-1}^{(k+1)} = \dfrac{s_{n-1} b_n}{s_n}$

Then using Eq. (B.17) and (B.18) we obtain

(B.19) $b_n^{(k+1)} b_{n-1}^{(k+1)} = -s_n s_{n-1}^2 c_{n-2} b_n b_{n-1}\left(\dfrac{b_n}{\bar{a}_{n-1}}\right)$

From Eq. (B.4) expressed as $|b_n^{(k)}/\bar{a}_{n-1}^{(k)}| \leq 1$ and Eq. (B.17) we have

(B.20) $$|b_n^{(k+1)}| \leq |b_{n-1}^{(k)}|, \qquad k = 1, 2, \ldots$$

From Eq. (B.19) we see that

(B.21) $$|b_n^{(k+1)} b_{n-1}^{(k+1)}| \leq |b_n^{(k)} b_{n-1}^{(k)}| \qquad k = 1, 2, \ldots$$

so that $|b_n^{(k)} b_{n-1}^{(k)}|$ converges to a limit $L \geq 0$ as $k \to \infty$.

(B.22) LEMMA   *The limit L is zero.*

Proof: Assume $L > 0$. Passing to the limit, as $k \to \infty$, in both members of Eq. (B.19) yields

(B.23) $$\left| \frac{b_n^{(k)}}{\bar{a}_{n-1}^{(k)}} \right| \to 1$$

(B.24) $$|s_n^{(k)}| \to 1$$

(B.25) $$|c_{n-2}^{(k)}| \to 1$$

(B.26) $$(s_{n-1}^{(k)})^2 \to 1$$

From Eq. (B.10) and (B.24) and the fact that the sequence $\{p_{n-1}^{(k)}\}$ is bounded [Lemma (B.2)], we obtain

(B.27) $$x_{n-1}^{(k)} \to 0$$

Thus from Eq. (B.13) and (B.14),

(B.28) $$\bar{a}_{n-1}^{(k)} c_{n-1}^{(k)} - b_{n-1}^{(k)} c_{n-2}^{(k)} s_{n-1}^{(k)} \to 0$$

Since Eq. (B.26) implies $c_{n-1}^{(k)} \to 0$, it follows from Eq. (B.28) that

(B.29) $$b_{n-1}^{(k)} c_{n-2}^{(k)} s_{n-1}^{(k)} \to 0$$

From Eq. (B.25), (B.26), and (B.29) we have $b_{n-1}^{(k)} \to 0$, but since the sequence $\{b_n^{(k)}\}$ is bounded, this implies $|b_{n-1}^{(k)} b_n^{(k)}| \to 0$, contradicting the assumption that $L > 0$. This completes the proof of Lemma (B.22).

(B.30) LEMMA

The sequence $\{|b_n^{(k)}|\}$, $k = 1, 2, \ldots$, contains arbitrarily small terms.

Proof: Let $\tau > 0$. By Lemma (B.2) the sequences $\{|b_n^{(k)}|\}$ and $\{|b_{n-1}^{(k)}|\}$ are bounded. Thus by Lemma (B.22) there exists an integer $\bar{k}$ for which either

(B.31) $$|b_{n-1}^{(k)}| < \tau$$

or

(B.32)
$$|b_n^{(k)}| < \tau$$

If Eq. (B.31) holds, then with Eq. (B.20) we have

$$|b_n^{(k+1)}| < \tau$$

It follows that for any $\tau > 0$ there exists an integer $\hat{k}$ depending upon τ such that

$$|b_n^{(\hat{k})}| < \tau$$

This completes the proof of Lemma (B.30).

(B.33)   LEMMA

> If $\tau > 0$ is sufficiently small and if $\hat{k}$ is such that $|b_n^{(\hat{k})}| < \tau$, then $|b_n^{(k)}| < \tau$ for all $k > \hat{k}$.

Proof: Define $\delta = \min\{|\lambda_i - \lambda_j| : i \neq j\}$. Note that $\delta > 0$ since the eigenvalues λ_i of A are distinct.

Define

(B.34)
$$f(t) = \frac{t[1 + t/(\delta - 3t)]}{\delta - 3t}$$

Let $\tau_0 > 0$ be a number small enough so that

(B.35)
$$\tau_0 < \frac{\delta}{3}$$

(B.36)
$$f(\tau_0) < 1$$

and

(B.37)
$$\frac{df(t)}{dt} > 0 \text{ for } 0 \leq t \leq \tau_0$$

Choose τ, $0 < \tau < \tau_0$, and let k be an integer such that

(B.38)
$$|b_n^{(k)}| < \tau$$

For notational convenience we introduce

(B.39)
$$\epsilon = b_n^{(k)}$$

If $b_n^{(k)} = 0$, the algorithm has converged. Otherwise without loss of generality we may assume $\epsilon > 0$.

Since $\bar{a}_n^{(k)}\bar{a}_{n-1}^{(k)} = \epsilon^2$ we can write

(B.40)
$$A_k - \sigma_k I_n = \begin{bmatrix} & & & & 0 \\ & & & \vdots & \cdot \\ & B & & \vdots & \cdot \\ & & & \vdots & \cdot \\ & & & \vdots & 0 \\ & & & \vdots & \epsilon \\ \hline 0 & \cdots & 0 & \epsilon & \dfrac{\epsilon^2}{\bar{a}_{n-1}^{(k)}} \end{bmatrix}$$

Let $\mu_1, \ldots, \mu_{n-1}$ be the roots of the $(n-1) \times (n-1)$ symmetric matrix B, and $\lambda'_1, \ldots, \lambda'_n$ be the roots of the shifted matrix $A_k - \sigma_k I_n$. By Theorem (5.1), the ordered eigenvalues of $A_k - \sigma_k I_n$ and those of

$$\begin{pmatrix} B & 0 \\ 0 & \dfrac{\epsilon^2}{\bar{a}_{n-1}^{(k)}} \end{pmatrix}$$

differ at most by ϵ. Thus with a possible reindexing of the λ'_i we have the inequalities

(B.41)
$$|\lambda'_i - \mu_i| \leq \epsilon \qquad i = 1, \ldots, n-1$$
$$\left| \lambda'_n - \frac{\epsilon^2}{\bar{a}_{n-1}^{(k)}} \right| \leq \epsilon$$

From the identity

$$\mu_i = \mu_i - \lambda'_i + \lambda'_i - \lambda'_n + \lambda'_n - \frac{\epsilon^2}{\bar{a}_{n-1}^{(k)}} + \frac{\epsilon^2}{\bar{a}_{n-1}^{(k)}}$$

we have

(B.42)
$$|\mu_i| \geq |\lambda'_i - \lambda'_n| - |\mu_i - \lambda'_i| - \left| \lambda'_n - \frac{\epsilon^2}{\bar{a}_{n-1}^{(k)}} \right| - \frac{\epsilon^2}{|\bar{a}_{n-1}^{(k)}|}$$
$$i = 1, \ldots, n-1$$

Now using Eq. (B.41) on the second and third terms of the right side of inequality (B.42), observing that $|\lambda'_i - \lambda'_n| = |\lambda_i - \lambda_n| \geq \delta$, followed by use of Eq. (B.4) with $b_n^{(k)} = \epsilon$ on the fourth term of the right side of inequality (B.42) we have

(B.43)
$$|\mu_i| \geq \delta - 3\epsilon \qquad i = 1, \ldots, n-1$$

At the next-to-last step in the forward triangularization of $A_k - \sigma_k I_n$ the configuration in the lower right 2×2 submatrix is seen by Eq. (B.3), (B.7),

(B.14), and (B.39) to be

(B.44)
$$\begin{pmatrix} x^{(k)}_{n-1} & \epsilon c^{(k)}_{n-1} \\ \epsilon & \dfrac{\epsilon^2}{\bar{a}^{(k)}_{n-1}} \end{pmatrix}$$

Note that $x^{(k)}_{n-1}$ of Eq. (B.44) is a diagonal element of the $(n-1) \times (n-1)$ upper triangular matrix resulting from premultiplication of the matrix B of Eq. (B.40) by $(n-2)$ rotation matrices. Thus by Eq. (6.3) and (B.43)

(B.45)
$$|x^{(k)}_{n-1}| \geq \min |\mu_i| \geq \delta - 3\epsilon$$

Completing the similarity transformation we have, from Eq. (B.3), (B.9), (B.13), and (B.17) with $y^{(k)}_{n-1} = c^{(k)}_{n-1} b^{(k)}_n = c^{(k)}_{n-1} \epsilon$,

(B.46)
$$\begin{aligned} b^{(k+1)}_n &= s^{(k)}_n p^{(k)}_n \\ &= (c^{(k)}_n \bar{a}^{(k)}_n - s^{(k)}_n y^{(k)}_{n-1}) s^{(k)}_n \\ &= \left(\frac{\epsilon^2 c^{(k)}_n}{\bar{a}^{(k)}_{n-1}} - \epsilon c^{(k)}_{n-1} s^{(k)}_n \right) s^{(k)}_n \end{aligned}$$

Now from Eq. (B.11) and the inequality (B.45)

(B.47)
$$|s^{(k)}_n| = \frac{\epsilon}{[(x^{(k)}_{n-1})^2 + \epsilon^2]^{1/2}} \leq \frac{\epsilon}{\delta - 3\epsilon}$$

Finally, using Eq. (B.46) and the inequality (B.47) we have

(B.48)
$$|b^{(k+1)}_n| \leq \frac{\epsilon^3}{|\bar{a}^{(k)}_{n-1}|(\delta - 3\epsilon)} + \frac{\epsilon^3}{(\delta - 3\epsilon)^2}$$

The inequality in Eq. (B.48) shows that if $|\bar{a}^{(k)}_{n-1}|$ is bounded away from zero, then the convergence ultimately is cubic. However, all we know generally from Eq. (B.4) is that $|\bar{a}^{(k)}_{n-1}| \geq \epsilon$, and therefore

(B.49)
$$|b^{(k+1)}_n| \leq \frac{\epsilon^2 + \epsilon^3/(\delta - 3\epsilon)}{\delta - 3\epsilon}$$

or, recalling Eq. (B.34) and (B.39),

(B.50)
$$|b^{(k+1)}_n| \leq [b^{(k)}_n]^2 \frac{1 + b^{(k)}_n/(\delta - 3b^{(k)}_n)}{\delta - 3b^{(k)}_n} = b^{(k)}_n f(b^{(k)}_n)$$

From conditions (B.36) to (B.38) we have $f(b^{(k)}_n) < 1$ and thus

(B.51)
$$|b^{(k+1)}_n| < |b^{(k)}_n|$$

By induction it follows that inequalities (B.50) and (B.51) with k replaced by $k + l$ hold for all $l = 0, 1, \ldots$. This completes the proof of Lemma (B.33).

Lemma (B.33) implies that $b_n^{(l)} \to 0$ as $l \to \infty$, which is conclusion (b) of Theorem (18.5). Since $b_n^{(l)} \to 0$ and the inequality in Eq. (B.50) holds for all sufficiently large k the quadratic convergence asserted in the final conclusion (c) of Theorem (18.5) is also established.

We conclude this appendix with these remarks:

1. The shifts σ_k do not need to be explicitly subtracted from each of the diagonal terms of A_k when forming A_{k+1}. Discussion of this as it applies to computation of the singular value decomposition is given in Chapter 18.

2. In practice the iteration procedure for each eigenvalue is terminated when the terms $b_n^{(k)}$ are "zero to working accuracy." This can be defined in a variety of ways. One criterion for this is given in Chapter 18 as it concerns the numerical aspects of the singular value decomposition.

3. The proof we have given assumes that $n \geq 3$. For $n = 2$ it is easy to see that if

$$A = A_1 = \begin{pmatrix} a_1^{(1)} & b_2^{(1)} \\ b_2^{(1)} & a_2^{(1)} \end{pmatrix}$$

then performing one shifted QR transformation as given in Eq. (18.1) to (18.4) gives

$$A_2 = \begin{pmatrix} a_1^{(2)} & b_2^{(2)} \\ b_2^{(2)} & a_2^{(2)} \end{pmatrix}$$

with $b_2^{(2)} = 0$. Thus A_2 is diagonal and the eigenvalues of A have been computed.

4. More generally if an actual eigenvalue λ of A instead of σ_k of Eq (18.4) were used as a shift, then the matrix A_{k+1} would break up after the next QR sweep. In fact, $b_n^{(k+1)} = 0$ and $a_n^{(k+1)} = \lambda$. To verify this note that since $A_k - \lambda I_n$ is rank deficient at least one of the diagonal terms $p_i^{(k)}$ of the triangular matrix R_k must be zero. From Eq. (B.8), $p_i^{(k)} > 0$ for $i = 1, \ldots, n - 1$ and therefore it must be $p_n^{(k)}$ that is zero. Upon completion of the similarity transformation and translation of Eq. (B.15) while using Eq. (B.17) we see that $b_n^{(k+1)} = s_n^{(k)} p_n^{(k)} = 0$, and therefore $a_n^{(k+1)} = \lambda$.

EXERCISE

(B.52)   Prove Lemma (B.2).

C DESCRIPTION AND USE OF FORTRAN CODES FOR SOLVING PROBLEM LS

Throughout most of this book attention has been focused primarily on the mathematical and algorithmic descriptions of various methods for obtaining solutions to Problem LS (Chapter 1) or its variants such as Problem LSE (20.1) and Problem LSI (23.1). These algorithms can be implemented in any of several well-known computer source languages such as Fortran, ALGOL, or PL/I. Accomplishing this can be expensive and time consuming. For this reason we have included Fortran source codes for some of the algorithms. We believe that our code conforms to ANSI Fortran [Ref: ANSI Subcommittee X3J3 (1971) and ASA Committee X3 (1964)].

The entire set of code and sample data presented occupies 2149 card images. Persons or organizations wishing to purchase this code and data in machine-readable form should direct their inquiry to International Mathematical and Statistical Libraries, Inc., Suite 510, 6200 Hillcroft, Houston, Texas 77036.

Tables C.1 and C.2 give a directory of these Fortran programs and subprograms. As an additional aid to setting up computer runs, Fig. C.1 indicates the **CALL** dependencies among the various programs and subprograms. For example, to set up a run of the Fortran main program **PROG4** one can follow the connecting lines downward from **PROG4** in Fig. C.1 and see that **PROG4** requires the data element **DATA4** and the subprograms **SVA**, **SVDRS**, **QRBD**, **H12**, **G1**, **G2**, **MFEOUT**, and **DIFF**.

These Fortran programs and subprograms have all been compiled and executed successfully on the UNIVAC 1108 and IBM 360/75 computer systems at the Jet Propulsion Laboratory, on the IBM 360/67 at Washington State University, on the CDC 6500 at Purdue University, on the CDC 6600 and 7600 at the National Center for Atmospheric Research (NCAR), and

Table C.1 DIRECTORY OF MAIN PROGRAMS AND THE DATA ELEMENT **DATA4**

Name of Program	Fortran Code Page Number	Purpose of Program
PROG1	279	Demonstrates Algorithms HFT, HS1, and COV of Chapters 11 and 12. Calls subprograms **H12** and **GEN**.
PROG2	281	Demonstrates Algorithms HFTI and COV of Chapters 14 and 12, respectively. Calls subprograms **HFTI** and **GEN**.
PROG3	283	Demonstrates the singular value decomposition algorithm of Chapter 18. Calls subprograms **SVDRS** and **GEN**.
PROG4	285	Demonstrates singular value analysis including computation of Levenberg–Marquardt solution norms and residual norms as described in Chapters 18, 25, and 26. Calls subroutine **SVA** and reads the data element **DATA4**.
DATA4	285	This is not a program. It is a set of 15 card images containing data to be read by **PROG4**. This example is discussed in Chapter 26.
PROG5	286	Demonstrates the band-limited sequential accumulation algorithm of Chapter 27, Sections 2 and 4. The algorithm is used to fit a cubic spline (with uniformly spaced breakpoints) to a table of data. Calls subroutines **BNDACC** and **BNDSOL**.
PROG6	288	Computes the constrained line-fitting problem given as an example in Chapter 23. The program illustrates a typical usage of the subroutine **LDP**, which in turn uses the subroutine **NNLS**. **PROG6** also calls **SVDRS**.

Table C.2 DIRECTORY OF SUBPROGRAMS

Name of Subprogram	User's Guide Page Number	Fortran Code Page Number	Purpose of Subprogram
HFTI	254	290	Implements Algorithm HFTI of Chapter 14. Calls subprograms **H12** and **DIFF**.
SVA	256	292	Implements singular value analysis and Levenberg–Marquardt analysis as described in Chapters 18 and 25. Produces printed output of quantities of interest. Calls subroutines **SVDRS** and **MFEOUT**.

Table C.2 (continued)

Name of Subprogram	User's Guide Page Number	Fortran Code Page Number	Purpose of Subprogram
SVDRS	260	295	Computes the singular value decomposition as described in Chapter 18. Calls subroutines **H12** and **QRBD**.
QRBD	262	298	Computes the singular value decomposition of a bidiagonal matrix as described in Chapter 18. Calls subprograms **G1**, **G2**, and **DIFF**.
BNDACC and **BNDSOL**	264	301 and 302	Implements the band-limited sequential accumulation algorithm of Chapter 27, Section 2. Calls subroutine **H12**.
LDP	267	303	Solves the least distance programming problem as described in Chapter 23. Calls subprograms **NNLS** and **DIFF**.
NNLS	269	304	Computes a least squares solution, subject to all variables being nonnegative, as described in Chapter 23. Calls subprograms **H12**, **G1**, **G2**, and **DIFF**.
H12	271	308	Constructs and applies a Householder transformation as described in Chapter 10.
G1 and **G2**	274	309	Constructs and applies a Givens rotation as described in Chapter 10.
MFEOUT	275	310	Prints a two-dimensional array in a choice of two pleasing formats.
GEN	277	311	Generates a sequence of numbers for use in constructing test data. Used by **PROG1, PROG2,** and **PROG3**.
DIFF	278	311	Computes the difference between two floating point arguments.

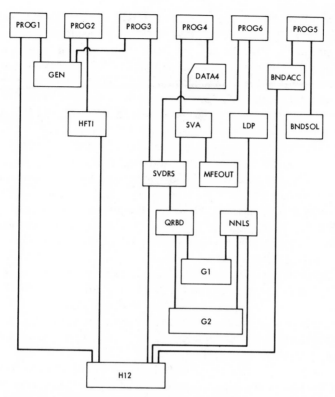

Fig. C.1 Network showing subprograms and data elements used by each main program. In addition the subprogram **DIFF** is called by the subprograms, **HFTI, QRBD, LDP,** and **NNLS**.

on the CDC 7600 at Aerospace Corporation. We wish to express our appreciation to Drs. T. J. Aird, Alan Cline, H. J. Wertz, and Mr. C. T. Verhey for their valuable cooperation in making these runs on CDC computers.

The six main programs, **PROG1**, . . . , **PROG6**, are designed to demonstrate various algorithms and the usage of the subprograms.

The program, **PROG1**, is an illustration of the simplest method of solving a least squares problem described in this book. It implements the Algorithms HFT, HS1, and COV of Chapters 11 and 12. This code may give unacceptable answers if the matrix is nearly rank-deficient. *So beware; no checking for nearly rank-deficient matrices is made with this first program.*

Methods coping with problems having these nearly rank-deficient matrices are illustrated by **PROG2** and **PROG3**, which demonstrate, respectively, Algorithms HFTI (Chapter 14) and SVD (Chapter 18). The reader who is solving least squares problems with near rank deficiencies (highly correlated variables) would be advised to use code in his program based on one of these more general programs. The program **PROG2** also implements Algorithm COV (Chapter 12) in the cases that are of full pseudorank.

The programs **PROG2** and **PROG3** each use an absolute tolerance parameter τ for pseudorank determination. In **PROG2**, τ is compared with diagonal elements of the triangular matrix resulting from Householder triangularization; while, in **PROG3**, τ is compared with the singular values.

Each of the three main programs, **PROG1**, **PROG2**, and **PROG3**, uses the function subprogram **GEN** to generate the same set of 36 test cases. For the first 18 cases the data consist of a sequence of integers between -500 and 500 with a period of 10. This short period causes certain of the matrices to be mathematically rank-deficient. All the generated matrices have norms of approximately 500.

For these first 18 cases **PROG2** sets the absolute tolerance $\tau = 0.5$. In **PROG3** a relative tolerance ρ is set to 10^{-3}. Then for each case the absolute tolerance is computed as $\tau = \rho s_1$ where s_1 is the largest singular value of the test matrix A. Recall that $s_1 = \|A\|$.

The second set of 18 cases are the same as the first 18 except that "noise," simulating data uncertainty, is added to all data values. The relative magnitude of this "noise" is approximately $v = 10^{-4}$. For these cases **PROG2** sets $\tau = 10 v \alpha$ where α is preset to 500 to represent the approximate norm of the test matrices. The program **PROG3** sets $\rho = 10v$ and then, for each test matrix A, computes $\tau = \rho \|A\|$.

This method of setting τ, particularly in the second 18 cases, is intended to stress the idea of choosing τ on the basis of a priori knowledge about the size of the data uncertainty.

If the reader executes these programs, he will find that in the pseudorank-deficient cases the solutions computed by **PROG2** and **PROG3** are similar and of order of magnitude unity, while those computed by **PROG1** contain very large numbers that would probably be unacceptable in practical applications. For example, Table C.3 lists the solutions and pseudoranks obtained for the test case $m = 7$, $n = 6$ by the three different programs with the two different noise levels.

Table C.3 SOLUTIONS, RESIDUAL NORM, AND PSEUDORANK FOR THE TEST CASE $m = 7$, $n = 6$

	Noise Level $= 0$			Relative Noise Level $= 10^{-4}$		
	PROG1	**PROG2**	**PROG3**	**PROG1**	**PROG2**	**PROG3**
x_1	0.237×10^8	0.1458	0.1458	-4544.0	0.1470	0.1456
x_2	-0.237×10^8	-0.0208	-0.0208	4544.7	0.0217	-0.0207
x_3	-0.565×10^8	0.0833	-0.0833	5064.5	-0.0846	0.0832
x_4	0.565×10^8	-0.0833	-0.0833	-5064.5	-0.0841	-0.0831
x_5	0.327×10^8	-0.1042	-0.1042	-521.2	-0.1030	-0.1044
x_6	-0.327×10^8	-0.2708	-0.2708	520.9	-0.2717	-0.2706
Pseudorank	6	4	4	6	4	4

PROG4 reads the 15 card images of data, identified as **DATA4**, and applies the subroutine **SVA** to that data. This example is discussed in Chapter 26.

PROG5 illustrates the use of subroutines **BNDACC** and **BNDSOL** by setting up and solving a sequence of six cubic spline curve-fitting programs as described in Chapter 27, Sections 2 and 4.

PROG6 illustrates the use of subroutines **LDP** and **NNLS** by solving the example of the constrained line-fitting problem given in Section 8 of Chapter 23.

Machine and Problem Dependent Tolerances

In order to avoid storing the machine precision constant η in the various subprograms tests of the general form "If $(|h| > \eta |x|)$" have been replaced by the test "If $((x + h) - x \neq 0)$." To obtain the intended result, it is essential that the sum $x + h$ be truncated (or rounded) to η-precision before computing the difference $(x + h) - x$. If the expression $(x + h) - x$ were written directly in Fortran code, there is the possibility that the intended result would not be obtained due to an optimizing compiler or the use of a double length accumulator. To circumvent these hazards a **FUNCTION** subprogram, **DIFF**, is used so that the test is coded as "If $(\mathbf{DIFF}(x + h, x) \neq 0)$."

The "**CALL**" statement for subroutine **HFTI** includes a parameter, **TAU**, which is used as an absolute tolerance in determining the pseudorank of the matrix. It is intended that the user set the value of this parameter to express information about the accuracy of the data rather than to express the machine's arithmetic precision.

Conversion to Double Precision

Users of these Fortran subprograms may wish to convert them for full double precision computation. We suggest the following approach to this conversion.

1. Do not change the **FUNCTION** subprogram **GEN** at all.

2. In all other subprograms, the variables and **FUNCTION** names which presently appear in **REAL, LOGICAL,** or **INTEGER** statements should remain so typed. All other variables whose names begin with **A, B, . . . , H,** or **O, P, . . . , Z** should be typed **DOUBLE PRECISION**.

3. Change the intrinsic Fortran functions as shown.

Single Precision Name	*Double Precision Name or Equivalent*
ABS(X)	**DABS(X)**
ALOG(X)	**DLOG(X)**
ALOG10(X)	**DLOG10(X)**

Single Precision Name	*Double Precision Name or Equivalent*
DBLE(X)	**X**
EXP(X)	**DEXP(X)**
SQRT(X)	**DSQRT(X)**
AMAX1(X, Y)	**DMAX1(X, Y)**
SIGN(X, Y)	**DSIGN(X, Y)**

One systematic way to accomplish this is to redefine these single precision "intrinsic function" names to denote "statement functions" which in turn are defined in terms of the appropriate double precision "intrinsic functions" [Sec. 8.1 and 8.2 of ASA Committee (1964)]. This can be done in subroutine **G1**, for example, as follows:

After line 12 insert

$$\textbf{ABS(X)} = \textbf{DABS(X)}$$
$$\textbf{SQRT(X)} = \textbf{DSQRT(X)}$$
$$\textbf{SIGN(X, Y)} = \textbf{DSIGN(X, Y)}$$

4. On some computer systems no changes in the **FORMAT** statements will be needed. On other systems, or to retain conformity with ANSI Fortran, the **"F"** and **"E"** specifications in **FORMAT**'s must be changed to **"D"**. **FORMAT** statements are grouped at the end of the listing in any subprograms in which they occur.

The remainder of this Appendix consists of User's Guides to the Fortran subprograms and listings of the Fortran code. See Tables C.1 and C.2 for a directory covering this material.

HFTI

USER'S GUIDE TO HFTI: SOLUTION OF THE LEAST SQUARES PROBLEM BY HOUSEHOLDER TRANSFORMATIONS

Subroutines Called **H12, DIFF**

Purpose

This subroutine solves a linear least squares problem or a set of linear least squares problems having the same matrix but different right-side vectors. The problem data consists of an **M** × **N** matrix A, an **M** × **NB** matrix B, and an absolute tolerance parameter τ. The **NB** column vectors of B represent right-side vectors b_j for **NB** distinct linear least squares problems.

$$Ax_j \cong b_j \qquad j = 1, \ldots, \text{NB}$$

This set of problems can also be written as the matrix least squares problem:

$$AX \cong B$$

where X is the $\text{N} \times \text{NB}$ matrix having column vectors x_j.

Note that if B is the $\text{M} \times \text{M}$ identity matrix, then X will be the pseudo-inverse of A.

Method

This subroutine first transforms the augmented matrix $[A : B]$ to a matrix $[R : C]$ using premultiplying Householder transformations with column interchanges. All subdiagonal elements in the matrix R are zero and its diagonal elements satisfy $|r_{ii}| \geq |r_{i+1,i+1}|$, $i = 1, \ldots, l - 1$, where $l = \min \{\text{M, N}\}$.

The subroutine will set the pseudorank **KRANK** equal to the number of diagonal elements of R exceeding τ in magnitude. Minimal length solution vectors $\hat{x}_j$, $j = 1, \ldots, \text{NB}$, will be computed for the problems defined by the first **KRANK** rows of $[R : C]$.

If the relative uncertainty in the data matrix B is ρ, it is suggested that τ be set approximately equal to $\rho \| A \|$.

For further algorithmic details, see Algorithm HFTI in Chapter 14.

Usage

DIMENSION A(MDA, n_1), {B(MDB, n_2) or B(m_1)}, RNORM(n_2), H(n_1), G(n_1)

INTEGER IP(n_1)

CALL HFTI (A, MDA, M, N, B, MDB, NB, TAU, KRANK, RNORM, H, G, IP)

The dimensioning parameters must satisfy $\text{MDA} \geq \text{M}$, $n_1 \geq \text{N}$, $\text{MDB} \geq \max \{\text{M, N}\}$, $m_1 \geq \max \{\text{M, N}\}$, and $n_2 \geq \text{NB}$.

The subroutine parameters are defined as follows:

A(,), MDA, M, N   The array **A(,)** initially contains the $\text{M} \times \text{N}$ matrix A of the least squares problem $AX \cong B$. The first dimensioning parameter of the array **A(,)** is **MDA**, which must satisfy $\text{MDA} \geq \text{M}$. Either $\text{M} \geq \text{N}$ or $\text{M} < \text{N}$ is permitted. There is no restriction on the rank of A. The contents of the array **A(,)** will be modified by the subroutine. See Fig. 14.1 for an example illustrating the final contents of **A(,)**.

B( ), MDB, NB   If $\text{NB} = 0$ the subroutine will make no references to the array **B( )**. If $\text{NB} > 0$ the array **B( )** must initially contain the $\text{M} \times \text{NB}$ matrix B of the least squares problem $AX \cong B$ and on return the array **B( )** will contain the **N** $\times$ **NB** solution matrix $\hat{X}$. If **NB** $\geq$ **2** the array **B( )** must be double subscripted with first dimensioning parameter **MDB** $\geq$ max $\{$**M, N**$\}$. If **NB** $=$ **1** the array **B( )** may be either doubly or singly subscripted. In the latter case the value of **MDB** is arbitrary but some Fortran compilers require that **MDB** be assigned a valid integer value, say **MDB** $=$ **1**.

TAU   Absolute tolerance parameter provided by user for pseudorank determination.

KRANK   Set by the subroutine to indicate the pseudorank of A.

RNORM( )   On exit, **RNORM(J)** will contain the euclidean norm of the residual vector for the problem defined by the jth column vector of the array **B(,)** for $j =$ **1, . . . , NB**.

H( ), G( )   Arrays of working space. See Fig. 14.1 for an example illustrating the final contents of these arrays.

IP( )   Array in which the subroutine records indices describing the permutation of column vectors. See Fig. 14.1 for an example illustrating the final contents of this array.

Example of Usage

See **PROG2** for an example of the usage of this subroutine.

SVA

USER'S GUIDE TO SVA: SINGULAR VALUE ANALYSIS

Subroutines Called **SVDRS, MFEOUT**

Purpose

The subroutine **SVA** uses subroutine **SVDRS** to obtain the singular value decomposition of a column scaled matrix $\tilde{A} = AD$ and the associated transformation of the right-side vector b of a least squares problem $Ax \cong b$. The subroutine **SVA** prints quantities derived from this decomposition to provide the user with information useful to the understanding of the given least squares problem.

Method

The user provides an $m \times n$ matrix A and an m-vector b, defining a least squares problem $Ax \cong b$. The user selects one of three options regarding column scaling described below in the definition of the parameter **ISCALE**. This selection defines an $n \times n$ diagonal matrix D.

Introducing the change of variables

$$x = Dy$$

the subroutine performs a singular value analysis of the least squares problem

$$\tilde{A}y \cong b$$

where

$$\tilde{A} = AD$$

The subroutine **SVA** uses the subroutine **SVDRS** to compute the singular value decomposition

$$\tilde{A} = USV^T$$

and the transformed right-side vector

$$g = U^T b$$

Denote the (ordered) singular values of $\tilde{A}$, i.e., the diagonal elements of S, by $s_1, \ldots, s_n$. Let $\bar{n}$ denote the index of the last nonzero singular value and define $\mu = \min (m, n)$.

Compute the vector

$$p = S^+ g$$

with components

$$p_i = \begin{cases} \dfrac{g_i}{s_i} & i = 1, \ldots, \bar{n} \\ 0 & i = \bar{n} + 1, \ldots, n \end{cases}$$

The kth candidate solution vector $y^{(k)}$ is defined as

$$y^{(0)} = 0$$

$$y^{(k)} = V \begin{bmatrix} I_k & 0 \\ 0 & 0 \end{bmatrix} p$$

$$= \sum_{j=1}^{k} p_j v_j \qquad k = 1, \ldots, \mu$$

where v_j denotes the jth column vector of V. Reverting to the original variables we have the corresponding candidate solutions

$$x^{(k)} = Dy^{(k)} \qquad k = 0, \ldots, \mu$$

The quantities

$$p_k^2 = \sum_{i=k+1}^{m} g_i^2 \qquad k = 0, \ldots, \mu$$

are computed. For $k \leq \bar{n}$ the quantity p_k^2 is the sum of squares of residuals associated with the kth candidate solution; i.e., $p_k^2 = \| b - Ax^{(k)} \|^2$.

It is possible that the $m \times (n + 1)$ data matrix $[A:b]$ provided to this subroutine may be a compressed representation of a least squares problem involving more than m rows. For example, $[A:b]$ may be the triangular matrix resulting from sequential Householder triangularization of a large set of data as described in Chapter 27. The user provides an integer **MDATA** that specifies the number of rows of data in the original problem. Of course, if $[A:b]$ is the original data, then **MDATA** and **M** should be set to the same value. The number **MDATA** is used in computing

$$\sigma_k = \left[\frac{p_k^2}{\max(1, \mathbf{MDATA} - k)} \right]^{1/2} \qquad k = 0, \ldots, \bar{n}$$

Under appropriate statistical hypotheses on the data $[A:b]$, the number σ_k can be interpreted as an unbiased estimate of the standard deviation of errors in the data vector b.

Adapting Eq. (25.47) and (25.48) to the notation of this Subroutine User's Guide, we compute the following quantities, which may be used for a Levenberg–Marquardt analysis:

$$v_\lambda^2 \equiv \| y(\lambda) \|^2 = \sum_{i=1}^{\bar{n}} p_i^2 \left(\frac{s_i^2}{s_i^2 + \lambda^2} \right)^2$$

$$\bar{\omega}_\lambda^2 \equiv \| r(\lambda) \|^2 = \sum_{i=\bar{n}+1}^{m} g_i^2 + \sum_{i=1}^{\bar{n}} g_i^2 \left(\frac{\lambda^2}{s_i^2 + \lambda^2} \right)^2$$

This subroutine prints the following quantities:

1. The quantities **M, N, MDATA**, identification of the scaling option used, and the diagonal elements of the scaling matrix D.
2. The matrix V, multiplied by 10^4 to facilitate scanning for large and small elements.
3. (a) The quantity s_k for $k = 1, \ldots, \mu$.
   (b) The quantities $p_k, s_k^{-1}, g_k, g_k^2$ for $k = 1, \ldots, \bar{n}$.
   (c) The quantities p_k^2 and σ_k for $k = 0, \ldots, \bar{n}$.
4. The quantities k, $\| y^{(k)} \|$, p_k, $\log_{10} \| y^{(k)} \|$, and $\log_{10} p_k$ for $k = 1, \ldots, \bar{n}$.
5. The quantities λ, v_λ, $\bar{\omega}_\lambda$, $\log_{10} \lambda$, $\log_{10} v_\lambda$, and $\log_{10} \bar{\omega}_\lambda$ for a sequence of 21 values of λ ranging from $10s_1$ to $s_{\bar{n}}/10$ in increments that are uniformly spaced in $\log_{10} \lambda$.

6. The candidate solutions $x^{(k)}$ for $k = 1, \ldots, \bar{n}$.

Usage

DIMENSION A(MDA, n_1), B(m_1), SING(n_2), NAMES(n_1), D(n_1)
CALL SVA (A, MDA, M, N, MDATA, B, SING, NAMES, ISCALE, D)

The dimensioning parameters must satisfy **MDA** $\geq$ max **(M, N)**, $n_1 \geq$ **N**, $m_1 \geq$ **M**, and $n_2 \geq$ **3*N**.

The subroutine parameters are defined as follows:

A(,), MDA, M, N The array **A(,)** initially contains the **M** $\times$ **N** matrix A of the least squares problem $Ax \cong b$. Either **M** $\geq$ **N** or **M** $<$ **N** is permitted. The first dimensioning parameter of the array **A(,)** is **MDA**, which must satisfy **MDA** $\geq$ max **(M, N)**. On output the kth candidate solution $x^{(k)}$ will be stored in the first **N** locations of the kth column of **A(,)** for $k = 1, \ldots,$ min **(M, N)**.

MDATA The number of rows in the original least squares problem, which may have had more rows than $[A : b]$. This number is used only in computing the numbers $\sigma_k, k = 0, \ldots, \bar{n}$.

B( ) The array **B( )** initially contains the **M**-vector b of the least squares problem $Ax \cong b$. On output the **M**-vector g is stored in the array **B( )**.

SING( ) The locations **(SING(I), I = 1, . . . , 3*N)** are used as temporary working space. On output the singular values of the scaled matrix $\tilde{A}$ are stored in **SING(I), I = 1, . . . ,** min **(M, N)**.

NAMES( ) In location **NAMES(I)** the user may store an alphanumeric name as an identifier for the I**th** component of the solution vector **(I = 1, . . . , N)**. These names will be printed with the appropriate components of the V matrix and with the candidate solutions. Each name is printed with an **A6** format. If **NAMES(1)** has the same contents as the quantity **IBLANK**, defined by

DATA IBLANK/1H /

then the remaining words of the array **NAMES( )** will be ignored and no identifying names will be printed.

ISCALE, D( ) The user sets **ISCALE = 1, 2,** or **3**. If **ISCALE = 1**, the subroutine functions as though the scaling matrix D were an **N** $\times$ **N** identity matrix.

In this case the subroutine makes no reference to the array **D( )**.

If **ISCALE = 2**, the subroutine computes $\|a_j\|$ where a_j denotes the jth column vector of A and sets

$$\mathbf{D}(j) = \begin{cases} \|a_j\|^{-1} & \text{if } \|a_j\| \neq 0 \\ 1 & \text{if } \|a_j\| = 0 \end{cases}$$

In this case all nonzero columns of the scaled matrix $\tilde{A} = AD$ will have unit euclidean length.

If **ISCALE = 3**, the subroutine will use the user-supplied values **D(J), J = 1, . . . , N** as the diagonal elements of the diagonal scaling matrix D. In this case the user must assign values to **D(J), J = 1, . . . , N**, and the subroutine will not modify these values. For example, the user might set **D(J)** equal to the a priori standard deviation of the **J**th component of the solution vector if such information is known. As a further example, the user might set **D(J)** equal to the reciprocal of the mean of the **M** components of the **J**th column vector's uncertainty. This type of scaling has the effect of making the uncertainty of every column the same magnitude in the matrix $\tilde{A}$.

Example of Usage

See **PROG4** for an example of the usage of this subroutine. Output printed by this subroutine is illustrated in Fig. 26.1.

SVDRS

USER'S GUIDE TO SVDRS: SINGULAR VALUE DECOMPOSITION OF PROBLEM LS

Subroutines Called **H12, QRBD**

Purpose

Given an **M** $\times$ **N** matrix A and an **M** $\times$ **NB** matrix B, this subroutine computes the singular values of A and also computes auxiliary quantities useful in analyzing and solving the matrix least squares problem $AX \cong B$. Denote the singular value decomposition of A by

$$A = USV^T$$

This subroutine computes S, V, and $G = U^T B$.

To complete the solution of the least squares problem $AX \cong B$, the user must first decide which small singular values are to be treated as zero. Let S^+ denote the matrix obtained by transposing S and reciprocating the significant singular values and setting the others to zero. Then the solution matrix X can be obtained by computing $P = S^+ G$ and $X = VP$.

Either $\mathbf{M} \geq \mathbf{N}$ or $\mathbf{M} < \mathbf{N}$ is permitted. Note that if $B = I$, then X is the pseudoinverse of A.

Input

The matrices A and B and their dimensioning parameters are input.

Output

The matrices V, $G = U^T B$, and the diagonal elements of S are output in the arrays $\mathbf{A(\ ,\ )}$, $\mathbf{B(\ ,\ )}$, and $\mathbf{S(\ \ )}$, respectively.

Usage

DIMENSION A(MDA, n_1), {B(MDB, n_2) or B(m_1)}, S(n_3)
CALL SVDRS (A, MDA, M, N, B, MDB, NB, S)

The dimensioning parameters must satisfy $\mathbf{MDA} \geq \max\{\mathbf{M, N}\}, n_1 \geq \mathbf{N}$, $\mathbf{MDB} \geq \mathbf{M}, n_2 \geq \mathbf{NB}, m_1 \geq \mathbf{M}$, and $n_3 \geq 3*\mathbf{N}$.

The subroutine parameters are defined as follows:

A(,), MDA, M, N   The array $\mathbf{A(\ ,\ )}$ is a doubly subscripted array with first dimensioning parameter equal to **MDA**. The array $\mathbf{A(\ ,\ )}$ initially contains the $\mathbf{M} \times \mathbf{N}$ matrix A with $\mathbf{A(I, J)} := a_{IJ}$. On output the array $\mathbf{A(\ ,\ )}$ contains the $\mathbf{N} \times \mathbf{N}$ matrix V with $\mathbf{A(I, J)} := v_{IJ}$. Either $\mathbf{M} \geq \mathbf{N}$ or $\mathbf{M} < \mathbf{N}$ is permitted.

B( ), MDB, NB   **NB** denotes the number of column vectors in the matrix B. If $\mathbf{NB} = \mathbf{0}$ the array $\mathbf{B(\ )}$ will not be referenced by this subroutine. If $\mathbf{NB} \geq \mathbf{2}$, the array **B** should be doubly subscripted with first dimensioning parameter equal to **MDB**. If $\mathbf{NB} = \mathbf{1}$, then **B** will be used as a singly subscripted array of length **M**. In this latter case the value of **MDB** is arbitrary but for successful functioning of some Fortran compilers it must be set to some acceptable integer value, say $\mathbf{MDB} = \mathbf{1}$. The contents of the array **B** are initially
$\mathbf{B(I, J)} := b_{IJ}, \mathbf{I} = \mathbf{1}, \ldots, \mathbf{M}, \mathbf{J} = \mathbf{1}, \ldots, \mathbf{NB}$ or
$\mathbf{B(I)} := b_I, \mathbf{I} = \mathbf{1}, \ldots, \mathbf{M}.$

At the conclusion

$B(I, J) := g_{IJ}, I = 1, \ldots , M, J = 1, \ldots , NB$ or

$B(I) := g_I, I = 1, \ldots , M$.

S( )   The array **S( )** is used as $3*N$ locations of temporary working space by the subroutine. On conclusion the first **N** locations of the array **S( )** contain the ordered singular values of the matrix $A : S(1) \geq S(2) \geq \cdots \geq S(N) \geq 0$.

Error Message

If convergence does not occur in subroutine **QRBD** then subroutine **SVDRS** prints the message

CONVERGENCE FAILURE IN QR BIDIAGONAL SVD ROUTINE

and executes the Fortran statement, **STOP**.

Example of Usage

See **PROG3** and the subroutine **SVA** for examples of usage of this subroutine.

QRBD

USER'S GUIDE TO QRBD: SINGULAR VALUE DECOMPOSITION OF A BIDIAGONAL MATRIX

Subroutines Called **G1**, **G2**, and **DIFF**

Purpose

Compute the singular value decomposition of an **N** × **N** bidiagonal matrix

$$B = \tilde{U}S\tilde{V}^T$$

This subroutine implements the special QR iteration described in Chapter 18. The output of the subroutine consists of the singular values of B (the diagonal elements of the matrix S) and the two matrix products $V\tilde{V}$ and $\tilde{U}^T C$ where V and C are given matrices of dimensions **NRV** × **N** and **N** × **NC**, respectively.

Usage

DIMENSION D(n_1), E(n_1), V(MDV, n_1), C(MDC, m_1)
CALL QRBD (IPASS, D, E, N, V, MDV, NRV, C, MDC, NC)

The dimension parameters must satisfy $n_1 \geq$ **N**, $m_1 \geq$ **NC, MDV** $\geq$ **NRV, MDC** $\geq$ **N**.

The subroutine parameters are defined as follows:

IPASS	This integer flag is returned with either of the values **1** or **2**. **IPASS** = **1** signifies that convergence of the QR algorithm did occur. A singular value decomposition of the matrix B was obtained. **IPASS** = **2** denotes an error condition. This indicates that convergence was not attained after **10∗N** QR sweeps. (Convergence usually occurs in about **2∗N** sweeps.)
D( )	The array **D( )** initially contains the diagonal elements of the matrix B. **D(I)** := b_{II}, **I** = **1, . . . , N** On return the array **D( )** contains the **N** (nonnegative) singular values of B in nonincreasing order.
E( )	The array **E( )** initially contains the superdiagonal elements of the matrix B. **E(1)** := arbitrary. **E(I)** := $b_{I-1,I}$, **I** = **2, . . . , N**. The contents of the array **E( )** are modified by the subroutine.
N	Order of the matrix B.
V(,), MDV, NRV	If **NRV** $\leq$ **0**, the parameters **V(,)** and **MDV** will not be used. If **NRV** $\geq$ **1**, then **MDV** must satisfy **MDV** $\geq$ **NRV**. The first dimensioning parameter of **V(,)** must be equal to **MDV**. The array **V(,)** initially contains an **NRV** $\times$ **N** matrix V and on return this matrix will be replaced by the **NRV** $\times$ **N** product matrix $V\tilde{V}$.
C(,), MDC, NC	If **NC** $\leq$ **0**, the parameters **C(,)** and **MDC** will not be used. If **NC** $\geq$ **2**, then **MDC** must satisfy **MDC** $\geq$ **N**. The first dimensioning parameter of the array **C(,)** must be equal to **MDC**.

If **NC = 1**, the parameter **MDC** can be assigned any value, say, **MDC = 1**. In this case the array **C(,)** may be either singly or doubly subscripted. When **NC ≥ 1**, the array **C(,)** must initially contain an **N × NC** matrix C. On output this matrix will be replaced by the **N × NC** product matrix $\tilde{U}^T C$.

Example of Usage

This subroutine is used by the subroutine **SVDRS**.

BNDACC, BNDSOL

USER'S GUIDE TO BNDACC AND BNDSOL: SEQUENTIAL PROCESSING OF BANDED LEAST SQUARES PROBLEMS

Subroutine Called **H12**

Purpose

These subroutines implement the algorithms given in Section 2 of Chapter 27 for solving $Ax \cong b$ where the matrix A has bandwidth **NB**. The four principal parts of these algorithms are obtained by the following **CALL** statements:

CALL BNDACC(. . .)   Introduce new rows of data.
CALL BNDSOL(1, . . .)   Compute solution vector and norm of residual vector.
CALL BNDSOL(2, . . .)   Given a vector d, solve $R^T y = d$ for y [see Eq. (27.22)].
CALL BNDSOL(3, . . .)   Given a vector w, solve $Rc = w$ for c [see Eq. (27.23)].

The dots indicate additional parameters that will be specified in the following paragraphs.

Usage of **BNDACC**

DIMENSION G(MDG, n_1)

The dimensioning parameters must satisfy **MDG** $\geq \mu$ [see Eq. (27.9) for the definition of μ] and $n_1 \geq$ **NB + 1**.

The user must set **IP = 1** and **IR = 1** before the first call to **BNDACC**.

The subroutine **BNDACC** is to be called once for each block of data $[C_t : b_t]$ to be introduced into the problem. See Eq. (27.14) and Fig. 27.2. For each block of data the calling program must assign values to **MT** and **JT** and copy the $[\mathbf{MT} \times (\mathbf{NB} + 1)]$-array of data $[C_t : b_t]$ into rows **IR** through **IR** + **MT** − **1** of the working array **G(,)** and then execute the statement

CALL BNDACC (G, MDG, NB, IP, IR, MT, JT)

The subroutine parameters are defined as follows:

G(,)   The working array. See Eq. (27.15) to (27.17) and adjacent text.

MDG   Set by user to indicate the number of rows in the array **G(,)**.

NB   Set by user to indicate the bandwidth of the data matrix A. See text following Eq. (27.14).

IP   Must be set to the value **1** before the first call to **BNDACC**. Its value is subsequently controlled by **BNDACC**.

IR   Index of first row of **G(,)** into which user is to place new data. The variable **IR** is initialized by the user to the value **1** and subsequently updated by **BNDACC**. It is not to be modified by the user.

MT   Set by user to indicate the number of new rows of data being introduced by the current call to **BNDACC**. See Eq. (27.14).

JT   Set by user to indicate the column of the submatrix A_t that is identified with the first column of the array C_t. This parameter **JT** has the same meaning as j_t of Eq. (27.14).

Usage of **BNDSOL**

This subroutine performs three distinct functions selected by the first parameter **MODE**, which may have the values **1**, **2**, or **3**.

The statement **CALL BNDSOL (1, . . .)** may be executed after one or more calls to **BNDACC** to compute a solution for the least squares problem whose data has been introduced and triangularized by the calls to **BNDACC**. The problem being solved is represented by Eq. (27.11). This statement also computes the number $|e|$ of Eq. (27.6), which has the interpretation given in Eq. (27.12). The number $|e|$ will be stored in **RNORM**.

The computation performed by **CALL BNDSOL (1, . . .)** does not alter the contents of the array **G(,)** or the value of **IR**. Thus, after executing **CALL BNDSOL (1, . . .)**, more calls can be made to **BNDACC** to introduce additional data.

The statement **CALL BNDSOL (2, . . .)** may be used to solve $R^T y = d$ and the statement **CALL BNDSOL (3, . . .)** may be used to solve $Rc = w$. These entries do not modify the contents of the array **G(,)** or the variable

IR. The variable **RNORM** is set to zero by each of these latter two statements.

The primary purpose of these latter two statements is to facilitate computation of columns of the covariance matrix $C = (R^T R)^{-1}$ as described in Chapter 27. To compute c_j, the jth column of C, one would set $d = e_j$, the jth column of the identity matrix. After solving $R^T y = d$, set $w = y$ and solve $Rc_j = w$. This type of usage is illustrated in the Fortran program **PROG5**.

The appropriate specification statement is

DIMENSION G(MDG, n_1), X(n_2)

The dimensioning parameters must satisfy **MDG** $\geq \mu$ [see Eq. (27.9) for the definition of μ], $n_1 \geq$ **NB** $+$ **1**, and $n_2 \geq$ **N**. The **CALL** statement is

CALL BNDSOL(MODE, G, MDG, NB, IP, IR, X, N, RNORM)

The subroutine parameters are defined as follows:

MODE	Set by the user to the value **1**, **2**, or **3** to select the desired function as described above.
G(,), MDG, NB, IP, IR	These parameters must contain values as they were defined upon the return from a preceding call to **BNDACC**.
X( )	On input, with **MODE = 2** or **3**, this array must contain the **N**-dimensional right-side vector of the system to be solved. On output with **MODE = 1**, **2**, or **3** this array will contain the **N**-dimensional solution vector of the appropriate system that has been solved.
N	Set by user to specify the dimensionality of the desired solution vector. This causes the subroutine **BNDSOL** to use only the leading **N** $\times$ **N** submatrix of the triangular matrix currently represented in columns **1** through **NB** of the array **G(,)** and (if **MODE = 1**) only the first **N** components of the right-side vector currently represented in column **NB** $+$ **1** of the array **G(,)**. If any of the first **N** diagonal elements are zero, this subroutine executes the Fortran statement **STOP** after printing an appropriate message. See the text following Algorithm (27.24) for a

discussion of some steps one might take to obtain a usable solution in this situation.

RNORM   If **MODE = 1, RNORM** is set by the subroutine to the value $|e|$ defined by Eq. (27.6). This number is computed as

$$\left[\sum_{I=N+1}^{IR-1} G(I, NB + 1)**2 \right]^{1/2}$$

If **MODE = 2** or **3, RNORM** is set to zero.

Example of Usage

See **PROG5** for an example of the usage of **BNDACC** and **BNDSOL**.

LDP

USER'S GUIDE TO LDP: LEAST DISTANCE PROGRAMMING

Subroutine Called **NNLS, DIFF**

Purpose

This subroutine computes the solution of the following constrained least squares problem:

Problem LDP: Minimize $\|x\|$ subject to $Gx \geq h$. Here G is an **M** $\times$ **N** matrix and h is an **M**-vector.

No initial feasible vector, x_0, is required. Thus the subroutine **LDP** can be used to obtain a solution to an arbitrary set of linear inequalities, $Gx \geq h$. If these inequalities have no solution the user is notified by means of a flag returned by the subroutine.

Method

This subroutine implements Algorithm LDP (23.27), which reduces Problem LDP (23.3) to a related Problem NNLS (23.2). Problem NNLS is solved with subroutine **NNLS** and a solution to Problem LDP is then obtained or it is noted that no solution exists.

Usage

> **DIMENSION G(MDG, n_1), H(m_1), X(n_1), W(n_2)**
> **INTEGER INDEX(m_1)**
> **CALL LDP (G, MDG, M, N, H, X, XNORM, W, INDEX, MODE)**

The dimensioning parameters must satisfy **MDG $\geq$ M**, $n_1 \geq$ **N**, $m_1 \geq$ **M**, and $n_2 \geq$ **(M + 2)∗(N + 1) + 2∗M**.

The subroutine parameters are defined as follows:

G(,), MDG, M, N   The array **G(,)** has first dimensioning parameter **MDG** and contains the **M $\times$ N** matrix G of Problem LDP. Either **M $\geq$ N** or **M < N** is permitted and there is no restriction on the rank of G. The contents of **G(,)** are not modified.

H( )   The array **H( )** contains the **M**-vector h of Problem LDP. The contents of **H( )** are not modified.

X( )   The contents of the array **X( )** do not need to be defined initially. On normal termination **(MODE = 1)** **X( )** contains the solution vector $\hat{x}$. On an error termination **(MODE $\geq$ 2)** the contents of **X( )** are set to zero.

XNORM   On normal termination **(MODE = 1) XNORM** contains $\|\hat{x}\|$. On an error termination **(MODE $\geq$ 2)** the value of **XNORM** is set to zero.

W( )   This array is used as temporary working storage.

INDEX( )   This array is used as temporary type **INTEGER** working storage.

MODE   This flag is set by the subroutine to indicate the status of the computation on completion. Its value indicates the reason for terminating:
1 = successful.
2 = bad dimension. The condition **N $\leq$ 0** was noted in subroutine **LDP**.
3 = maximum number **(3∗M)** of iterations exceeded in subroutine **NNLS**.
4 = inequality constraints $Gx \geq h$ are incompatible.

Example 1

See program **PROG6** for one example of the usage of this subroutine.

Example 2

Suppose a 5 $\times$ 15 matrix G and a 5-dimensional vector h are given. The following code will solve Problem LDP for this data:

```
     DIMENSION G(5, 15), H(5), X(15), W(122)
     INTEGER INDEX (5)
     DO 10 I = 1, 5
     H(I) = h_i
     DO 10 J = 1, 15
 10  G(I, J) = g_ij
     CALL LDP (G, 5, 5, 15, H, X, XNORM, W, INDEX, MODE)
     GO TO (20, 30, 40, 50), MODE
 20  [Successful return. The solution vector x̂ is in X( ).]
 30  [Bad dimension. N ≤ 0 was noted.]
 40  [Excessive iterations (more than 3*M) in subroutine NNLS.]
 50  [The inequalities Gx ≥ h are incompatible.]
```

NNLS

USER'S GUIDE TO NNLS: NONNEGATIVE LINEAR LEAST SQUARES

Subroutines Called H12, G1, G2, DIFF

Purpose

This subroutine computes a solution vector, $\hat{x}$, for the following constrained least squares problem:

Problem NNLS: Solve $Ax \cong b$ subject to $x \geq 0$.

Here A is a given M × N matrix and b is a given m-vector. This problem always has a solution but it is nonunique if the rank of A is less than N.

Method

Problem NNLS is solved using Algorithm NNLS (23.10). It can be proved that this algorithm converges in a finite number of iterations. The technique for adding and removing vectors in the QR decomposition of A is that described in Method 3 of Chapter 24.

Usage

```
     DIMENSION A(MDA, n_1), B(m_1), X(n_1), W(n_1), Z(m_1)
     INTEGER INDEX(n_1)
     CALL NNLS (A, MDA, M, N, B, X, RNORM, W, Z, INDEX, MODE)
```

The dimensioning parameters must satisfy $\mathbf{MDA} \geq \mathbf{M}$, $n_1 \geq \mathbf{N}$, $m_1 \geq \mathbf{M}$. The subroutine parameters are defined as follows:

A(,), MDA, M, N The array **A(,)** has first dimensioning parameter **MDA** and initially contains the **M** × **N** matrix A. Either **M** $\geq$ **N** or **M** $<$ **N** is permissible. There is no restriction on the rank of A. On termination **A(,)** contains $\tilde{A} = QA$, where Q is an **M** × **M** orthogonal matrix implicitly generated by this subroutine.

B( ) The array **B( )** initially contains the **M**-vector, b. On termination **B( )** contains Qb.

X( ) The initial content of the array **X( )** is arbitrary. On termination with **MODE = 1 X( )** contains a solution vector, $\hat{x} \geq 0$.

RNORM On termination **RNORM** contains the euclidean norm of the final residual vector **RNORM** := $\|b - A\hat{x}\|$.

W( ), Z( ) The initial contents of these two arrays are arbitrary. On termination **W( )** will contain the **N**-dimensional dual vector $w = A^T(b - A\hat{x})$. The array **Z( )** is working space.

INDEX( ) The initial content of this array is arbitrary. The array **INDEX( )** is integer working space.

MODE This flag is set by the subroutine to indicate the status of the computation on completion. Its value indicates the reason for terminating:
1 = successful.
2 = bad dimensions. One of the conditions **M** $\leq$ **0** or **N** $\leq$ **0** has occurred.
3 = maximum number **(3∗N)** of iterations has been exceeded.

Error Message

The subroutine will write the error message

NNLS QUITTING ON ITERATION COUNT

and exit with a feasible (approximate solution), $\hat{x}$, if the iteration count exceeds **3∗N**.

Example of Usage

Solve the least squares problem $Ax \cong b$ subject to $x \geq 0$ where A is a 20 × 10 matrix and b is a 20-dimensional vector.

```
      DIMENSION A(20, 10), B(20), X(10), W(10), Z(20), INDEX(10)
      DO 10 I = 1, 20
      B(I) = bᵢ
      DO 10 J = 1,10
10    A(I, J) = aᵢⱼ
      CALL NNLS (A, 20, 20, 10, B, X, RNORM, W, Z, INDEX, MODE)
```

H12

USER'S GUIDE TO H12: CONSTRUCTION AND APPLICATION OF A HOUSEHOLDER TRANSFORMATION

Purpose

This subroutine implements Algorithms H1 and H2 (10.22). Given an m-vector v and integers l_p and l_1, this subroutine computes an m-vector u and a number s such that the $m \times m$ (Householder) symmetric orthogonal matrix $Q = I + (uu^T)/(su_{l_p})$ satisfies $Qv = w$ where $w_i = v_i$ for $i < l_p$, $|w_{l_p}| = (v_{l_p}^2 + \sum_{i=l_1}^{m} v_i^2)^{1/2}$, $w_i = v_i$ for $l_p < i < l_1$, and $w_i = 0$ for $l_1 \leq i \leq m$. Optionally this matrix Q may be multiplied times a given set of **NCV** m-vectors.

Usage

CALL H12 (MODE, LPIVOT, L1, M, U, IUE, UP, C, ICE, ICV, NCV)

The parameters are defined as follows:

MODE   Set by the user to the value **1** or **2**. If **MODE = 1**, the subroutine executes Algorithm H1 (10.22), which computes a Householder transformation and, if **NCV > 0**, multiplies it times the set of **NCV** m-vectors stored in the array **C**. If **MODE = 2**, the subroutine executes Algorithm H2 (10.22), which assumes that a Householder transformation has already been defined by a previous call with **MODE = 1** (Algorithm H1) and, if **NCV > 0**, multiplies it times the set of **NCV** m-vectors stored in the array **C**.

LPIVOT, L1, M   If these integers satisfy

$$1 \leq \textbf{LPIVOT} < \textbf{L1} \leq \textbf{M}$$

then they constitute the quantities l_p, l_1, and m defined above under *Purpose*. If these inequalities are not all satisfied, then the subroutine returns without doing any computation. This implicitly effects an identity transformation.

U( ), IUE, UP

The array **U( )** contains **M** elements with a positive storage increment of **IUE** between elements. If **MODE = 1**, the array **U( )** must initially contain the *m*-vector v stored as **U(1 + $(j - 1)$*IUE)** $= v_j$, $j = 1, \ldots, m$. The subroutine will compute s and u_{l_p} (see *Purpose*), storing s in place of v_{l_p} and storing u_{l_p} in **UP**. The other elements of **U( )** remain unchanged; however, the elements v_j for $l_1 \leq j \leq m$ will be regarded on exit as constituting elements u_j since $u_j = v_j$ for $l_1 \leq j \leq m$. If **MODE = 2**, the contents of **U( )** and **UP** must contain the results produced by a previous call to the subroutine with **MODE = 1**.

Example 1: If v is stored in a single subscripted Fortran array called **W** with **W(I)** $:= v_i$, then the parameters **"U, IUE"** should be written as **"W, 1"**.

Example 2: If v is stored as the Jth column of a doubly subscripted Fortran array called **A** with **A(I, J)** $:= v_i$, then the parameters **"U,IUE"** should be written as **"A(1, J), 1"**.

Example 3: If v is stored as the Ith row of a doubly subscripted Fortran array called **A** with **A(I, J)** $:= v_j$ and **A** is dimensioned **A(50, 40)**, then the parameters **"U, IUE"** should be written as **"A(I, 1), 50"**.

C( ), ICE, ICV, NCV

If **NCV** $\leq$ **0**, no action is taken involving the parameters **C( )**, **ICE**, and **ICV**. If **NCV** > **0**, the array **C( )** must initially contain a set of **NCV** *m*-vectors stored with a positive storage increment of **ICE** between elements of a vector and a storage increment of **ICV** between vectors. Thus if z_{ij} denotes the *i*th element of the *j*th vector, then **C(1 +** $(i - 1)$***ICE +** $(j - 1)$***ICV)** $= z_{ij}$. On output, **C( )** contains the set of **NCV** *m*-vectors resulting from multiplying the given vectors by Q. The following examples illustrate the two most typical ways in which the parameters **C, ICE, ICV,** and **NCV** would be used.

Example 1: Compute $E = QF$ where F is a 50 × 30 matrix stored in a Fortran array **F** dimensioned as **F(60, 40)**. Here **M = 50** and the parameters "**C, ICE, ICV, NCV**" should be written as "**F, 1, 60, 30**".

Example 2: Compute $G = FQ$ where F is a 20 × 60 matrix stored in a Fortran array **F** dimensioned as **F(30, 70)**. Here **M = 60** and the parameters "**C, ICE, ICV, NCV**" should be written as "**F, 30, 1, 20**".

Usage Examples

A. Reduce an $n \times n$ real matrix A to upper Hessenberg form using Householder transformations. This is often used as a first step in computing eigenvalues of an unsymmetric matrix.

$$QAQ^T = H$$

$$H = \begin{bmatrix} h_{11} & h_{12} & \cdots & h_{1n} \\ h_{21} & \cdot & & \cdot \\ & \cdot & \cdot & \cdot \\ & & \cdot & \cdot \\ 0 & & h_{n,n-1} & h_{nn} \end{bmatrix}$$

```
DIMENSION A(50, 50),  UP(50)
IF(N.LE.2) GO TO 20
DO 10 I = 3, N
    CALL H12(1,I−1,I,N,A(1,I−2),1,UP(I−2),A(1,I−1),1,50,N−I+2)
10 CALL H12(2, I−1, I, N, A(1, I−2), 1, UP(I−2), A, 50, 1, N)
20 CONTINUE
```

Note that the matrix H occupies the upper triangular and first subdiagonal part of the array named **A** on output. The data defining Q occupies the remaining part of the array **A** and the array **UP**.

B. Suppose that we have the least squares problem $Ax \cong b$ with the special form

$$A = \begin{bmatrix} R_{n \times n} \\ 0_{1 \times n} \\ Y_{k \times n} \end{bmatrix}, \qquad b = \begin{bmatrix} d_{n \times 1} \\ e_{1 \times 1} \\ z_{k \times 1} \end{bmatrix}$$

where R is upper triangular. Use Householder transformations to reduce this system to upper triangular form.

The augmented matrix $[A:b]$ occupies the $(N + K + 1) \times (N + 1)$ -array named **A**. This is an example of block by block sequential accumulation as described in Algorithm SEQHT (27.10).

```
DIMENSION A(50, 30)
NP1 = N + 1
DO 10 J = 1, NP1
10 CALL H12(1,J,N+2,N+K+1,A(1,J),1,T,A(1,J+1),1,50,N+1−J)
```

Notice that this code does not explicitly zero the lower triangular part of the **A**-array in storage.

G1, G2

USER'S GUIDE TO G1 AND G2: CONSTRUCTION AND APPLICATION OF ROTATION MATRICES

Purpose

Given the numbers x_1 and x_2, the subroutine **G1** computes data that defines the 2×2 orthogonal rotation matrix G such that

$$G\begin{bmatrix} x_1 \\ x_2 \end{bmatrix} \equiv \begin{pmatrix} c & s \\ -s & c \end{pmatrix}\begin{bmatrix} x_1 \\ x_2 \end{bmatrix} = \begin{bmatrix} (x_1^2 + x_2^2)^{1/2} \\ 0 \end{bmatrix} \equiv \begin{bmatrix} r \\ 0 \end{bmatrix}$$

The subroutine **G2** performs the matrix–vector product

$$G\begin{bmatrix} z_1 \\ z_2 \end{bmatrix} = \begin{bmatrix} \tilde{z}_1 \\ \tilde{z}_2 \end{bmatrix}$$

See Algorithms G1 (10.25) and G2 (10.26) for further details.

Input

(a) The numbers x_1 and x_2 are input to **G1**.
(b) The numbers c, s, z_1, and z_2 are input to **G2**.

Output

(a) The numbers c, s, and r are output by **G1**.

(b) The matrix product $\begin{bmatrix} \tilde{z}_1 \\ \tilde{z}_2 \end{bmatrix}$ is output by **G2** following a previous execution of **G1** that defined the appropriate c and s of the G matrix. The output quantities $\tilde{z}_1$ and $\tilde{z}_2$ replace the input quantities z_1 and z_2 in storage.

Usage

CALL G1(X1, X2, C, S, R)
CALL G2(C, S, Z1, Z2)

Usage Example

Suppose that we have the least squares problem $Ax \cong b$ with the special form

$$A = \begin{bmatrix} R_{n \times n} \\ 0_{1 \times n} \\ y_{1 \times n} \end{bmatrix}, \qquad b = \begin{bmatrix} d_{n \times 1} \\ e_{1 \times 1} \\ z_{1 \times 1} \end{bmatrix}$$

where R is upper triangular.

Use rotation matrices to reduce this system to upper triangular form and reduce the right side so that only its first $n + 1$ components are nonzero. The augmented matrix $[A : b]$ occupies the **(N + 2) $\times$ (N + 1)**-array named **A**.

This is an example of a row by row "sequential accumulator." See the remarks following Algorithm SEQHT (27.10).

```
      NP1 = N + 1
      DO 10 I = 1, NP1
      CALL  G1 (A(I, I), A(N+2, I), C, S, A(I, I))
      A(N + 2, I) = 0.
      IF (I.GT.N) GO TO 20
      DO 10 J = I, N
10    CALL  G2 (C, S, A(I, J+1), A(N+2, J+1))
20    CONTINUE
```

MFEOUT

USER'S GUIDE TO MFEOUT: MATRIX OUTPUT SUBROUTINE

Purpose

This subroutine prints an **M $\times$ N** matrix A using an eight-column format. Let A be partitioned as $[A_1 : A_2 : A_3 : \cdots]$ where each submatrix except possibly the last has dimensions **M** $\times$ **8**. The last submatrix may have fewer than eight columns. These submatrices will be printed in the form

$$\begin{bmatrix} A_1 \\ A_2 \\ A_3 \\ \cdot \\ \cdot \\ \cdot \end{bmatrix}$$

Identifying row and column numbers will be printed. Additionally, rows can be given individual alphanumeric labels if provided by the user in the array **NAMES( )**.

Because of the intended use of this subroutine by the subroutine **SVA**, a parameter **MODE** is provided by which the calling program selects one of two choices of main headings, column headings, and numerical formats.

Usage

> **DIMENSION A(MDA, n_1), NAMES(m_1)**
> **CALL MFEOUT (A, MDA, M, N, NAMES, MODE)**

The dimensioning parameters must satisfy **MDA** $\geq$ **M**, $n_1 \geq$ **N**, and $m_1 \geq$ **M**.

The subroutine parameters are defined as follows:

A(,)   Array containing an **M** $\times$ **N** matrix to be printed.

MDA   Dimension of first subscript of the array **A(,)**.

M, N   Number of rows and columns, respectively, in the matrix to be printed.

NAMES( )   **NAMES(I)** contains an alphanumeric label to be output in association with row **I** of the matrix using an **A6** format. If **NAMES(1)** matches the variable **IBLANK** defined by **DATA IBLANK/1H /**, then the subroutine will ignore the remainder of the **NAMES( )** array and print blank labels for all rows.

MODE   Set by user to either **1** or **2**.
            MODE = 1 provides headings appropriate for the V matrix of the singular value decomposition and selects a numerical format of **4PF15.0**.
            MODE = 2 provides headings appropriate for the array of candidate solutions resulting from singular value analysis and selects a numerical format of **E15.8**.

GEN

USER'S GUIDE TO GEN: DATA GENERATION FUNCTION

Purpose

This **FUNCTION** is used by **PROG1**, **PROG2**, and **PROG3** to generate data for test cases. By basing the generation method on small integers it is intended that the same test cases can be generated on virtually all computers.

Method

This **FUNCTION** generates two sequences of integers:

$$I_0 = 5$$
$$I_k = (891*I_{k-1}) \bmod (1000) \qquad k = 1, 2, \ldots$$

and

$$J_0 = 7$$
$$J_l = (457*J_{l-1}) \bmod (997) \qquad l = 1, 2, \ldots$$

The sequence $\{I_k\}$ has period 10, while the sequence $\{J_l\}$ has period 332. On the kth call after initialization, **GEN** produces the **REAL** output value

FLOAT $(I_k - 500)$ + **ANOISE**$*$**FLOAT** $(J_l - 498)$

The next member of the sequence $\{J_l\}$ is produced only when **ANOISE** > **0**. No claim is made that this sequence has any particular pseudorandom properties.

Usage

This **FUNCTION** must be initialized by a statement of the form

X = GEN(A)

where **A** < **0**. In this case the value assigned to **X** is zero.

The next number in the sequence defined above is produced by a statement of the form

X = GEN(ANOISE)

where the input parameter **ANOISE** is nonnegative.

DIFF

USER'S GUIDE TO DIFF

Purpose

This **FUNCTION** is used by the subroutines **HFTI**, **QRBD**, **LDP** and **NNLS** to make the test "If $((x + h) - x) \neq 0$" which is used in place of the test "If $(|h| > \eta |x|)$" where η is the relative precision of the machine floating point arithmetic.

Method

In the intended usage of this **FUNCTION** the intermediate sum $z = x + h$ is computed in the calling program using η-precision arithmetic. Then the difference $d = z - x$ is computed by this **FUNCTION** using η-precision arithmetic.

Usage

This **FUNCTION** can be used for the test described in *Purpose* with an arithmetic **IF** statement of the form

$$\text{IF(DIFF(X+H, X))} \ \ 10, 20, 10$$

The statement numbered **10** corresponds to the condition $|h| > \eta |x|$. The statement numbered **20** corresponds to the condition $|h| \leq \eta |x|$.

PROG1

```
 1   C     PROG1
 2   C     C.L.LAWSON AND R.J.HANSON, JET PROPULSION LABORATORY, 1973 JUN 12
 3   C     TO APPEAR IN 'SOLVING LEAST SQUARES PROBLEMS', PRENTICE-HALL, 1974
 4   C        DEMONSTRATE ALGORITHMS HFT AND HS1 FOR SOLVING LEAST SQUARES
 5   C     PROBLEMS AND ALGORITHM COV FOR COMPUTING THE ASSOCIATED COVARIANCE
 6   C     MATRICES.
 7   C
 8         DIMENSION A(8,8),H(8),B(8)
 9         REAL GEN,ANOISE
10         DOUBLE PRECISION SM
11         DATA MDA/8/
12   C
13             DO 180 NOISE=1,2
14             ANOISE=0.
15             IF (NOISE.EQ.2) ANOISE=1.E-4
16             WRITE (6,230)
17             WRITE (6,240) ANOISE
18   C     INITIALIZE THE DATA GENERATION FUNCTION
19   C       ..
20             DUMMY=GEN(-1.)
21               DO 180 MN1=1,6,5
22               MN2=MN1+2
23                 DO 180 M=MN1,MN2
24                 DO 180 N=MN1,MN2
25                 NP1=N+1
26                 WRITE (6,250) M,N
27   C     GENERATE DATA
28   C       ..
29                     DO 10 I=1,M
30                       DO 10 J=1,N
31   10                    A(I,J)=GEN(ANOISE)
32                     DO 20 I=1,M
33   20                  B(I)=GEN(ANOISE)
34                   IF(M .LT. N) GO TO 180
35   C
36   C     ****** BEGIN ALGORITHM HFT ******
37   C       ..
38                     DO 30 J=1,N
39   30                  CALL H12 (1,J,J+1,M,A(1,J),1,H(J),A(1,J+1),1,MDA,N-J)
40   C       ..
41   C     THE ALGORITHM 'HFT' IS COMPLETED.
42   C
43   C     ****** BEGIN ALGORITHM HS1 ******
44   C     APPLY THE TRANSFORMATIONS   Q(N)...Q(1)=Q TO B
45   C     REPLACING THE PREVIOUS CONTENTS OF THE ARRAY, B .
46   C       ..
47                     DO 40 J=1,N
48   40                  CALL H12 (2,J,J+1,M,A(1,J),1,H(J),B,1,1,1)
49   C     SOLVE THE TRIANGULAR SYSTEM FOR THE SOLUTION X.
50   C     STORE X IN THE ARRAY, B .
51   C       ..
52                     DO 80 K=1,N
53                     I=NP1-K
54                     SM=0.D0
55                     IF (I.EQ.N) GO TO 60
56                     IP1=I+1
57                       DO 50 J=IP1,N
58   50                    SM=SM+A(I,J)*DBLE(B(J))
59   60                  IF (A(I,I)) 80,70,80
60   70                  WRITE (6,260)
61                       GO TO 180
62   80                  B(I)=(B(I)-SM)/A(I,I)
```

PROG1 (Continued)

```
 63   C        COMPUTE LENGTH OF RESIDUAL VECTOR.
 64   C    ..
 65                        SRSMSQ=0.
 66                        IF (N.EQ.M) GO TO 100
 67                        MMN=M-N
 68                        DO 90 J=1,MMN
 69                        NPJ=N+J
 70     90                 SRSMSQ=SRSMSQ+B(NPJ)**2
 71                        SRSMSQ=SQRT(SRSMSQ)
 72   C    ****** BEGIN ALGORITHM  COV ******
 73   C        COMPUTE UNSCALED COVARIANCE MATRIX    ((A**T)*A)**(-1)
 74   C    ..
 75    100                 DO 110 J=1,N
 76    110                 A(J,J)=1./A(J,J)
 77                        IF (N.EQ.1) GO TO 140
 78                        NM1=N-1
 79                        DO 130 I=1,NM1
 80                        IP1=I+1
 81                            DO 130 J=IP1,N
 82                            JM1=J-1
 83                            SM=0.D0
 84                                DO 120 L=I,JM1
 85    120                         SM=SM+A(I,L)*DBLE(A(L,J))
 86    130                 A(I,J)=-SM*A(J,J)
 87   C    ..
 88   C        THE UPPER TRIANGLE OF A HAS BEEN INVERTED
 89   C        UPON ITSELF.
 90    140                 DO 160 I=1,N
 91                        DO 160 J=I,N
 92                        SM=0.D0
 93                            DO 150 L=J,N
 94    150                     SM=SM+A(I,L)*DBLE(A(J,L))
 95    160                 A(I,J)=SM
 96   C    ..
 97   C        THE UPPER TRIANGULAR PART OF THE
 98   C        SYMMETRIC MATRIX (A**T*A)**(-1) HAS
 99   C        REPLACED THE UPPER TRIANGULAR PART OF
100   C        THE A ARRAY.
101                        WRITE (6,200) (I,B(I),I=1,N)
102                        WRITE (6,190) SRSMSQ
103                        WRITE (6,210)
104                        DO 170 I=1,N
105    170                 WRITE (6,220) (I,J,A(I,J),J=I,N)
106    180                 CONTINUE
107             STOP
108    190 FORMAT (1H0,8X,17HRESIDUAL LENGTH =,E12.4)
109    200 FORMAT (1H0,8X,34HESTIMATED PARAMETERS,  X=A**(+)*B,,22H COMPUTED
110       1BY 'HFT,HS1'///(9X,I6,E16.8,I6,E16.8,I6,E16.8,I6,E16.8,I6,E16.8))
111    210 FORMAT (1H0,8X,31HCOVARIANCE MATRIX (UNSCALED) OF,22H ESTIMATED PA
112       1RAMETERS,,19H COMPUTED BY 'COV'./1X)
113    220 FORMAT (9X,2I3,E16.8,2I3,E16.8,2I3,E16.8,2I3,E16.8,2I3,E16.8)
114    230 FORMAT (52H1 PROG1.    THIS PROGRAM DEMONSTRATES THE ALGORITHMS,19
115       1H HFT, HS1, AND COV.//40H CAUTION.. 'PROG1' DOES NO CHECKING FOR ,
116       252HNEAR RANK DEFICIENT MATRICES.  RESULTS IN SUCH CASES,20H MAY BE
117       3 MEANINGLESS.,/34H SUCH CASES ARE HANDLED BY 'PROG2',11H OR 'PROG3
118       4')
119    240 FORMAT (1H0,54HTHE RELATIVE NOISE LEVEL OF THE GENERATED DATA WILL
120       1 BE,E16.4)
121    250 FORMAT (1H0////9H0    M    N/1X,2I4)
122    260 FORMAT (1H0,8X,36H******  TERMINATING THIS CASE DUE TO,37H A DIVIS
123       1OR BEING EXACTLY ZERO.  ******)
124            END
```

PROG2

```
1    C        PROG2
2    C        C.L.LAWSON AND R.J.HANSON, JET PROPULSION LABORATORY, 1973 JUN 12
3    C        TO APPEAR IN 'SOLVING LEAST SQUARES PROBLEMS', PRENTICE-HALL, 1974
4    C           DEMONSTRATE ALGORITHM  HFTI   FOR SOLVING LEAST SQUARES PROBLEMS
5    C        AND ALGORITHM  COV  FOR COMPUTING THE ASSOCIATED UNSCALED
6    C        COVARIANCE MATRIX.
7    C
8             DIMENSION A(8,8),H(8),B(8),G(8)
9             REAL GEN,ANOISE
10            INTEGER IP(8)
11            DOUBLE PRECISION SM
12            DATA MDA /8/
13   C
14            DO 180 NOISE=1,2
15            ANORM=500.
16            ANOISE=0.
17            TAU=.5
18            IF (NOISE.EQ.1) GO TO 10
19            ANOISE=1.E-4
20            TAU=ANORM*ANOISE*10.
21    10      CONTINUE
22   C        INITIALIZE THE DATA GENERATION FUNCTION
23   C        ..
24            DUMMY=GEN(-1.)
25            WRITE (6,230)
26            WRITE (6,240) ANOISE,ANORM,TAU
27   C
28                DO 180 MN1=1,6,5
29                MN2=MN1+2
30                    DO 180 M=MN1,MN2
31                        DO 180 N=MN1,MN2
32                        WRITE (6,250) M,N
33   C        GENERATE DATA
34   C        ..
35                            DO 20 I=1,M
36                                DO 20 J=1,N
37    20                          A(I,J)=GEN(ANOISE)
38                            DO 30 I=1,M
39    30                      B(I)=GEN(ANOISE)
40   C
41   C        ****** CALL HFTI  ******
42   C
43                            CALL HFTI(A,MDA,M,N,B,1,1,TAU,KRANK,SRSMSQ,H,G,IP)
44   C
45   C
46                            WRITE (6,260) KRANK
47                            WRITE (6,200) (I,B(I),I=1,N)
48                            WRITE (6,190) SRSMSQ
49                            IF (KRANK.LT.N) GO TO 180
50   C        ****** ALGORITHM COV BIGINS HERE ******
51   C        ..
52                            DO 40 J=1,N
53    40                      A(J,J)=1./A(J,J)
54                            IF (N.EQ.1) GO TO 70
55                            NM1=N-1
56                            DO 60 I=1,NM1
57                            IP1=I+1
58                                DO 60 J=IP1,N
59                                JM1=J-1
60                                SM=0.D0
61                                    DO 50 L=I,JM1
62    50                              SM=SM+A(I,L)*DBLE(A(L,J))
63    60                          A(I,J)=-SM*A(J,J)
64   C        ..
65   C        THE UPPER TRIANGLE OF A HAS BEEN INVERTED
66   C        UPON ITSELF.
```

PROG2 (Continued)

```
 67     70                        DO 90 I=1,N
 68                                 DO 90 J=I,N
 69                                 SM=0.D0
 70                                   DO 80 L=J,N
 71     80                             SM=SM+A(I,L)*DBLE(A(J,L))
 72     90                           A(I,J)=SM
 73                          IF (N.LT.2) GO TO 160
 74                              DO 150 II=2,N
 75                              I=N+1-II
 76                              IF (IP(I).EQ.I) GO TO 150
 77                              K=IP(I)
 78                              TMP=A(I,I)
 79                              A(I,I)=A(K,K)
 80                              A(K,K)=TMP
 81                              IF (I.EQ.1) GO TO 110
 82                                DO 100 L=2,I
 83                                TMP=A(L-1,I)
 84                                A(L-1,I)=A(L-1,K)
 85    100                         A(L-1,K)=TMP
 86    110                       IP1=I+1
 87                              KM1=K-1
 88                              IF (IP1.GT.KM1) GO TO 130
 89                                DO 120 L=IP1,KM1
 90                                TMP=A(I,L)
 91                                A(I,L)=A(L,K)
 92    120                         A(L,K)=TMP
 93    130                       IF (K.EQ.N) GO TO 150
 94                              KP1=K+1
 95                                DO 140 L=KP1,N
 96                                TMP=A(I,L)
 97                                A(I,L)=A(K,L)
 98    140                         A(K,L)=TMP
 99    150                      CONTINUE
100    160                   CONTINUE
101  C        ..
102  C        COVARIANCE HAS BEEN COMPUTED AND REPERMUTED.
103  C        THE UPPER TRIANGULAR PART OF THE
104  C        SYMMETRIC MATRIX (A**T*A)**(-1) HAS
105  C        REPLACED THE UPPER TRIANGULAR PART OF
106  C        THE A ARRAY.
107                           WRITE (6,210)
108                             DO 170 I=1,N
109    170                     WRITE (6,220) (I,J,A(I,J),J=I,N)
110    180                   CONTINUE
111            STOP
112    190 FORMAT (1H0,8X,17HRESIDUAL LENGTH =,E12.4)
113    200 FORMAT (1H0,8X,34HESTIMATED PARAMETERS,  X=A**(+)*B,,22H COMPUTED
114        1BY 'HFTI'   //(9X,I6,E16.8,I6,E16.8,I6,E16.8,I6,E16.8,I6,E16.8))
115    210 FORMAT (1H0,8X,31HCOVARIANCE MATRIX (UNSCALED) OF,22H ESTIMATED PA
116        1RAMETERS.,19H COMPUTED BY 'COV'./1X)
117    220 FORMAT (9X,2I3,E16.8,2I3,E16.8,2I3,E16.8,2I3,E16.8,2I3,E16.8)
118    230 FORMAT (52H1 PROG2.       THIS PROGRAM DEMONSTATES THE ALGORITHMS,16
119        1H HFTI  AND  COV.)
120    240 FORMAT (1H0,54HTHE RELATIVE NOISE LEVEL OF THE GENERATED DATA WILL
121        1 BE,E16.4/33H0THE MATRIX NORM IS APPROXIMATELY,E12.4/43H0THE ABSOL
122        2UTE PSEUDORANK TOLERANCE, TAU, IS,E12.4)
123    250 FORMAT (1H0////9H0   M    N/1X,2I4)
124    260 FORMAT (1H0,8X,12HPSEUDORANK =,I4)
125            END
```

PROG3

```
 1   C       PROG3
 2   C       C.L.LAWSON AND R.J.HANSON, JET PROPULSION LABORATORY, 1973 JUN 12
 3   C       TO APPEAR IN 'SOLVING LEAST SQUARES PROBLEMS', PRENTICE-HALL, 1974
 4   C          DEMONSTRATE THE USE OF SUBROUTINE   SVDRS   TO COMPUTE THE
 5   C       SINGULAR VALUE DECOMPOSITION OF A MATRIX, A, AND SOLVE A LEAST
 6   C       SQUARES PROBLEM,  A*X=B.
 7   C
 8   C       THE S.V.D.  A= U*(S,0)**T*V**T IS
 9   C       COMPUTED SO THAT..
10   C       (1)  U**T*B REPLACES THE M+1 ST. COL. OF B.
11   C
12   C       (2)  U**T   REPLACES THE M BY M IDENTITY IN
13   C       THE FIRST M COLS. OF  B.
14   C
15   C       (3) V REPLACES THE FIRST N ROWS AND COLS.
16   C       OF A.
17   C
18   C       (4) THE DIAGONAL ENTRIES OF THE S MATRIX
19   C       REPLACE THE FIRST N ENTRIES OF THE ARRAY S.
20   C
21   C       THE ARRAY S( ) MUST BE DIMENSIONED AT LEAST 3*N .
22   C
23           DIMENSION A(8,8), B(8,9), S(24), X(8), AA(8,8)
24           REAL GEN,ANOISE
25           DOUBLE PRECISION SM
26           DATA MDA,MDB /8,8/
27   C
28               DO 160 NOISE=1,2
29               ANOISE=0.
30               RHO=1.E-3
31               IF (NOISE.EQ.1) GO TO 10
32               ANOISE=1.E-4
33               RHO=10.*ANOISE
34       10      CONTINUE
35               WRITE (6,240)
36               WRITE (6,250) ANOISE,RHO
37   C       INITIALIZE DATA GENERATION FUNCTION
38   C       ..
39               DUMMY=GEN(-1.)
40   C
41               DO 160 MN1=1,6,5
42               MN2=MN1+2
43                   DO 160 M=MN1,MN2
44                   DO 160 N=MN1,MN2
45                   WRITE (6,170) M,N
46                       DO 30 I=1,M
47                           DO 20 J=1,M
48       20              B(I,J)=0.
49                       B(I,I)=1.
50                       DO 30 J=1,N
51                       A(I,J)=GEN(ANOISE)
52       30              AA(I,J)=A(I,J)
53                   DO 40 I=1,M
54       40          B(I,M+1)=GEN(ANOISE)
55   C
56   C       THE ARRAYS ARE NOW FILLED IN..
57   C       COMPUTE THE S.V.D.
58   C       ********************************************************
59                   CALL SVDRS (A,MDA,M,N,B,MDB,M+1,S)
60   C       ********************************************************
61   C
62               WRITE (6,180)
63               WRITE (6,230) (I,S(I),I=1,N)
64               WRITE (6,190)
65               WRITE (6,230) (I,B(I,M+1),I=1,M)
```

PROG3 (Continued)

```
66   C
67   C        TEST FOR DISPARITY OF RATIO OF SINGULAR VALUES.
68   C        ..
69                                 KRANK=N
70                                 TAU=RHO*S(1)
71                                    DO 50 I=1,N
72                                    IF (S(I).LE.TAU) GO TO 60
73        50                         CONTINUE
74                                 GO TO 70
75        60                       KRANK=I-1
76        70                       WRITE (6,260) TAU,KRANK
77   C        COMPUTE SOLUTION VECTOR ASSUMING PSEUDORANK IS KRANK
78   C        ..
79                                    DO 80 I=1,KRANK
80        80                          B(I,M+1)=B(I,M+1)/S(I)
81                                    DO 100 I=1,N
82                                 SM=0.D0
83                                       DO 90 J=1,KRANK
84        90                             SM=SM+A(I,J)*DBLE(B(J,M+1))
85        100                      X(I)=SM
86   C        COMPUTE PREDICTED NORM OF RESIDUAL VECTOR.
87   C        ..
88                                 SRSMSQ=0.
89                                 IF (KRANK.EQ.M) GO TO 120
90                                 KP1=KRANK+1
91                                    DO 110 I=KP1,M
92        110                         SRSMSQ=SRSMSQ+B(I,M+1)**2
93                                 SRSMSQ=SQRT(SRSMSQ)
94   C
95        120                      CONTINUE
96                                 WRITE (6,200)
97                                 WRITE (6,230) (I,X(I),I=1,N)
98                                 WRITE (6,210) SRSMSQ
99   C        COMPUTE THE FROBENIUS NORM OF  A**T- V*(S,0)*U**T.
100  C
101  C        COMPUTE   V*S  FIRST.
102  C
103                                 MINMN=MINO(M,N)
104                                    DO 130 J=1,MINMN
105                                    DO 130 I=1,N
106       130                          A(I,J)=A(I,J)*S(J)
107                                 DN=0.
108                                    DO 150 J=1,M
109                                    DO 150 I=1,N
110                                 SM=0.D0
111                                       DO 140 L=1,MINMN
112       140                             SM=SM+A(I,L)*DBLE(B(L,J))
113  C        COMPUTED DIFFERENCE OF (I,J) TH ENTRY
114  C        OF  A**T-V*(S,0)*U**T.
115  C        ..
116                                 T=AA(J,I)-SM
117       150                       DN=DN+T**2
118                                 DN=SQRT(DN)/(SQRT(FLOAT(N))*S(1))
119                                 WRITE (6,220) DN
120       160                      CONTINUE
121            STOP
122       170 FORMAT (1H0////9H0    M    N/1X,2I4)
123       180 FORMAT (1H0,8X,25HSINGULAR VALUES OF MATRIX)
124       190 FORMAT (1H0,8X,30HTRANSFORMED RIGHT SIDE, U**T*B)
125       200 FORMAT (1H0,8X,33HESTIMATED PARAMETERS,  X=A**(+)*B,21H COMPUTED B
126          1Y 'SVDRS' )
127       210 FORMAT (1H0,8X,24HRESIDUAL VECTOR LENGTH =,E12.4)
128       220 FORMAT (1H0,8X,43HFROBENIUS NORM (A-U*(S,0)**T*V**T)/(SQRT(N),22H*
129          1SPECTRAL NORM OF A) =,E12.4)
130       230 FORMAT (9X,I5,E16.8,I5,E16.8,I5,E16.8,I5,E16.8,I5,E16.8)
131       240 FORMAT (51H1PROG3.       THIS PROGRAM DEMONSTRATES THE ALGORITHM,9H.
132          1 SVDRS. )
133       250 FORMAT (55H0THE RELATIVE NOISE LEVEL OF THE GENERATED DATA WILL BE
134          1,E16.4/44H0THE RELATIVE TOLERANCE, RHO, FOR PSEUDORANK,17H DETERMI
135          2NATION IS,E16.4)
136       260 FORMAT (1H0,8X,36HABSOLUTE PSEUDORANK TOLERANCE, TAU =,E12.4,10X,1
137          12HPSEUDORANK =,I5)
138  C
139            END
```

PROG4

```
 1   C        PROG4
 2   C        C.L.LAWSON AND R.J.HANSON, JET PROPULSION LABORATORY, 1973 JUN 12
 3   C        TO APPEAR IN 'SOLVING LEAST SQUARES PROBLEMS', PRENTICE-HALL, 1974
 4   C            DEMONSTRATE SINGULAR VALUE ANALYSIS.
 5   C
 6            DIMENSION A(15,5),B(15),SING(15)
 7            DATA NAMES/1H /
 8   C
 9            READ (5,10) ((A(I,J),J=1,5),B(I),I=1,15)
10            WRITE (6,20)
11            WRITE (6,30) ((A(I,J),J=1,5),B(I),I=1,15)
12            WRITE (6,40)
13   C
14            CALL SVA (A,15,15,5,15,B,SING,NAMES,1,D)
15   C
16            STOP
17       10 FORMAT (6F12.0)
18       20 FORMAT (46H1PROG4.        DEMONSTRATE SINGULAR VALUE ANALYSIS/53H LIST
19          1ING OF INPUT MATRIX, A, AND VECTOR, B, FOLLOWS..)
20       30 FORMAT (1H /(5F12.8,F20.4))
21       40 FORMAT (1H1)
22            END
```

DATA4

```
 1   -.13405547   -.20162827   -.16930778   -.18971990   -.17387234   -.4361
 2   -.10379475   -.15766336   -.13346256   -.14848550   -.13597690   -.3437
 3   -.08779597   -.12883867   -.10683007   -.12011796   -.10932972   -.2657
 4    .02058554    .00335331   -.01641270    .00078606    .00271659   -.0392
 5   -.03248093   -.01876799    .00410639   -.01405894   -.01384391    .0193
 6    .05967662    .06667714    .04352153    .05740438    .05024962    .0747
 7    .06712457    .07352437    .04489770    .06471862    .05876455    .0935
 8    .08687186    .09368296    .05672327    .08141043    .07302320    .1079
 9    .02149662    .06222662    .07213486    .06200069    .05570931    .1930
10    .06687407    .10344506    .09153849    .09508223    .08393667    .2058
11    .15879069    .18088339    .11540692    .16160727    .14796479    .2606
12    .17642887    .20361830    .13057860    .18385729    .17005549    .3142
13    .11414080    .17259611    .14816471    .16007466    .14374096    .3529
14    .07846038    .14669563    .14365800    .14003842    .12571177    .3615
15    .10803175    .16994623    .14971519    .15885312    .14301547    .3647
```

PROG5

```
 1    C     PROG5
 2    C     C.L.LAWSON AND R.J.HANSON, JET PROPULSION LABORATORY, 1973 JUN 12
 3    C     TO APPEAR IN 'SOLVING LEAST SQUARES PROBLEMS', PRENTICE-HALL, 1974
 4    C          EXAMPLE OF THE USE OF SUBROUTINES BNDACC AND BNDSOL TO SOLVE
 5    C     SEQUENTIALLY THE BANDED LEAST SQUARES PROBLEM WHICH ARISES IN
 6    C     SPLINE CURVE FITTING.
 7    C
 8          DIMENSION X(12),Y(12),B(10),G(12,5),C(12),Q(4),COV(12)
 9    C
10    C        DEFINE FUNCTIONS P1 AND P2 TO BE USED IN CONSTRUCTING
11    C        CUBIC SPLINES OVER UNIFORMLY SPACED BREAKPOINTS.
12    C
13          P1(T)=.25*T**2*T
14          P2(T)=-(1.-T)**2*(1.+T)*.75+1.
15          ZERO=0.
16    C
17          WRITE (6,230)
18          MDG=12
19          NBAND=4
20          M=12
21    C                                    SET ORDINATES OF DATA
22          Y(1)=2.2
23          Y(2)=4.0
24          Y(3)=5.0
25          Y(4)=4.6
26          Y(5)=2.8
27          Y(6)=2.7
28          Y(7)=3.8
29          Y(8)=5.1
30          Y(9)=6.1
31          Y(10)=6.3
32          Y(11)=5.0
33          Y(12)=2.0
34    C                                    SET ABCISSAS OF DATA
35             DO 10 I=1,M
36    10       X(I)=2*I
37    C
38    C     BEGIN LOOP THRU CASES USING INCREASING NOS OF BREAKPOINTS.
39    C
40             DO 150 NBP=5,10
41          NC=NBP+2
42    C                                    SET BREAKPOINTS
43          B(1)=X(1)
44          B(NBP)=X(M)
45          H=(B(NBP)-B(1))/FLOAT(NBP-1)
46          IF (NBP.LE.2) GO TO 30
47             DO 20 I=3,NBP
48    20       B(I-1)=B(I-2)+H
49    30       CONTINUE
50          WRITE (6,240) NBP,(B(I),I=1,NBP)
51    C
52    C     INITIALIZE IR AND IP BEFORE FIRST CALL TO BNDACC.
53    C
54             IR=1
55             IP=1
56             I=1
57             JT=1
58    40       MT=0
59    50       CONTINUE
60          IF (X(I).GT.B(JT+1)) GO TO 60
61    C
62    C                              SET ROW   FOR ITH DATA POINT
63    C
64          U=(X(I)-B(JT))/H
65          IG=IR+MT
66          G(IG,1)=P1(1.-U)
67          G(IG,2)=P2(1.-U)
68          G(IG,3)=P2(U)
69          G(IG,4)=P1(U)
70          G(IG,5)=Y(I)
71          MT=MT+1
72          IF (I.EQ.M) GO TO 60
73          I=I+1
74          GO TO 50
```

PROG5 (Continued)

```
 75    C
 76    C                         SEND BLOCK OF DATA TO PROCESSOR
 77    C
 78        60        CONTINUE
 79                  CALL BNDACC (G,MDG,NBAND,IP,IR,MT,JT)
 80                  IF (I.EQ.M) GO TO 70
 81                  JT=JT+1
 82                  GO TO 40
 83    C                         COMPUTE SOLUTION C()
 84        70        CONTINUE
 85                  CALL BNDSOL (1,G,MDG,NBAND,IP,IR,C,NC,RNORM)
 86    C                   WRITE SOLUTION COEFFICIENTS
 87                  WRITE (6,180) (C(L),L=1,NC)
 88                  WRITE (6,210) RNORM
 89    C
 90    C             COMPUTE AND PRINT X,Y,YFIT,R=Y-YFIT
 91    C
 92                  WRITE (6,160)
 93                  JT=1
 94                  DO 110 I=1,M
 95        80        IF (X(I).LE.B(JT+1)) GO TO 90
 96                  JT=JT+1
 97                  GO TO 80
 98    C
 99        90        U=(X(I)-B(JT))/H
100                  Q(1)=P1(1.-U)
101                  Q(2)=P2(1.-U)
102                  Q(3)=P2(U)
103                  Q(4)=P1(U)
104                  YFIT=ZERO
105                  DO 100 L=1,4
106                  IC=JT-1+L
107       100        YFIT=YFIT+C(IC)*Q(L)
108                  R=Y(I)-YFIT
109                  WRITE (6,170) I,X(I),Y(I),YFIT,R
110       110        CONTINUE
111    C
112    C       COMPUTE RESIDUAL VECTOR NORM.
113    C
114                  IF (M.LE.NC) GO TO 150
115                  SIGSQ=(RNORM**2)/FLOAT(M-NC)
116                  SIGFAC=SQRT(SIGSQ)
117                  WRITE (6,220) SIGFAC
118                  WRITE (6,200)
119    C
120    C       COMPUTE AND PRINT COLS. OF COVARIANCE.
121    C
122                  DO 140 J=1,NC
123                  DO 120 I=1,NC
124       120        COV(I)=ZERO
125                  COV(J)=1.
126                  CALL BNDSOL (2,G,MDG,NBAND,IP,IR,COV,NC,RDUMMY)
127                  CALL BNDSOL (3,G,MDG,NBAND,IP,IR,COV,NC,RDUMMY)
128    C
129    C       COMPUTE THE JTH COL. OF THE COVARIANCE MATRIX.
130                  DO 130 I=1,NC
131       130        COV(I)=COV(I)*SIGSQ
132       140        WRITE (6,190) (L,J,COV(L),L=1,NC)
133       150        CONTINUE
134        STOP
135       160 FORMAT (4H0  I,8X,1HX,10X,1HY,6X,4HYFIT,4X,8HR=Y-YFIT/1X)
136       170 FORMAT (1X,I3,4X,F6.0,4X,F6.2,4X,F6.2,4X,F8.4)
137       180 FORMAT (4H0C =,10F10.5/(4X,10F10.5))
138       190 FORMAT (4(02X,2I5,E15.7))
139       200 FORMAT (46H0COVARIANCE MATRIX OF THE SPLINE COEFFICIENTS.)
140       210 FORMAT (9H0RNORM  =,E15.8)
141       220 FORMAT (9H0SIGFAC =,E15.8)
142       230 FORMAT (50H1PROG5.        EXECUTE A SEQUENCE OF CUBIC SPLINE FITS,27H
143         1TO A DISCRETE SET OF DATA.)
144       240 FORMAT (1X////,11H0NEW CASE...,/47H0NUMBER OF BREAKPOINTS, INCLUDIN
145         1G ENDPOINTS, IS,I5/14H0BREAKPOINTS =,10F10.5,/(14X,10F10.5))
146        END
```

PROG6

```
 1   C      PROG6
 2   C      C.L.LAWSON AND R.J.HANSON, JET PROPULSION LABORATORY, 1973 JUN 15
 3   C      TO APPEAR IN 'SOLVING LEAST SQUARES PROBLEMS', PRENTICE-HALL, 1974
 4   C            DEMONSTRATE THE USE OF THE SUBROUTINE   LDP   FOR LEAST
 5   C      DISTANCE PROGRAMMING BY SOLVING THE CONSTRAINED LINE DATA FITTING
 6   C      PROBLEM OF CHAPTER 23.
 7   C
 8          DIMENSION E(4,2), F(4), G(3,2), H(3), G2(3,2), H2(3), X(2), Z(2),
 9         1W(4)
10          DIMENSION WLDP(21), S(6), T(4)
11          INTEGER INDEX(3)
12   C
13          WRITE (6,110)
14          MDE=4
15          MDGH=3
16   C
17          ME=4
18          MG=3
19          N=2
20   C                  DEFINE THE LEAST SQUARES AND CONSTRAINT MATRICES.
21          T(1)=0.25
22          T(2)=0.5
23          T(3)=0.5
24          T(4)=0.8
25   C
26          W(1)=0.5
27          W(2)=0.6
28          W(3)=0.7
29          W(4)=1.2
30   C
31              DO 10 I=1,ME
32              E(I,1)=T(I)
33              E(I,2)=1.
34   10     F(I)=W(I)
35   C
36          G(1,1)=1.
37          G(1,2)=0.
38          G(2,1)=0.
39          G(2,2)=1.
40          G(3,1)=-1.
41          G(3,2)=-1.
42   C
43          H(1)=0.
44          H(2)=0.
45          H(3)=-1.
46   C
47   C      COMPUTE THE SINGULAR VALUE DECOMPOSITION OF THE MATRIX, E.
48   C
49          CALL SVDRS (E,MDE,ME,N,F,1,1,S)
50   C
51          WRITE (6,120) ((E(I,J),J=1,N),I=1,N)
52          WRITE (6,130) F,(S(J),J=1,N)
53   C
54   C      GENERALLY RANK DETERMINATION AND LEVENBERG-MARQUARDT
55   C      STABILIZATION COULD BE INSERTED HERE.
56   C
```

PROG6 (Continued)

```
57    C           DEFINE THE CONSTRAINT MATRIX FOR THE Z COORDINATE SYSTEM.
58                DO 30 I=1,MG
59                    DO 30 J=1,N
60                    SM=0.
61                        DO 20 L=1,N
62    20                 SM=SM+G(I,L)*E(L,J)
63    30            G2(I,J)=SM/S(J)
64    C           DEFINE CONSTRAINT RT SIDE FOR THE Z COORDINATE SYSTEM.
65                DO 50 I=1,MG
66                SM=0.
67                    DO 40 J=1,N
68    40            SM=SM+G2(I,J)*F(J)
69    50        H2(I)=H(I)-SM
70    C
71            WRITE (6,140) ((G2(I,J),J=1,N),I=1,MG)
72            WRITE (6,150) H2
73    C
74    C                           SOLVE THE CONSTRAINED PROBLEM IN Z-COORDINATES.
75    C
76            CALL LDP (G2,MDGH,MG,N,H2,Z,ZNORM,WLDP,INDEX,MODE)
77    C
78            WRITE (6,200) MODE,ZNORM
79            WRITE (6,160) Z
80    C
81    C                           TRANSFORM BACK FROM Z-COORDINATES TO X-COORDINATES.
82                DO 60 J=1,N
83    60        Z(J)=(Z(J)+F(J))/S(J)
84            DO 80 I=1,N
85            SM=0.
86                DO 70 J=1,N
87    70            SM=SM+E(I,J)*Z(J)
88    80        X(I)=SM
89            RES=ZNORM**2
90            NP1=N+1
91                DO 90 I=NP1,ME
92    90        RES=RES+F(I)**2
93            RES=SQRT(RES)
94    C                           COMPUTE THE RESIDUALS.
95                DO 100 I=1,ME
96    100       F(I)=W(I)-X(1)*T(I)-X(2)
97            WRITE (6,170) (X(J),J=1,N)
98            WRITE (6,180) (I,F(I),I=1,ME)
99            WRITE (6,190) RES
100           STOP
101   C
102   110 FORMAT (46H0PROG6.. EXAMPLE OF CONSTRAINED CURVE FITTING,26H USIN
103       1G THE SUBROUTINE LDP.,/43H0RELATED INTERMEDIATE QUANTITIES ARE GIV
104       2EN.)
105   120 FORMAT (10H0V =       ,2F10.5/(10X,2F10.5))
106   130 FORMAT (10H0F TILDA =,4F10.5/10H0S =       ,2F10.5)
107   140 FORMAT (10H0G TILDA =,2F10.5/(10X,2F10.5))
108   150 FORMAT (10H0H TILDA =,3F10.5)
109   160 FORMAT (10H0Z =       ,2F10.5)
110   170 FORMAT (52H0THE COEFICIENTS OF THE FITTED LINE F(T)=X(1)*T+X(2),12
111       1H ARE X(1) = ,F10.5,14H  AND X(2) = ,F10.5)
112   180 FORMAT (30H0THE CONSECUTIVE RESIDUALS ARE/1X,4(I10,F10.5))
113   190 FORMAT (23H0THE RESIDUALS NORM IS ,F10.5)
114   200 FORMAT (18H0MODE (FROM LDP) =,I3,2X,7HZNORM =,F10.5)
115   C
116           END
```

HFTI

```
1              SUBROUTINE HFTI (A,MDA,M,N,B,MDB,NB,TAU,KRANK,RNORM,H,G,IP)
2      C       C.L.LAWSON AND R.J.HANSON, JET PROPULSION LABORATORY, 1973 JUN 12
3      C       TO APPEAR IN 'SOLVING LEAST SQUARES PROBLEMS', PRENTICE-HALL, 1974
4      C           SOLVE LEAST SQUARES PROBLEM USING ALGORITHM, HFTI.
5      C
6              DIMENSION A(MDA,N),B(MDB,NB),H(N),G(N),RNORM(NB)
7              INTEGER IP(N)
8              DOUBLE PRECISION SM,DZERO
9              SZERO=0.
10             DZERO=0.D0
11             FACTOR=0.001
12     C
13             K=0
14             LDIAG=MIN0(M,N)
15             IF (LDIAG.LE.0) GO TO 270
16                 DO 80 J=1,LDIAG
17                 IF (J.EQ.1) GO TO 20
18     C
19     C       UPDATE SQUARED COLUMN LENGTHS AND FIND LMAX
20     C         ..
21                 LMAX=J
22                     DO 10 L=J,N
23                     H(L)=H(L)-A(J-1,L)**2
24                     IF (H(L).GT.H(LMAX)) LMAX=L
25        10          CONTINUE
26                 IF(DIFF(HMAX+FACTOR*H(LMAX),HMAX)) 20,20,50
27     C
28     C       COMPUTE SQUARED COLUMN LENGTHS AND FIND LMAX
29     C         ..
30        20     LMAX=J
31                     DO 40 L=J,N
32                     H(L)=0.
33                         DO 30 I=J,M
34        30              H(L)=H(L)+A(I,L)**2
35                     IF (H(L).GT.H(LMAX)) LMAX=L
36        40          CONTINUE
37             HMAX=H(LMAX)
38     C         ..
39     C       LMAX HAS BEEN DETERMINED
40     C
41     C       DO COLUMN INTERCHANGES IF NEEDED.
42     C         ..
43        50     CONTINUE
44             IP(J)=LMAX
45             IF (IP(J).EQ.J) GO TO 70
46                 DO 60 I=1,M
47                 TMP=A(I,J)
48                 A(I,J)=A(I,LMAX)
49        60      A(I,LMAX)=TMP
50             H(LMAX)=H(J)
51     C
52     C       COMPUTE THE J-TH TRANSFORMATION AND APPLY IT TO A AND B.
53     C         ..
54        70     CALL H12 (1,J,J+1,M,A(1,J),1,H(J),A(1,J+1),1,MDA,N-J)
55        80     CALL H12 (2,J,J+1,M,A(1,J),1,H(J),B,1,MDB,NB)
56     C
57     C       DETERMINE THE PSEUDORANK, K, USING THE TOLERANCE, TAU.
58     C         ..
59                 DO 90 J=1,LDIAG
60                 IF (ABS(A(J,J)).LE.TAU) GO TO 100
61        90      CONTINUE
62             K=LDIAG
63             GO TO 110
64       100 K=J-1
65       110 KP1=K+1
```

HFTI (Continued)

```
 66   C
 67   C          COMPUTE THE NORMS OF THE RESIDUAL VECTORS.
 68   C
 69              IF (NB.LE.0) GO TO 140
 70                   DO 130 JB=1,NB
 71                   TMP=SZERO
 72                   IF (KP1.GT.M) GO TO 130
 73                        DO 120 I=KP1,M
 74   120                  TMP=TMP+B(I,JB)**2
 75   130             RNORM(JB)=SQRT(TMP)
 76   140 CONTINUE
 77   C                                                  SPECIAL FOR PSEUDORANK = 0
 78              IF (K.GT.0) GO TO 160
 79              IF (NB.LE.0) GO TO 270
 80                   DO 150 JB=1,NB
 81                        DO 150 I=1,N
 82   150                  B(I,JB)=SZERO
 83              GO TO 270
 84   C
 85   C          IF THE PSEUDORANK IS LESS THAN N COMPUTE HOUSEHOLDER
 86   C          DECOMPOSITION OF FIRST K ROWS.
 87   C    ..
 88   160 IF (K.EQ.N) GO TO 180
 89              DO 170 II=1,K
 90              I=KP1-II
 91   170        CALL H12 (1,I,KP1,N,A(I,1),MDA,G(I),A,MDA,1,I-1)
 92   180 CONTINUE
 93   C
 94   C
 95              IF (NB.LE.0) GO TO 270
 96                   DO 260 JB=1,NB
 97   C
 98   C          SOLVE THE K BY K TRIANGULAR SYSTEM.
 99   C    ..
100                   DO 210 L=1,K
101                   SM=DZERO
102                   I=KP1-L
103                   IF (I.EQ.K) GO TO 200
104                   IP1=I+1
105                        DO 190 J=IP1,K
106   190                  SM=SM+A(I,J)*DBLE(B(J,JB))
107   200             SM1=SM
108   210             B(I,JB)=(B(I,JB)-SM1)/A(I,I)
109   C
110   C          COMPLETE COMPUTATION OF SOLUTION VECTOR.
111   C    ..
112                   IF (K.EQ.N) GO TO 240
113                   DO 220 J=KP1,N
114   220             B(J,JB)=SZERO
115                   DO 230 I=1,K
116   230             CALL H12 (2,I,KP1,N,A(I,1),MDA,G(I),B(1,JB),1,MDB,1)
117   C
118   C          RE-ORDER THE SOLUTION VECTOR TO COMPENSATE FOR THE
119   C          COLUMN INTERCHANGES.
120   C    ..
121   240             DO 250 JJ=1,LDIAG
122                   J=LDIAG+1-JJ
123                   IF (IP(J).EQ.J) GO TO 250
124                   L=IP(J)
125                   TMP=B(L,JB)
126                   B(L,JB)=B(J,JB)
127                   B(J,JB)=TMP
128   250             CONTINUE
129   260        CONTINUE
130   C    ..
131   C          THE SOLUTION VECTORS, X, ARE NOW
132   C          IN THE FIRST  N  ROWS OF THE ARRAY B(,).
133   C
134   270 KRANK=K
135              RETURN
136              END
```

SVA

```
 1          SUBROUTINE SVA (A,MDA,M,N,MDATA,B,SING,NAMES,ISCALE,D)
 2    C     C.L.LAWSON AND R.J.HANSON, JET PROPULSION LABORATORY, 1973 JUN 12
 3    C     TO APPEAR IN 'SOLVING LEAST SQUARES PROBLEMS', PRENTICE-HALL, 1974
 4    C            SINGULAR VALUE ANALYSIS PRINTOUT
 5    C
 6    C     ISCALE          SET BY USER TO 1, 2, OR 3 TO SELECT COLUMN SCALING
 7    C                     OPTION.
 8    C                     1    SUBR WILL USE IDENTITY SCALING AND IGNORE THE D()
 9    C                          ARRAY.
10    C                     2    SUBR WILL SCALE NONZERO COLS TO HAVE UNIT EUCLID-
11    C                          EAN LENGTH AND WILL STORE RECIPROCAL LENGTHS OF
12    C                          ORIGINAL NONZERO COLS IN D().
13    C                     3    USER SUPPLIES COL SCALE FACTORS IN D(). SUBR
14    C                          WILL MULT COL J BY D(J) AND REMOVE THE SCALING
15    C                          FROM THE SOLN AT THE END.
16    C
17          DIMENSION  A(MDA,N), B(M), SING(01),D(N)
18    C     SING(3*N)
19          DIMENSION NAMES(N)
20          LOGICAL TEST
21          DOUBLE PRECISION    SB, DZERO
22          DZERO=0.D0
23          ONE=1.
24          ZERO=0.
25          IF (M.LE.0 .OR. N.LE.0) RETURN
26          NP1=N+1
27          WRITE (6,270)
28          WRITE (6,260) M,N,MDATA
29          GO TO (60,10,10), ISCALE
30    C
31    C     APPLY COLUMN SCALING TO A
32    C
33    10        DO 50 J=1,N
34              A1=D(J)
35              GO TO (20,20,40), ISCALE
36    20        SB=DZERO
37                  DO 30 I=1,M
38    30                SB=SB+DBLE(A(I,J))**2
39              A1=DSQRT(SB)
40              IF (A1.EQ.ZERO) A1=ONE
41              A1=ONE/A1
42              D(J)=A1
43    40            DO 50 I=1,M
44    50                A(I,J)=A(I,J)*A1
45          WRITE (6,280) ISCALE,(D(J),J=1,N)
46          GO TO 70
47    60    CONTINUE
48          WRITE (6,290)
49    70    CONTINUE
50    C
51    C     OBTAIN SING. VALUE DECOMP. OF SCALED MATRIX.
52    C
53    C*********************************************************************
54          CALL SVDRS (A,MDA,M,N,B,1,1,SING)
55    C*********************************************************************
56    C
```

SVA (Continued)

```
 57   C       PRINT THE V MATRIX.
 58   C
 59           CALL MFEOUT (A,MDA,N,N,NAMES,1)
 60           IF (ISCALE.EQ.1) GO TO 90
 61   C
 62   C       REPLACE V BY D*V IN THE ARRAY A()
 63   C
 64               DO 80 I=1,N
 65                   DO 80 J=1,N
 66        80            A(I,J)=D(I)*A(I,J)
 67   C
 68   C       G NOW IN  B ARRAY.  V NOW IN A ARRAY.
 69   C
 70   C
 71   C       OBTAIN SUMMARY OUTPUT.
 72   C
 73        90 CONTINUE
 74           WRITE (6,220)
 75   C
 76   C       COMPUTE CUMULATIVE SUMS OF SQUARES OF COMPONENTS OF
 77   C       G AND STORE THEM IN SING(I), I=MINMN+1,...,2*MINMN+1
 78   C
 79           SB=DZERO
 80           MINMN=MIN0(M,N)
 81           MINMN1=MINMN+1
 82           IF (M.EQ.MINMN) GO TO 110
 83               DO 100 I=MINMN1,M
 84       100       SB=SB+DBLE(B(I))**2
 85       110 SING(2*MINMN+1)=SB
 86               DO 120 JJ=1,MINMN
 87           J=MINMN+1-JJ
 88           SB=SB+DBLE(B(J))**2
 89           JS=MINMN+J
 90       120    SING(JS)=SB
 91           A3=SING(MINMN+1)
 92           A4=SQRT(A3/FLOAT(MAX0(1,MDATA)))
 93           WRITE (6,230) A3,A4
 94   C
 95           NSOL=0
 96   C
 97   C
 98   C
 99               DO 160 K=1,MINMN
100               IF (SING(K).EQ.ZERO) GO TO 130
101               NSOL=K
102               PI=B(K)/SING(K)
103               A1=ONE/SING(K)
104               A2=B(K)**2
105               A3=SING(MINMN1+K)
106               A4=SQRT(A3/FLOAT(MAX0(1,MDATA-K)))
107               TEST=SING(K).GE.100..OR.SING(K).LT..001
108               IF (TEST) WRITE (6,240) K,SING(K),PI,A1,B(K),A2,A3,A4
109               IF (.NOT.TEST) WRITE (6,250) K,SING(K),PI,A1,B(K),A2,A3,A4
110               GO TO 140
111       130 WRITE (6,240) K,SING(K)
112       140 PI=ZERO
113       140     DO 150 I=1,N
114               A(I,K)=A(I,K)*PI
115       150     IF (K.GT.1) A(I,K)=A(I,K)+A(I,K-1)
116       160     CONTINUE
117   C
```

SVA (Continued)

```
118   C       COMPUTE AND PRINT VALUES OF YNORM AND RNORM.
119   C
120           WRITE (6,300)
121           J=0
122           YSQ=ZERO
123           GO TO 180
124   170     J=J+1
125           YSQ=YSQ+(B(J)/SING(J))**2
126   180     YNORM=SQRT(YSQ)
127           JS=NP1+J
128           RNORM=SQRT(SING(JS))
129           YL=-1000.
130           IF (YNORM.GT.0.) YL=ALOG10(YNORM)
131           RL=-1000.
132           IF (RNORM.GT.0.) RL=ALOG10(RNORM)
133           WRITE (6,310) J,YNORM,RNORM,YL,RL
134           IF (J.LT.NSOL) GO TO 170
135   C
136   C       COMPUTE VALUES OF XNORM AND RNORM FOR A SEQUENCE OF VALUES OF
137   C       THE LEVENBERG-MARQUARDT PARAMETER.
138   C
139           IF (SING(1).EQ.ZERO) GO TO 210
140           EL=ALOG10(SING(1))+ONE
141           EL2=ALOG10(SING(NSOL))-ONE
142           DEL=(EL2-EL)/20.
143           TEN=10.
144           ALN10=ALOG(TEN)
145           WRITE (6,320)
146           DO 200 IE=1,21
147   C       COMPUTE       ALAMB=10.**EL
148           ALAMB=EXP(ALN10*EL)
149           YS=0.
150           JS=NP1+NSOL
151           RS=SING(JS)
152           DO 190 I=1,N
153           SL=SING(I)**2+ALAMB**2
154           YS=YS+(B(I)*SING(I)/SL)**2
155           RS=RS+(B(I)*(ALAMB**2)/SL)**2
156   190     CONTINUE
157           YNORM=SQRT(YS)
158           RNORM=SQRT(RS)
159           RL=-1000.
160           IF (RNORM.GT.ZERO) RL=ALOG10(RNORM) .
161           YL=-1000.
162           IF (YNORM.GT.ZERO) YL=ALOG10(YNORM)
163           WRITE (6,330) ALAMB,YNORM,RNORM,EL,YL,RL
164           EL=EL+DEL
165   200     CONTINUE
166   C
167   C       PRINT CANDIDATE SOLUTIONS.
168   C
169   210     IF (NSOL.GE.1) CALL MFEOUT (A,MDA,N,NSOL,NAMES,2)
170           RETURN
171   220     FORMAT (42H0 INDEX  SING. VALUE        P COEF          ,48HRECIP. S.
172          1V.          G COEF          G**2 ,39H          C.S.S.
173          2    N.S.R.C.S.S.)
174   230     FORMAT (1H ,5X,1H0,88X,1PE15.4,1PE17.4)
175   240     FORMAT (1H ,I6,E12.4,1P(E15.4,4X,E15.4,4X,E15.4,4X,E15.4,
176          12X,E15.4))
177   250     FORMAT (1H ,I6,F12.4,1P(E15.4,4X,E15.4,4X,E15.4,4X,E15.4,
178          12X,E15.4))
179   260     FORMAT (5H0M = ,I6,8H,   N = ,I4,12H,    MDATA = ,I8)
180   270     FORMAT (45H0SINGULAR VALUE ANALYSIS OF THE LEAST SQUARES,42H PROBL
181          1EM,  A*X=B,  SCALED AS  (A*D)*Y=B  .)
182   280     FORMAT (19H0SCALING OPTION NO.,I2,18H.  D IS A DIAGONAL,46H MATRIX
183          1 WITH THE FOLLOWING DIAGONAL ELEMENTS../(5X,10E12.4))
184   290     FORMAT (50H0SCALING OPTION NO. 1.   D IS THE IDENTITY MATRIX./1X)
185   300     FORMAT (6H0INDEX,13X,28H          YNORM            RNORM,14X,28H  LOG1
186          10(YNORM)   LOG10(RNORM)/1X)
187   310     FORMAT (1H ,I4,14X,2E14.5,14X,2F14.5)
188   320     FORMAT (54H0NORMS OF SOLUTION AND RESIDUAL VECTORS FOR A RANGE OF,
189          144H VALUES OF THE LEVENBERG-MARQUARDT PARAMETER,9H, LAMBDA./1H0,4X
190          2,42H          LAMBDA           YNORM            RNORM,42H LOG10(LAMBDA)
191          3LOG10(YNORM)  LOG10(RNORM))
192   330     FORMAT (5X,3E14.5,3F14.5)
193           END
```

SVDRS

```
 1  C      SUBROUTINE SVDRS (A,MDA,MM,NN,B,MDB,NB,S)
 2  C      C.L.LAWSON AND R.J.HANSON, JET PROPULSION LABORATORY, 1974 MAR 1
 3  C      TO APPEAR IN 'SOLVING LEAST SQUARES PROBLEMS', PRENTICE-HALL, 1974
 4  C           SINGULAR VALUE DECOMPOSITION ALSO TREATING RIGHT SIDE VECTOR.
 5  C
 6  C      THE ARRAY S OCCUPIES 3*N CELLS.
 7  C      A OCCUPIES M*N CELLS
 8  C      B OCCUPIES M*NB CELLS.
 9  C
10  C      SPECIAL SINGULAR VALUE DECOMPOSITION SUBROUTINE.
11  C      WE HAVE THE M X N MATRIX A AND THE SYSTEM A*X=B TO SOLVE.
12  C      EITHER M .GE. N  OR  M .LT. N IS PERMITTED.
13  C                  THE SINGULAR VALUE DECOMPOSITION
14  C      A = U*S*V**(T) IS MADE IN SUCH A WAY THAT ONE GETS
15  C         (1) THE MATRIX V IN THE FIRST N ROWS AND COLUMNS OF A.
16  C         (2) THE DIAGONAL MATRIX OF ORDERED SINGULAR VALUES IN
17  C             THE FIRST N CELLS OF THE ARRAY S(IP), IP .GE. 3*N.
18  C         (3) THE MATRIX PRODUCT U**(T)*B=G GETS PLACED BACK IN B.
19  C         (4) THE USER MUST COMPLETE THE SOLUTION AND DO HIS OWN
20  C             SINGULAR VALUE ANALYSIS.
21  C      *******
22  C      GIVE SPECIAL
23  C      TREATMENT TO ROWS AND COLUMNS WHICH ARE ENTIRELY ZERO.   THIS
24  C      CAUSES CERTAIN ZERO SING. VALS. TO APPEAR AS EXACT ZEROS RATHER
25  C      THAN AS ABOUT ETA TIMES THE LARGEST SING. VAL.   IT SIMILARLY
26  C      CLEANS UP THE ASSOCIATED COLUMNS OF U AND V.
27  C      METHOD..
28  C         1. EXCHANGE COLS OF A TO PACK NONZERO COLS TO THE LEFT.
29  C            SET N = NO. OF NONZERO COLS.
30  C            USE LOCATIONS A(1,NN),A(1,NN-1),....,A(1,N+1) TO RECORD THE
31  C            COL PERMUTATIONS.
32  C         2. EXCHANGE ROWS OF A TO PACK NONZERO ROWS TO THE TOP.
33  C            QUIT PACKING IF FIND N NONZERO ROWS.  MAKE SAME ROW EXCHANGES
34  C            IN B.  SET M SO THAT ALL NONZERO ROWS OF THE PERMUTED A
35  C            ARE IN FIRST M ROWS.  IF M .LE. N THEN ALL M ROWS ARE
36  C            NONZERO.  IF M .GT. N THEN      THE FIRST N ROWS ARE KNOWN
37  C            TO BE NONZERO,AND ROWS N+1 THRU M MAY BE ZERO OR NONZERO.
38  C         3. APPLY ORIGINAL ALGORITHM TO THE M BY N PROBLEM.
39  C         4. MOVE PERMUTATION RECORD FROM A(,) TO S(I),I=N+1,....,NN.
40  C         5. BUILD V UP FROM  N BY N  TO  NN BY NN  BY PLACING ONES ON
41  C            THE DIAGONAL AND ZEROS ELSEWHERE.  THIS IS ONLY PARTLY DONE
42  C            EXPLICITLY.  IT IS COMPLETED DURING STEP 6.
43  C         6. EXCHANGE ROWS OF V TO COMPENSATE FOR COL EXCHANGES OF STEP 2.
44  C         7. PLACE ZEROS IN  S(I),I=N+1,NN  TO REPRESENT ZERO SING VALS.
45  C
46         SUBROUTINE SVDRS (A,MDA,MM,NN,B,MDB,NB,S)
47         DIMENSION A(MDA,NN),B(MDB,NB),S(NN,3)
48         ZERO=0.
49         ONE=1.
50  C
51  C                              BEGIN.. SPECIAL FOR ZERO ROWS AND COLS.
52  C
53  C                              PACK THE NONZERO COLS TO THE LEFT
54  C
55         N=NN
56         IF (N.LE.0.OR.MM.LE.0) RETURN
57         J=N
58      10 CONTINUE
59            DO 20 I=1,MM
60            IF (A(I,J)) 50,20,50
61      20    CONTINUE
62  C
63  C         COL J  IS ZERO. EXCHANGE IT WITH COL N.
64  C
65         IF (J.EQ.N) GO TO 40
66            DO 30 I=1,MM
67      30       A(I,J)=A(I,N)
68      40 CONTINUE
69         A(1,N)=J
70         N=N-1
71      50 CONTINUE
72         J=J-1
73         IF (J.GE.1) GO TO 10
```

SVDRS (Continued)

```
74    C                                         IF N=0 THEN A IS ENTIRELY ZERO AND SVD
75    C                                         COMPUTATION CAN BE SKIPPED
76          NS=0
77          IF (N.EQ.0) GO TO 240
78    C                                         PACK NONZERO ROWS TO THE TOP
79    C                                         QUIT PACKING IF FIND N NONZERO ROWS
80          I=1
81          M=MM
82     60 IF (I.GT.N.OR.I.GE.M) GO TO 150
83        IF (A(I,I)) 90,70,90
84     70    DO 80 J=1,N
85             IF (A(I,J)) 90,80,90
86     80     CONTINUE
87        GO TO 100
88     90 I=I+1
89        GO TO 60
90    C                                         ROW I IS ZERO
91    C                                         EXCHANGE ROWS I AND M
92    100 IF(NB.LE.0) GO TO 115
93             DO 110 J=1,NB
94             T=B(I,J)
95             B(I,J)=B(M,J)
96    110      B(M,J)=T
97    115    DO 120 J=1,N
98    120    A(I,J)=A(M,J)
99        IF (M.GT.N) GO TO 140
100           DO 130 J=1,N
101   130     A(M,J)=ZERO
102   140 CONTINUE
103   C                               EXCHANGE IS FINISHED
104        M=M-1
105        GO TO 60
106   C
107   150 CONTINUE
108   C                                         END.. SPECIAL FOR ZERO ROWS AND COLUMNS
109   C                                         BEGIN.. SVD ALGORITHM
110   C     METHOD..
111   C     (1)      REDUCE THE MATRIX TO UPPER BIDIAGONAL FORM WITH
112   C     HOUSEHOLDER TRANSFORMATIONS.
113   C          H(N)...H(1)AQ(1)...Q(N-2) = (D**T,0)**T
114   C     WHERE D IS UPPER BIDIAGONAL.
115   C
116   C     (2)      APPLY H(N)...H(1) TO B.  HERE H(N)...H(1)*B REPLACES B
117   C     IN STORAGE.
118   C
119   C     (3)      THE MATRIX PRODUCT W= Q(1)...Q(N-2) OVERWRITES THE FIRST
120   C     N ROWS OF A IN STORAGE.
121   C
122   C     (4)      AN SVD FOR D IS COMPUTED.  HERE K ROTATIONS RI AND PI ARE
123   C     COMPUTED SO THAT
124   C          RK...R1*D*P1**(T)...PK**(T) = DIAG(S1,...,SM)
125   C     TO WORKING ACCURACY.  THE SI ARE NONNEGATIVE AND NONINCREASING.
126   C     HERE RK...R1*B OVERWRITES B IN STORAGE WHILE
127   C     A*P1**(T)...PK**(T)  OVERWRITES A IN STORAGE.
128   C
129   C     (5)      IT FOLLOWS THAT,WITH THE PROPER DEFINITIONS,
130   C     U**(T)*B OVERWRITES B, WHILE V OVERWRITES THE FIRST N ROW AND
131   C     COLUMNS OF A.
132   C
```

SVDRS (Continued)

```
133           L=MIN0(M,N)
134    C                   THE FOLLOWING LOOP REDUCES A TO UPPER BIDIAGONAL AND
135    C                   ALSO APPLIES THE PREMULTIPLYING TRANSFORMATIONS TO B.
136    C
137           DO 170 J=1,L
138           IF (J.GE.M) GO TO 160
139           CALL H12 (1,J,J+1,M,A(1,J),1,T,A(1,J+1),1,MDA,N-J)
140           CALL H12 (2,J,J+1,M,A(1,J),1,T,B,1,MDB,NB)
141    160    IF (J.GE.N-1) GO TO 170
142           CALL H12 (1,J+1,J+2,N,A(J,1),MDA,S(J,3),A(J+1,1),MDA,1,M-J)
143    170    CONTINUE
144    C
145    C      COPY THE BIDIAGONAL MATRIX INTO THE ARRAY S() FOR QRBD.
146    C
147           IF (N.EQ.1) GO TO 190
148           DO 180 J=2,N
149           S(J,1)=A(J,J)
150    180    S(J,2)=A(J-1,J)
151    190 S(1,1)=A(1,1)
152    C
153           NS=N
154           IF (M.GE.N) GO TO 200
155           NS=M+1
156           S(NS,1)=ZERO
157           S(NS,2)=A(M,M+1)
158    200 CONTINUE
159    C
160    C      CONSTRUCT THE EXPLICIT N BY N PRODUCT MATRIX, W=Q1*Q2*...*QL*I
161    C      IN THE ARRAY A().
162    C
163           DO 230 K=1,N
164           I=N+1-K
165           IF (I.GE.N-1) GO TO 210
166           CALL H12 (2,I+1,I+2,N,A(I,1),MDA,S(I,3),A(1,I+1),1,MDA,N-I)
167    210        DO 220 J=1,N
168    220        A(I,J)=ZERO
169    230    A(I,I)=ONE
170    C
171    C             COMPUTE THE SVD OF THE BIDIAGONAL MATRIX
172    C
173           CALL QRBD (IPASS,S(1,1),S(1,2),NS,A,MDA,N,B,MDB,NB)
174    C
175           GO TO (240,310), IPASS
176    240 CONTINUE
177           IF (NS.GE.N) GO TO 260
178           NSP1=NS+1
179           DO 250 J=NSP1,N
180    250    S(J,1)=ZERO
181    260 CONTINUE
182           IF (N.EQ.NN) RETURN
183           NP1=N+1
184    C                                   MOVE RECORD OF PERMUTATIONS
185    C                                   AND STORE ZEROS
186           DO 280 J=NP1,NN
187           S(J,1)=A(1,J)
188           DO 270 I=1,N
189    270    A(I,J)=ZERO
190    280    CONTINUE
191    C                           PERMUTE ROWS AND SET ZERO SINGULAR VALUES.
192           DO 300 K=NP1,NN
193           I=S(K,1)
194           S(K,1)=ZERO
195           DO 290 J=1,NN
196           A(K,J)=A(I,J)
197    290    A(I,J)=ZERO
198           A(I,K)=ONE
199    300    CONTINUE
200    C                           END.. SPECIAL FOR ZERO ROWS AND COLUMNS
201           RETURN
202    310 WRITE (6,320)
203           STOP
204    320 FORMAT (49H CONVERGENCE FAILURE IN QR BIDIAGONAL SVD ROUTINE)
205           END
```

QRBD

```
 1    C       SUBROUTINE QRBD (IPASS,Q,E,NN,V,MDV,NRV,C,MDC,NCC)
 2    C       C.L.LAWSON AND R.J.HANSON, JET PROPULSION LABORATORY, 1973 JUN 12
 3    C       TO APPEAR IN 'SOLVING LEAST SQUARES PROBLEMS', PRENTICE-HALL, 1974
 4    C       QR ALGORITHM FOR SINGULAR VALUES OF A BIDIAGONAL MATRIX.
 5    C
 6    C       THE BIDIAGONAL MATRIX
 7    C
 8    C                         (Q1,E2,0...      )
 9    C                         (   Q2,E3,0...   )
10    C             D=          (        .       )
11    C                         (          .   0 )
12    C                         (            .EN )
13    C                         (           0,QN )
14    C
15    C                    IS PRE AND POST MULTIPLIED BY
16    C                    ELEMENTARY ROTATION MATRICES
17    C                    RI AND PI SO THAT
18    C
19    C                    RK...R1*D*P1**(T)...PK**(T) = DIAG(S1,...,SN)
20    C
21    C                    TO WITHIN WORKING ACCURACY.
22    C
23    C       1. EI AND QI OCCUPY E(I) AND Q(I) AS INPUT.
24    C
25    C       2. RM...R1*C REPLACES 'C' IN STORAGE AS OUTPUT.
26    C
27    C       3. V*P1**(T)...PM**(T) REPLACES 'V' IN STORAGE AS OUTPUT.
28    C
29    C       4. SI OCCUPIES Q(I) AS OUTPUT.
30    C
31    C       5. THE SI'S ARE NONINCREASING AND NONNEGATIVE.
32    C
33    C          THIS CODE IS BASED ON THE PAPER AND 'ALGOL' CODE..
34    C REF..
35    C       1. REINSCH,C.H. AND GOLUB,G.H. 'SINGULAR VALUE DECOMPOSITION
36    C          AND LEAST SQUARES SOLUTIONS' (NUMER. MATH.), VOL. 14,(1970).
37    C
38            SUBROUTINE QRBD (IPASS,Q,E,NN,V,MDV,NRV,C,MDC,NCC)
39            LOGICAL WNTV ,HAVERS,FAIL
40            DIMENSION Q(NN),E(NN),V(MDV,NN),C(MDC,NCC)
41            ZERO=0.
42            ONE=1.
43            TWO=2.
44    C
45            N=NN
46            IPASS=1
47            IF (N.LE.0) RETURN
48            N10=10*N
49            WNTV=NRV.GT.0
50            HAVERS=NCC.GT.0
51            FAIL=.FALSE.
52            NQRS=0
53            E(1)=ZERO
54            DNORM=ZERO
55                DO 10 J=1,N
56         10     DNORM=AMAX1(ABS(Q(J))+ABS(E(J)),DNORM)
```

QRBD (Continued)

```
57                  DO 200 KK=1,N
58                  K=N+1-KK
59      C
60      C       TEST FOR SPLITTING OR RANK DEFICIENCIES..
61      C           FIRST MAKE TEST FOR LAST DIAGONAL TERM, Q(K), BEING SMALL.
62          20      IF(K.EQ.1) GO TO 50
63                  IF(DIFF(DNORM+Q(K),DNORM)) 50,25,50
64      C
65      C       SINCE Q(K) IS SMALL WE WILL MAKE A SPECIAL PASS TO
66      C       TRANSFORM E(K) TO ZERO.
67      C
68          25      CS=ZERO
69                  SN=-ONE
70                      DO 40 II=2,K
71                      I=K+1-II
72                      F=-SN*E(I+1)
73                      E(I+1)=CS*E(I+1)
74                      CALL G1 (Q(I),F,CS,SN,Q(I))
75      C           TRANSFORMATION CONSTRUCTED TO ZERO POSITION (I,K).
76      C
77                      IF (.NOT.WNTV) GO TO 40
78                      DO 30 J=1,NRV
79          30          CALL G2 (CS,SN,V(J,I),V(J,K))
80      C           ACCUMULATE RT. TRANSFORMATIONS IN V.
81      C
82          40      CONTINUE
83      C
84      C       THE MATRIX IS NOW BIDIAGONAL, AND OF LOWER ORDER
85      C       SINCE E(K) .EQ. ZERO..
86      C
87          50      DO 60 LL=1,K
88                  L=K+1-LL
89                  IF(DIFF(DNORM+E(L),DNORM)) 55,100,55
90          55      IF(DIFF(DNORM+Q(L-1),DNORM)) 60,70,60
91          60      CONTINUE
92      C       THIS LOOP CAN'T COMPLETE SINCE E(1) = ZERO.
93      C
94                  GO TO 100
95      C
96      C       CANCELLATION OF E(L), L.GT.1.
97          70      CS=ZERO
98                  SN=-ONE
99                      DO 90 I=L,K
100                     F=-SN*E(I)
101                     E(I)=CS*E(I)
102                     IF(DIFF(DNORM+F,DNORM)) 75,100,75
103         75          CALL G1 (Q(I),F,CS,SN,Q(I))
104                     IF (.NOT.HAVERS) GO TO 90
105                     DO 80 J=1,NCC
106         80          CALL G2 (CS,SN,C(I,J),C(L-1,J))
107         90      CONTINUE
108     C
109     C       TEST FOR CONVERGENCE..
110        100      Z=Q(K)
111                 IF (L.EQ.K) GO TO 170
112     C
113     C       SHIFT FROM BOTTOM 2 BY 2 MINOR OF B**(T)*B.
114                 X=Q(L)
115                 Y=Q(K-1)
116                 G=E(K-1)
117                 H=E(K)
118                 F=((Y-Z)*(Y+Z)+(G-H)*(G+H))/(TWO*H*Y)
119                 G=SQRT(ONE+F**2)
120                 IF (F.LT.ZERO) GO TO 110
121                 T=F+G
122                 GO TO 120
123        110      T=F-G
124        120      F=((X-Z)*(X+Z)+H*(Y/T-H))/X
125     C
126     C       NEXT QR SWEEP..
127                 CS=ONE
128                 SN=ONE
129                 LP1=L+1
```

QRBD (Continued)

```
130                         DO 160 I=LP1,K
131                         G=E(I)
132                         Y=Q(I)
133                         H=SN*G
134                         G=CS*G
135                         CALL G1 (F,H,CS,SN,E(I-1))
136                         F=X*CS+G*SN
137                         G=-X*SN+G*CS
138                         H=Y*SN
139                         Y=Y*CS
140                         IF (.NOT.WNTV) GO TO 140
141     C
142     C                   ACCUMULATE ROTATIONS (FROM THE RIGHT) IN 'V'
143                             DO 130 J=1,NRV
144        130                  CALL G2 (CS,SN,V(J,I-1),V(J,I))
145        140              CALL G1 (F,H,CS,SN,Q(I-1))
146                         F=CS*G+SN*Y
147                         X=-SN*G+CS*Y
148                         IF (.NOT.HAVERS) GO TO 160
149                             DO 150 J=1,NCC
150        150                  CALL G2 (CS,SN,C(I-1,J),C(I,J))
151     C                   APPLY ROTATIONS FROM THE LEFT TO
152     C                   RIGHT HAND SIDES IN 'C'..
153     C
154        160              CONTINUE
155                     E(L)=ZERO
156                     E(K)=F
157                     Q(K)=X
158                     NQRS=NQRS+1
159                     IF (NQRS.LE.N10) GO TO 20
160     C               RETURN TO 'TEST FOR SPLITTING'.
161     C
162                     FAIL=.TRUE.
163     C           . .
164     C           CUTOFF FOR CONVERGENCE FAILURE. 'NQRS' WILL BE 2*N USUALLY.
165        170      IF (Z.GE.ZERO) GO TO 190
166                 Q(K)=-Z
167                 IF (.NOT.WNTV) GO TO 190
168                     DO 180 J=1,NRV
169        180          V(J,K)=-V(J,K)
170        190      CONTINUE
171     C           CONVERGENCE. Q(K) IS MADE NONNEGATIVE..
172     C
173        200      CONTINUE
174             IF (N.EQ.1) RETURN
175             DO 210 I=2,N
176             IF (Q(I).GT.Q(I-1)) GO TO 220
177        210  CONTINUE
178         IF (FAIL) IPASS=2
179         RETURN
180     C       . .
181     C       EVERY SINGULAR VALUE IS IN ORDER..
182        220      DO 270 I=2,N
183                 T=Q(I-1)
184                 K=I-1
185                     DO 230 J=I,N
186                     IF (T.GE.Q(J)) GO TO 230
187                     T=Q(J)
188                     K=J
189        230          CONTINUE
190                 IF (K.EQ.I-1) GO TO 270
191                 Q(K)=Q(I-1)
192                 Q(I-1)=T
193                 IF (.NOT.HAVERS) GO TO 250
194                     DO 240 J=1,NCC
195                     T=C(I-1,J)
196                     C(I-1,J)=C(K,J)
197        240          C(K,J)=T
198        250      IF (.NOT.WNTV) GO TO 270
199                     DO 260 J=1,NRV
200                     T=V(J,I-1)
201                     V(J,I-1)=V(J,K)
202        260          V(J,K)=T
203        270      CONTINUE
204     C           END OF ORDERING ALGORITHM.
205     C
206         IF (FAIL) IPASS=2
207         RETURN
208         END
```

BNDACC

```
 1        SUBROUTINE BNDACC (G,MDG,NB,IP,IR,MT,JT)
 2   C    C.L.LAWSON AND R.J.HANSON, JET PROPULSION LABORATORY, 1973 JUN 12
 3   C    TO APPEAR IN 'SOLVING LEAST SQUARES PROBLEMS', PRENTICE-HALL, 1974
 4   C       SEQUENTIAL ALGORITHM FOR BANDED LEAST SQUARES PROBLEM..
 5   C       ACCUMULATION PHASE.        FOR SOLUTION PHASE USE BNDSOL.
 6   C
 7   C    THE CALLING PROGRAM MUST SET IR=1 AND IP=1 BEFORE THE FIRST CALL
 8   C    TO BNDACC FOR A NEW CASE.
 9   C
10   C    THE SECOND SUBSCRIPT OF G( ) MUST BE DIMENSIONED AT LEAST
11   C    NB+1 IN THE CALLING PROGRAM.
12        DIMENSION G(MDG,1)
13        ZERO=0.
14   C
15   C             ALG. STEPS 1-4 ARE PERFORMED EXTERNAL TO THIS SUBROUTINE.
16   C
17        NBP1=NB+1
18        IF (MT.LE.0) RETURN
19   C                                          ALG. STEP 5
20        IF (JT.EQ.IP) GO TO 70
21   C                                          ALG. STEPS 6-7
22        IF (JT.LE.IR) GO TO 30
23   C                                          ALG. STEPS 8-9
24        DO 10 I=1,MT
25           IG1=JT+MT-I
26           IG2=IR+MT-I
27           DO 10 J=1,NBP1
28   10      G(IG1,J)=G(IG2,J)
29   C                                          ALG. STEP 10
30        IE=JT-IR
31        DO 20 I=1,IE
32           IG=IR+I-1
33           DO 20 J=1,NBP1
34   20      G(IG,J)=ZERO
35   C                                          ALG. STEP 11
36        IR=JT
37   C                                          ALG. STEP 12
38   30   MU=MIN0(NB-1,IR-IP-1)
39        IF (MU.EQ.0) GO TO 60
40   C                                          ALG. STEP 13
41        DO 50 L=1,MU
42   C                                          ALG. STEP 14
43           K=MIN0(L,JT-IP)
44   C                                          ALG. STEP 15
45           LP1=L+1
46           IG=IP+L
47           DO 40 I=LP1,NB
48              JG=I-K
49   40         G(IG,JG)=G(IG,I)
50   C                                          ALG. STEP 16
51           DO 50 I=1,K
52           JG=NBP1-I
53   50      G(IG,JG)=ZERO
54   C                                          ALG. STEP 17
55   60   IP=JT
56   C                                          ALG. STEPS 18-19
57   70   MH=IR+MT-IP
58        KH=MIN0(NBP1,MH)
59   C                                          ALG. STEP 20
60        DO 80 I=1,KH
61   80      CALL H12 (1,I,MAX0(I+1,IR-IP+1),MH,G(IP,I),1,RHO,
62        1             G(IP,I+1),1,MDG,NBP1-I)
63   C                                          ALG. STEP 21
64        IR=IP+KH
65   C                                          ALG. STEP 22
66        IF (KH.LT.NBP1) GO TO 100
67   C                                          ALG. STEP 23
68        DO 90 I=1,NB
69   90      G(IR-1,I)=ZERO
70   C                                          ALG. STEP 24
71   100  CONTINUE
72   C                                          ALG. STEP 25
73        RETURN
74        END
```

BNDSOL

```
 1          SUBROUTINE BNDSOL (MODE,G,MDG,NB,IP,IR,X,N,RNORM)
 2    C     C.L.LAWSON AND R.J.HANSON, JET PROPULSION LABORATORY, 1973 JUN 12
 3    C     TO APPEAR IN 'SOLVING LEAST SQUARES PROBLEMS', PRENTICE-HALL, 1974
 4    C        SEQUENTIAL SOLUTION OF A BANDED LEAST SQUARES PROBLEM..
 5    C        SOLUTION PHASE.    FOR THE ACCUMULATION PHASE USE BNDACC.
 6    C
 7    C     MODE = 1       SOLVE R*X=Y   WHERE R AND Y ARE IN THE G( ) ARRAY
 8    C                    AND X WILL BE STORED IN THE X( ) ARRAY.
 9    C            2       SOLVE (R**T)*X=Y   WHERE R IS IN G( ),
10    C                    Y IS INITIALLY IN X( ), AND X REPLACES Y IN X( ),
11    C            3       SOLVE R*X=Y   WHERE R IS IN G( ).
12    C                    Y IS INITIALLY IN X( ), AND X REPLACES Y IN X( ).
13    C
14    C     THE SECOND SUBSCRIPT OF G( ) MUST BE DIMENSIONED AT LEAST
15    C     NB+1 IN THE CALLING PROGRAM.
16          DIMENSION G(MDG,1),X(N)
17          ZERO=0.
18    C
19          RNORM=ZERO
20          GO TO (10,90,50), MODE               ****************** MODE = 1
21    C
22    C                                          ALG. STEP 26
23    10      DO 20 J=1,N
24    20      X(J)=G(J,NB+1)
25          RSQ=ZERO
26          NP1=N+1
27          IRM1=IR-1
28          IF (NP1.GT.IRM1) GO TO 40
29          DO 30 J=NP1,IRM1
30    30      RSQ=RSQ+G(J,NB+1)**2
31          RNORM=SQRT(RSQ)
32    40 CONTINUE                                ****************** MODE = 3
33    C
34    C                                          ALG. STEP 27
35    50      DO 80 II=1,N
36          I=N+1-II
37    C                                          ALG. STEP 28
38          S=ZERO
39          L=MAX0(0,I-IP)
40    C                                          ALG. STEP 29
41          IF (I.EQ.N) GO TO 70
42    C                                          ALG. STEP 30
43          IE=MIN0(N+1-I,NB)
44              DO 60 J=2,IE
45              JG=J+L
46              IX=I-1+J
47    60          S=S+G(I,JG)*X(IX)
48    C                                          ALG. STEP 31
49    70      IF (G(I,L+1)) 80,130,80
50    80      X(I)=(X(I)-S)/G(I,L+1)
51    C                                          ALG. STEP 32
52        RETURN
53    C                                          ****************** MODE = 2
54    90      DO 120 J=1,N
55          S=ZERO
56          IF (J.EQ.1) GO TO 110
57          I1=MAX0(1,J-NB+1)
58          I2=J-1
59              DO 100 I=I1,I2
60              L=J-I+1+MAX0(0,I-IP)
61    100         S=S+X(I)*G(I,L)
62    110     L=MAX0(0,J-IP)
63          IF (G(J,L+1)) 120,130,120
64    120     X(J)=(X(J)-S)/G(J,L+1)
65        RETURN
66    C
67    130 WRITE (6,140) MODE,I,J,L
68        STOP
69    140 FORMAT (30H0ZERO DIAGONAL TERM IN BNDSOL..12H MODE,I,J,L=,4I6)
70        END
```

LDP

```
 1          SUBROUTINE LDP (G,MDG,M,N,H,X,XNORM,W,INDEX,MODE)
 2      C   C.L.LAWSON AND R.J.HANSON, JET PROPULSION LABORATORY, 1974 MAR 1
 3      C   TO APPEAR IN 'SOLVING LEAST SQUARES PROBLEMS', PRENTICE-HALL, 1974
 4      C
 5      C        ********** LEAST DISTANCE PROGRAMMING **********
 6      C
 7          INTEGER INDEX(M)
 8          DIMENSION G(MDG,N), H(M), X(N), W(1)
 9          ZERO=0.
10          ONE=1.
11          IF (N.LE.0) GO TO 120
12             DO 10 J=1,N
13       10       X(J)=ZERO
14          XNORM=ZERO
15          IF (M.LE.0) GO TO 110
16      C
17      C   THE DECLARED DIMENSION OF W() MUST BE AT LEAST (N+1)*(M+2)+2*M.
18      C
19      C   FIRST (N+1)*M LOCS OF W()  - MATRIX E FOR PROBLEM NNLS.
20      C      NEXT      N+1 LOCS OF W()  - VECTOR F FOR PROBLEM NNLS.
21      C      NEXT      N+1 LOCS OF W()  - VECTOR Z FOR PROBLEM NNLS.
22      C      NEXT       M LOCS OF W()  - VECTOR Y FOR PROBLEM NNLS.
23      C      NEXT       M LOCS OF W()  - VECTOR WDUAL FOR PROBLEM NNLS.
24      C   COPY G**T INTO FIRST N ROWS AND M COLUMNS OF E.
25      C   COPY H**T INTO ROW N+1 OF E.
26      C
27          IW=0
28             DO 30 J=1,M
29                DO 20 I=1,N
30                   IW=IW+1
31       20           W(IW)=G(J,I)
32                IW=IW+1
33       30          W(IW)=H(J)
34          IF=IW+1
35      C                           STORE N ZEROS FOLLOWED BY A ONE INTO F.
36             DO 40 I=1,N
37                IW=IW+1
38       40          W(IW)=ZERO
39          W(IW+1)=ONE
40      C
41          NP1=N+1
42          IZ=IW+2
43          IY=IZ+NP1
44          IWDUAL=IY+M
45      C
46          CALL NNLS (W,NP1,NP1,M,W(IF),W(IY),RNORM,W(IWDUAL),W(IZ),INDEX,
47         *           MODE)
48      C                   USE THE FOLLOWING RETURN IF UNSUCCESSFUL IN NNLS.
49          IF (MODE.NE.1) RETURN
50          IF (RNORM) 130,130,50
51       50 FAC=ONE
52          IW=IY-1
53             DO 60 I=1,M
54                IW=IW+1
55      C                           HERE WE ARE USING THE SOLUTION VECTOR Y.
56       60          FAC=FAC-H(I)*W(IW)
57      C
58          IF (DIFF(ONE+FAC,ONE)) 130,130,70
59       70 FAC=ONE/FAC
60             DO 90 J=1,N
61                IW=IY-1
62                DO 80 I=1,M
63                   IW=IW+1
64      C                           HERE WE ARE USING THE SOLUTION VECTOR Y.
65       80           X(J)=X(J)+G(I,J)*W(IW)
66       90          X(J)=X(J)*FAC
67             DO 100 J=1,N
68      100       XNORM=XNORM+X(J)**2
69          XNORM=SQRT(XNORM)
70      C                           SUCCESSFUL RETURN.
71      110 MODE=1
72          RETURN
73      C                           ERROR RETURN.      N .LE. 0.
74      120 MODE=2
75          RETURN
76      C                           RETURNING WITH CONSTRAINTS NOT COMPATIBLE.
77      130 MODE=4
78          RETURN
79          END
```

NNLS

```
1   C       SUBROUTINE NNLS   (A,MDA,M,N,B,X,RNORM,W,ZZ,INDEX,MODE)
2   C       C.L.LAWSON AND R.J.HANSON, JET PROPULSION LABORATORY, 1973 JUNE 15
3   C       TO APPEAR IN 'SOLVING LEAST SQUARES PROBLEMS', PRENTICE-HALL, 1974
4   C
5   C       **********   NONNEGATIVE LEAST SQUARES   **********
6   C
7   C       GIVEN AN M BY N MATRIX, A, AND AN M-VECTOR, B,  COMPUTE AN
8   C       N-VECTOR, X, WHICH SOLVES THE LEAST SQUARES PROBLEM
9   C
10  C                    A * X = B   SUBJECT TO X .GE. 0
11  C
12  C       A(),MDA,M,N      MDA IS THE FIRST DIMENSIONING PARAMETER FOR THE
13  C                        ARRAY, A().   ON ENTRY A() CONTAINS THE M BY N
14  C                        MATRIX, A.           ON EXIT A() CONTAINS
15  C                        THE PRODUCT MATRIX, Q*A , WHERE Q IS AN
16  C                        M BY M ORTHOGONAL MATRIX GENERATED IMPLICITLY BY
17  C                        THIS SUBROUTINE.
18  C       B()        ON ENTRY B() CONTAINS THE M-VECTOR, B.   ON EXIT B() CON-
19  C                  TAINS Q*B.
20  C       X()        ON ENTRY X() NEED NOT BE INITIALIZED.  ON EXIT X() WILL
21  C                  CONTAIN THE SOLUTION VECTOR.
22  C       RNORM      ON EXIT RNORM CONTAINS THE EUCLIDEAN NORM OF THE
23  C                  RESIDUAL VECTOR.
24  C       W()        AN N-ARRAY OF WORKING SPACE.   ON EXIT W() WILL CONTAIN
25  C                  THE DUAL SOLUTION VECTOR.   W WILL SATISFY W(I) = 0.
26  C                  FOR ALL I IN SET P  AND W(I) .LE. 0. FOR ALL I IN SET Z
27  C       ZZ()       AN M-ARRAY OF WORKING SPACE.
28  C       INDEX()         AN INTEGER WORKING ARRAY OF LENGTH AT LEAST N.
29  C                       ON EXIT THE CONTENTS OF THIS ARRAY DEFINE THE SETS
30  C                       P AND Z AS FOLLOWS..
31  C
32  C                       INDEX(1)    THRU INDEX(NSETP) = SET P.
33  C                       INDEX(IZ1)  THRU INDEX(IZ2)   = SET Z.
34  C                       IZ1 = NSETP + 1 = NPP1
35  C                       IZ2 = N
36  C       MODE       THIS IS A SUCCESS-FAILURE FLAG WITH THE FOLLOWING
37  C                  MEANINGS.
38  C                  1       THE SOLUTION HAS BEEN COMPUTED SUCCESSFULLY.
39  C                  2       THE DIMENSIONS OF THE PROBLEM ARE BAD.
40  C                          EITHER M .LE. 0 OR N .LE. 0.
41  C                  3       ITERATION COUNT EXCEEDED.  MORE THAN 3*N ITERATIONS.
42  C
43          SUBROUTINE NNLS  (A,MDA,M,N,B,X,RNORM,W,ZZ,INDEX,MODE)
44          DIMENSION A(MDA,N), B(M), X(N), W(N), ZZ(M)
45          INTEGER INDEX(N)
46          ZERO=0.
47          ONE=1.
48          TWO=2.
49          FACTOR=0.01
50  C
51          MODE=1
52          IF (M.GT.0.AND.N.GT.0) GO TO 10
53          MODE=2
54          RETURN
55       10 ITER=0
56          ITMAX=3*N
57  C
58  C                          INITIALIZE THE ARRAYS INDEX() AND X().
59  C
60             DO 20 I=1,N
61             X(I)=ZERO
62       20    INDEX(I)=I
63  C
64          IZ2=N
65          IZ1=1
66          NSETP=0
67          NPP1=1
```

NNLS (Continued)

```
68   C                              •••••• MAIN LOOP BEGINS HERE ••••••
69      30 CONTINUE
70   C                    QUIT IF ALL COEFFICIENTS ARE ALREADY IN THE SOLUTION.
71   C                    OR IF M COLS OF A HAVE BEEN TRIANGULARIZED.
72   C
73          IF (IZ1.GT.IZ2.OR.NSETP.GE.M) GO TO 350
74   C
75   C          COMPUTE COMPONENTS OF THE DUAL (NEGATIVE GRADIENT) VECTOR W().
76   C
77          DO 50 IZ=IZ1,IZ2
78          J=INDEX(IZ)
79          SM=ZERO
80              DO 40 L=NPP1,M
81      40          SM=SM+A(L,J)*B(L)
82      50  W(J)=SM
83   C                              FIND LARGEST POSITIVE W(J).
84      60 WMAX=ZERO
85          DO 70 IZ=IZ1,IZ2
86          J=INDEX(IZ)
87          IF (W(J).LE.WMAX) GO TO 70
88          WMAX=W(J)
89          IZMAX=IZ
90      70      CONTINUE
91   C
92   C              IF WMAX .LE. 0. GO TO TERMINATION.
93   C              THIS INDICATES SATISFACTION OF THE KUHN-TUCKER CONDITIONS.
94   C
95          IF (WMAX) 350,350,80
96      80 IZ=IZMAX
97          J=INDEX(IZ)
98   C
99   C       THE SIGN OF W(J) IS OK FOR J TO BE MOVED TO SET P.
100  C       BEGIN THE TRANSFORMATION AND CHECK NEW DIAGONAL ELEMENT TO AVOID
101  C       NEAR LINEAR DEPENDENCE.
102  C
103         ASAVE=A(NPP1,J)
104         CALL H12 (1,NPP1,NPP1+1,M,A(1,J),1,UP,DUMMY,1,1,0)
105         UNORM=ZERO
106         IF (NSETP.EQ.0) GO TO 100
107             DO 90 L=1,NSETP
108     90          UNORM=UNORM+A(L,J)**2
109    100 UNORM=SQRT(UNORM)
110         IF (DIFF(UNORM+ABS(A(NPP1,J))*FACTOR,UNORM)) 130,130,110
111  C
112  C       COL J IS SUFFICIENTLY INDEPENDENT.  COPY B INTO ZZ, UPDATE ZZ AND
113  C       SOLVE FOR ZTEST ( = PROPOSED NEW VALUE FOR X(J) ).
114  C
115    110     DO 120 L=1,M
116    120     ZZ(L)=B(L)
117         CALL H12 (2,NPP1,NPP1+1,M,A(1,J),1,UP,ZZ,1,1,1)
118         ZTEST=ZZ(NPP1)/A(NPP1,J)
119  C
120  C                              SEE IF ZTEST IS POSITIVE
121  C
122         IF (ZTEST) 130,130,140
123  C
124  C       REJECT J AS A CANDIDATE TO BE MOVED FROM SET Z TO SET P.
125  C       RESTORE A(NPP1,J), SET W(J)=0., AND LOOP BACK TO TEST DUAL
126  C       COEFFS AGAIN.
127  C
128    130 A(NPP1,J)=ASAVE
129         W(J)=ZERO
130         GO TO 60
```

NNLS (Continued)

```
131   C
132   C        THE INDEX  J=INDEX(IZ)  HAS BEEN SELECTED TO BE MOVED FROM
133   C        SET Z TO SET P.     UPDATE B,  UPDATE INDICES,   APPLY HOUSEHOLDER
134   C        TRANSFORMATIONS TO COLS IN NEW SET Z,  ZERO SUBDIAGONAL ELTS IN
135   C        COL J,   SET W(J)=0.
136   C
137     140      DO 150 L=1,M
138     150      B(L)=ZZ(L)
139   C
140              INDEX(IZ)=INDEX(IZ1)
141              INDEX(IZ1)=J
142              IZ1=IZ1+1
143              NSETP=NPP1
144              NPP1=NPP1+1
145   C
146              IF (IZ1.GT.IZ2) GO TO 170
147                  DO 160 JZ=IZ1,IZ2
148                  JJ=INDEX(JZ)
149     160          CALL H12 (2,NSETP,NPP1,M,A(1,J),1,UP,A(1,JJ),1,MDA,1)
150     170 CONTINUE
151   C
152              IF (NSETP.EQ.M) GO TO 190
153                  DO 180 L=NPP1,M
154     180          A(L,J)=ZERO
155     190 CONTINUE
156   C
157              W(J)=ZERO
158   C                                        SOLVE THE TRIANGULAR SYSTEM.
159   C                                        STORE THE SOLUTION TEMPORARILY IN ZZ().
160              ASSIGN 200 TO NEXT
161              GO TO 400
162     200 CONTINUE
163   C
164   C                        ****** SECONDARY LOOP BEGINS HERE ******
165   C
166   C                                 ITERATION COUNTER.
167   C
168     210 ITER=ITER+1
169              IF (ITER.LE.ITMAX) GO TO 220
170              MODE=3
171              WRITE (6,440)
172              GO TO 350
173     220 CONTINUE
174   C
175   C                        SEE IF ALL NEW CONSTRAINED COEFFS ARE FEASIBLE.
176   C                                 IF NOT COMPUTE ALPHA.
177   C
178              ALPHA=TWO
179                  DO 240 IP=1,NSETP
180                  L=INDEX(IP)
181                  IF (ZZ(IP)) 230,230,240
182   C
183     230      T=-X(L)/(ZZ(IP)-X(L))
184              IF (ALPHA.LE.T) GO TO 240
185              ALPHA=T
186              JJ=IP
187     240      CONTINUE
188   C
189   C                IF ALL NEW CONSTRAINED COEFFS ARE FEASIBLE THEN ALPHA WILL
190   C                STILL = 2.    IF SO EXIT FROM SECONDARY LOOP TO MAIN LOOP.
191   C
192              IF (ALPHA.EQ.TWO) GO TO 330
193   C
194   C                OTHERWISE USE ALPHA WHICH WILL BE BETWEEN 0. AND 1. TO
195   C                INTERPOLATE BETWEEN THE OLD X AND THE NEW ZZ.
196   C
197                  DO 250 IP=1,NSETP
198                  L=INDEX(IP)
199     250      X(L)=X(L)+ALPHA*(ZZ(IP)-X(L))
200   C
```

NNLS (Continued)

```
201   C          MODIFY A AND B AND THE INDEX ARRAYS TO MOVE COEFFICIENT I
202   C          FROM SET P TO SET Z.
203   C
204              I=INDEX(JJ)
205      260  X(I)=ZERO
206   C
207              IF (JJ.EQ.NSETP) GO TO 290
208              JJ=JJ+1
209                 DO 280 J=JJ,NSETP
210                 II=INDEX(J)
211                 INDEX(J-1)=II
212                 CALL G1 (A(J-1,II),A(J,II),CC,SS,A(J-1,II))
213                 A(J,II)=ZERO
214                    DO 270 L=1,N
215                    IF (L.NE.II) CALL G2 (CC,SS,A(J-1,L),A(J,L))
216      270        CONTINUE
217      280        CALL G2 (CC,SS,B(J-1),B(J))
218      290 NPP1=NSETP
219              NSETP=NSETP-1
220              IZ1=IZ1-1
221              INDEX(IZ1)=I
222   C
223   C          SEE IF THE REMAINING COEFFS IN SET P ARE FEASIBLE.  THEY SHOULD
224   C          BE BECAUSE OF THE WAY ALPHA WAS DETERMINED.
225   C          IF ANY ARE INFEASIBLE IT IS DUE TO ROUND-OFF ERROR.  ANY
226   C          THAT ARE NONPOSITIVE WILL BE SET TO ZERO
227   C          AND MOVED FROM SET P TO SET Z.
228   C
229              DO 300 JJ=1,NSETP
230              I=INDEX(JJ)
231              IF (X(I)) 260,260,300
232      300    CONTINUE
233   C
234   C          COPY B( ) INTO ZZ( ).  THEN SOLVE AGAIN AND LOOP BACK.
235   C
236              DO 310 I=1,M
237      310  ZZ(I)=B(I)
238              ASSIGN 320 TO NEXT
239              GO TO 400
240      320 CONTINUE
241              GO TO 210
242   C                         ****** END OF SECONDARY LOOP ******
243   C
244      330      DO 340 IP=1,NSETP
245              I=INDEX(IP)
246      340      X(I)=ZZ(IP)
247   C          ALL NEW COEFFS ARE POSITIVE.  LOOP BACK TO BEGINNING.
248              GO TO 30
249   C
250   C                         ****** END OF MAIN LOOP ******
251   C
252   C                   COME TO HERE FOR TERMINATION.
253   C                   COMPUTE THE NORM OF THE FINAL RESIDUAL VECTOR.
254   C
255      350 SM=ZERO
256              IF (NPP1.GT.M) GO TO 370
257                 DO 360 I=NPP1,M
258      360        SM=SM+B(I)**2
259              GO TO 390
260      370      DO 380 J=1,N
261      380      W(J)=ZERO
262      390 RNORM=SQRT(SM)
263              RETURN
264   C
265   C          THE FOLLOWING BLOCK OF CODE IS USED AS AN INTERNAL SUBROUTINE
266   C          TO SOLVE THE TRIANGULAR SYSTEM, PUTTING THE SOLUTION IN ZZ( ).
267   C
268      400      DO 430 L=1,NSETP
269              IP=NSETP+1-L
270              IF (L.EQ.1) GO TO 420
271                 DO 410 II=1,IP
272      410        ZZ(II)=ZZ(II)-A(II,JJ)*ZZ(IP+1)
273      420      JJ=INDEX(IP)
274      430      ZZ(IP)=ZZ(IP)/A(IP,JJ)
275              GO TO NEXT, (200,320)
276      440 FORMAT (35H0 NNLS QUITTING ON ITERATION COUNT.)
277              END
```

H12

```
 1   C      SUBROUTINE H12 (MODE,LPIVOT,L1,M,U,IUE,UP,C,ICE,ICV,NCV)
 2   C      C.L.LAWSON AND R.J.HANSON, JET PROPULSION LABORATORY, 1973 JUN 12
 3   C      TO APPEAR IN 'SOLVING LEAST SQUARES PROBLEMS', PRENTICE-HALL, 1974
 4   C
 5   C      CONSTRUCTION AND/OR APPLICATION OF A SINGLE
 6   C      HOUSEHOLDER TRANSFORMATION..      Q = I + U*(U**T)/B
 7   C
 8   C      MODE    = 1 OR 2    TO SELECT ALGORITHM  H1  OR  H2 .
 9   C      LPIVOT IS THE INDEX OF THE PIVOT ELEMENT.
10   C      L1,M    IF L1 .LE. M    THE TRANSFORMATION WILL BE CONSTRUCTED TO
11   C              ZERO ELEMENTS INDEXED FROM L1 THROUGH M.    IF L1 GT. M
12   C              THE SUBROUTINE DOES AN IDENTITY TRANSFORMATION.
13   C      U(),IUE,UP    ON ENTRY TO H1 U() CONTAINS THE PIVOT VECTOR.
14   C              IUE IS THE STORAGE INCREMENT BETWEEN ELEMENTS.
15   C                             ON EXIT FROM H1 U() AND UP
16   C              CONTAIN QUANTITIES DEFINING THE VECTOR U OF THE
17   C              HOUSEHOLDER TRANSFORMATION.   ON ENTRY TO H2 U()
18   C              AND UP SHOULD CONTAIN QUANTITIES PREVIOUSLY COMPUTED
19   C              BY H1.   THESE WILL NOT BE MODIFIED BY H2.
20   C      C()     ON ENTRY TO H1 OR H2 C() CONTAINS A MATRIX WHICH WILL BE
21   C              REGARDED AS A SET OF VECTORS TO WHICH THE HOUSEHOLDER
22   C              TRANSFORMATION IS TO BE APPLIED.   ON EXIT C() CONTAINS THE
23   C              SET OF TRANSFORMED VECTORS.
24   C      ICE     STORAGE INCREMENT BETWEEN ELEMENTS OF VECTORS IN C().
25   C      ICV     STORAGE INCREMENT BETWEEN VECTORS IN C().
26   C      NCV     NUMBER OF VECTORS IN C() TO BE TRANSFORMED. IF NCV .LE. 0
27   C              NO OPERATIONS WILL BE DONE ON C().
28   C
29          SUBROUTINE H12 (MODE,LPIVOT,L1,M,U,IUE,UP,C,ICE,ICV,NCV)
30          DIMENSION U(IUE,M), C(1)
31          DOUBLE PRECISION SM,B
32          ONE=1.
33   C
34          IF (0.GE.LPIVOT.OR.LPIVOT.GE.L1.OR.L1.GT.M) RETURN
35          CL=ABS(U(1,LPIVOT))
36          IF (MODE.EQ.2) GO TO 60
37   C                            ****** CONSTRUCT THE TRANSFORMATION. ******
38              DO 10 J=L1,M
39   10         CL=AMAX1(ABS(U(1,J)),CL)
40          IF (CL) 130,130,20
41   20     CLINV=ONE/CL
42          SM=(DBLE(U(1,LPIVOT))*CLINV)**2
43              DO 30 J=L1,M
44   30         SM=SM+(DBLE(U(1,J))*CLINV)**2
45   C                              CONVERT DBLE. PREC. SM TO SNGL. PREC. SM1
46          SM1=SM
47          CL=CL*SQRT(SM1)
48          IF (U(1,LPIVOT)) 50,50,40
49   40     CL=-CL
50   50     UP=U(1,LPIVOT)-CL
51          U(1,LPIVOT)=CL
52          GO TO 70
53   C                 ****** APPLY THE TRANSFORMATION  I+U*(U**T)/B  TO C. ******
54   C
55   60     IF (CL) 130,130,70
56   70     IF (NCV.LE.0) RETURN
57          B=DBLE(UP)*U(1,LPIVOT)
58   C                           B  MUST BE NONPOSITIVE HERE.   IF B = 0., RETURN.
59   C
60          IF (B) 80,130,130
61   80     B=ONE/B
62          I2=1-ICV+ICE*(LPIVOT-1)
63          INCR=ICE*(L1-LPIVOT)
64              DO 120 J=1,NCV
65              I2=I2+ICV
66              I3=I2+INCR
67              I4=I3
68              SM=C(I2)*DBLE(UP)
69                  DO 90 I=L1,M
70                  SM=SM+C(I3)*DBLE(U(1,I))
71   90             I3=I3+ICE
72              IF (SM) 100,120,100
73   100        SM=SM*B
74              C(I2)=C(I2)+SM*DBLE(UP)
75                  DO 110 I=L1,M
76                  C(I4)=C(I4)+SM*DBLE(U(1,I))
77   110            I4=I4+ICE
78   120        CONTINUE
79   130 RETURN
80          END
```

G1

```
 1          SUBROUTINE G1 (A,B,COS,SIN,SIG)
 2    C     C.L.LAWSON AND R.J.HANSON, JET PROPULSION LABORATORY, 1973 JUN 12
 3    C     TO APPEAR IN 'SOLVING LEAST SQUARES PROBLEMS', PRENTICE-HALL, 1974
 4    C
 5    C
 6    C     COMPUTE ORTHOGONAL ROTATION MATRIX..
 7    C     COMPUTE..  MATRIX   (C, S) SO THAT (C, S)(A) = (SQRT(A**2+B**2))
 8    C                         (-S,C)         (-S,C)(B)  (    0         )
 9    C     COMPUTE SIG = SQRT(A**2+B**2)
10    C         SIG IS COMPUTED LAST TO ALLOW FOR THE POSSIBILITY THAT
11    C         SIG MAY BE IN THE SAME LOCATION AS A OR B .
12    C
13          ZERO=0.
14          ONE=1.
15          IF (ABS(A).LE.ABS(B)) GO TO 10
16          XR=B/A
17          YR=SQRT(ONE+XR**2)
18          COS=SIGN(ONE/YR,A)
19          SIN=COS*XR
20          SIG=ABS(A)*YR
21          RETURN
22       10 IF (B) 20,30,20
23       20 XR=A/B
24          YR=SQRT(ONE+XR**2)
25          SIN=SIGN(ONE/YR,B)
26          COS=SIN*XR
27          SIG=ABS(B)*YR
28          RETURN
29       30 SIG=ZERO
30          COS=ZERO
31          SIN=ONE
32          RETURN
33          END
```

G2

```
 1          SUBROUTINE G2    (COS,SIN,X,Y)
 2    C     C.L.LAWSON AND R.J.HANSON, JET PROPULSION LABORATORY, 1972 DEC 15
 3    C     TO APPEAR IN 'SOLVING LEAST SQUARES PROBLEMS', PRENTICE-HALL, 1974
 4    C         APPLY THE ROTATION COMPUTED BY G1 TO (X,Y).
 5          XR=COS*X+SIN*Y
 6          Y=-SIN*X+COS*Y
 7          X=XR
 8          RETURN
 9          END
```

MFEOUT

```
 1          SUBROUTINE MFEOUT (A,MDA,M,N,NAMES,MODE)
 2   C      C.L.LAWSON AND R.J.HANSON, JET PROPULSION LABORATORY, 1973 JUN 12
 3   C      TO APPEAR IN 'SOLVING LEAST SQUARES PROBLEMS', PRENTICE-HALL, 1974
 4   C          SUBROUTINE FOR MATRIX OUTPUT WITH LABELING.
 5   C
 6   C      A( )            MATRIX TO BE OUTPUT
 7   C                      MDA     FIRST DIMENSION OF A ARRAY
 8   C                      M           NO. OF ROWS IN A MATRIX
 9   C                      N           NO. OF COLS IN A MATRIX
10   C      NAMES()         ARRAY OF NAMES.  IF NAMES(1) = 1H , THE REST
11   C                      OF THE NAMES() ARRAY WILL BE IGNORED.
12   C      MODE            =1    FOR    4P8F15.0  FORMAT   FOR V MATRIX.
13   C                      =2    FOR    8E15.8  FORMAT   FOR CANDIDATE SOLUTIONS.
14   C
15          DIMENSION     A(MDA,01)
16          INTEGER NAMES(M),IHEAD(2)
17          LOGICAL    NOTBLK
18          DATA  MAXCOL/8/, IBLANK/1H /,IHEAD(1)/4H COL/,IHEAD(2)/4HSOLN/
19   C
20          IF (M.LE.0.OR.N.LE.0) RETURN
21          NOTBLK=NAMES(1).NE.IBLANK
22   C
23          IF (MODE.EQ.2) GO TO 10
24          WRITE (6,70)
25          GO TO 20
26       10 WRITE (6,80)
27       20 CONTINUE
28   C
29          NBLOCK=N/MAXCOL
30          LAST=N-NBLOCK*MAXCOL
31          NCOL=MAXCOL
32          J1=1
33   C
34   C                              MAIN LOOP STARTS HERE
35   C
36       30 IF (NBLOCK.GT.0) GO TO 40
37          IF (LAST.LE.0) RETURN
38          NCOL=LAST
39          LAST=0
40   C
41       40 J2=J1+NCOL-1
42          WRITE (6,90) (IHEAD(MODE),J,J=J1,J2)
43   C
44              DO 60 I=1,M
45              NAME=IBLANK
46              IF (NOTBLK) NAME=NAMES(I)
47   C
48              IF (MODE.EQ.2) GO TO 50
49              WRITE (6,100) I,NAME,(A(I,J),J=J1,J2)
50              GO TO 60
51       50     WRITE (6,110) I,NAME,(A(I,J),J=J1,J2)
52       60     CONTINUE
53   C
54          J1=J1+MAXCOL
55          NBLOCK=NBLOCK-1
56          GO TO 30
57   C
58       70 FORMAT (45H0V-MATRIX OF THE SINGULAR VALUE DECOMPOSITION,
59        * 8H OF A*D./47H (ELEMENTS OF V SCALED UP BY A FACTOR OF 10**4))
60       80 FORMAT (35H0SEQUENCE OF CANDIDATE SOLUTIONS, X)
61       90 FORMAT (1H0,11X,8(6X,A4,I4,1X)/1X)
62      100 FORMAT (1X,I3,1X,A6,1X,4P8F15.0)
63      110 FORMAT (1X,I3,1X,A6,1X,8E15.8)
64          END
```

GEN

```
1          FUNCTION    GEN(ANOISE)
2     C    C.L.LAWSON AND R.J.HANSON, JET PROPULSION LABORATORY, 1972 DEC 15
3     C    TO APPEAR IN 'SOLVING LEAST SQUARES PROBLEMS', PRENTICE-HALL, 1974
4     C        GENERATE NUMBERS FOR CONSTRUCTION OF TEST CASES.
5          IF (ANOISE) 10,30,20
6     10 MI=891
7          MJ=457
8          I=5
9          J=7
10         AJ=0.
11         GEN=0.
12         RETURN
13    C
14    C    THE SEQUENCE OF VALUES OF J  IS BOUNDED BETWEEN 1 AND 996
15    C    IF INITIAL J = 1,2,3,4,5,6,7,8, OR 9, THE PERIOD IS 332
16    20 J=J*MJ
17         J=J-997*(J/997)
18         AJ=J-498
19    C    THE SEQUENCE OF VALUES OF I  IS BOUNDED BETWEEN 1 AND 999
20    C    IF INITIAL I = 1,2,3,6,7, OR 9,  THE PERIOD WILL BE 50
21    C    IF INITIAL I = 4 OR 8    THE PERIOD WILL BE 25
22    C    IF INITIAL I = 5         THE PERIOD WILL BE 10
23    30 I=I*MI
24         I=I-1000*(I/1000)
25         AI=I-500
26         GEN=AI+AJ*ANOISE
27         RETURN
28         END
```

DIFF

```
1          FUNCTION DIFF(X,Y)
2     C    C.L.LAWSON AND R.J.HANSON, JET PROPULSION LABORATORY, 1973 JUNE 7
3     C    TO APPEAR IN 'SOLVING LEAST SQUARES PROBLEMS', PRENTICE-HALL, 1974
4          DIFF=X-Y
5          RETURN
6          END
```

BIBLIOGRAPHY

ALBERT, ARTHUR (1972) *Regression and the Moore–Penrose Pseudoinverse.* Academic Press, New York and London, 180 pp.

ANSI Subcommittee X3J3 (1971) "Clarification of Fortran Standards— Second Report," *Comm. ACM*, **14**, No. 10, 628–642.

ASA Committee X3 (1964) "Fortran vs. Basic Fortran," *Comm. ACM*, **7**, No. 10, 591–625.

AUTONNE, L. (1915) "Sur les Matrices Hypohermitienes et sur les Matrices Unitaries," *Ann. Univ. Lyon*, (2), **38**, 1–77.

BARD, YONATHAN (1970) "Comparison of Gradient Methods for the Solution of Nonlinear Parameter Estimation Problems," *SIAM J. Numer. Anal.*, **7**, No. 1, 157–186.

BARTELS, R. H., GOLUB, G. H., AND SAUNDERS, M. A. (1970) "Numerical Techniques in Mathematical Programming," in *Nonlinear Programming.* Academic Press, New York, 123–176.

BARTELS, RICHARD H. (1971) "A Stabilization of the Simplex Method," *Numer. Math.*, **16**, 414–434.

BAUER, F. L. (1963) "Optimal Scaling of Matrices and the Importance of the Minimal Condition," *Proc. IFIP Congress 1962*, North-Holland, 198–201.

BAUER, F. L., see Wilkinson, J. H.

BEN–ISRAEL, ADI (1965) "A Modified Newton–Raphson Method for the Solution of Systems of Equations," *Israel J. Math.*, **3**, 94–98.

BEN–ISRAEL, ADI (1966) "On Error Bounds for Generalized Inverses," *SIAM J. Numer. Anal.*, **3**, No. 4, 585–592.

BENNETT, JOHN M. (1965) "Triangular Factors of Modified Matrices," *Numer. Math.*, **7**, 217–221.

312

BJÖRCK, ÅKE (1967a) "Solving Linear Least Squares Problems by Gram-Schmidt Orthogonalization," *BIT*, **7**, 1–21.

BJÖRCK, ÅKE (1967b) "Iterative Refinement of Linear Least Squares Solutions I," *BIT*, **7**, 257–278.

BJÖRCK, ÅKE AND GOLUB, G. H. (1967) "Iterative Refinement of Linear Least Squares Solutions by Householder Transformation," *BIT*, **7**, 322–337.

BJÖRCK, ÅKE AND GOLUB, G. H. (1973) "Numerical Methods for Computing Angles between Linear Subspaces," *Math. Comp.*, **27**, 579–594.

BOGGS, D. H. (1972) "A Partial-Step Algorithm for the Nonlinear Estimation Problem," *AIAA Journal*, **10**, No. 5, 675–679.

BOULLION, T. AND ODELL, P. (1971) *Generalized Inverse Matrices*. Wiley (Interscience), New York.

BROWN, KENNETH M. (1971) "Computer Solution of Nonlinear Algebraic Equations and Nonlinear Optimization Problems," *Proc. Share*, **XXXVII**, New York, 12 pp.

BUCHANAN, J. E. AND THOMAS, D. H. (1968) "On Least-Squares Fitting of Two-Dimensional Data with a Special Structure," *SIAM J. Numer. Anal.*, **5**, No. 2, 252–257.

BUSINGER, P. A. (1967) "Matrix Scaling with Respect to the Maximum-Norm, the Sum-Norm, and the Euclidean Norm," Thesis TNN71, The University of Texas, Austin, 119 pp.

BUSINGER, P. A. (1970a) "MIDAS—Solution of Linear Algebraic Equations," *Bell Telephone Laboratories, Numerical Mathematics Computer Programs* **3**, Issue 1. Murray Hill, N.J.

BUSINGER, P. A. (1970b) "Updating a Singular Value Decomposition," *BIT*, **10**, 376–385.

BUSINGER, P. A., see Golub, G. H.

BUSINGER, P. A. AND GOLUB, G. H. (1967) "An Algol Procedure for Computing the Singular Value Decomposition," *Stanford Univ. Report No. CS-73*, Stanford, Calif.

BUSINGER, P. A. AND GOLUB, G. H. (1969) "Singular Value Decomposition of a Complex Matrix," *Comm. ACM*, **12**, No. 10, 564–565.

CARASSO, C. AND LAURENT, P. J. (1968) "On the Numerical Construction and the Practical Use of Interpolating Spline-Functions," *Proc. IFIP Congress 1968*, North-Holland Pub. Co., Amsterdam.

CHAMBERS, JOHN M. (1971) "Regression Updating," *J. Amer. Statist. Assn.*, **66**, 744–748.

CHENEY, E. W. AND GOLDSTEIN, A. A. (1966) "Mean-Square Approximation by Generalized Rational Functions," *Boeing Scientific Research Lab. Math. Note*, No. 465, p. 17, Seattle.

CODY, W. J. (1971) "Software for the Elementary Functions," *Mathematical Software*, ed. by J. R. Rice. Academic Press, New York, 171–186.

COWDEN, DUDLEY J. (1958) "A Procedure for Computing Regression Coefficients," *J. Amer. Statist. Assn.*, **53**, 144–150.

COX, M. G. (1971) "The Numerical Evaluation of B-Splines," Report No. NPL-DNAC-4, *National Physical Lab.*, Teddington, Middlesex, England, 34 pp.

CRUISE, D. R. (1971) "Optimal Regression Models for Multivariate Analysis (Factor Analysis)," *Naval Weapons Center Report NWC TP 5103*, China Lake, Calif., 60 pp.

DAVIDON, WILLIAM C. (1968) "Variance Algorithm for Minimization," *Optimization* (Sympos. Keele, 1968) Acad. Press London, 13–20.

DAVIS, CHANDLER AND KAHAN, W. M. (1970) "The Rotation of Eigenvectors by a Perturbation, III," *SIAM J. Numer. Anal.*, **7**, No. 1, 1–46.

DE BOOR, CARL (1971) "Subroutine Package for Calculating with B-Splines," Report No. LA-4728-MS, *Los Alamos Scientific Lab*, 12 pp.

DE BOOR, CARL (1972) "On Calculating with B-Splines," *J. Approximation Theory*, **6**, No. 1, 50–62.

DE BOOR, CARL AND RICE, JOHN R. (1968a) "Least Squares Cubic Spline Approximation I—Fixed Knots," Purdue Univ., CSD TR 20, Lafayette, Ind., 30 pp.

DE BOOR, CARL AND RICE, JOHN R. (1968b) "Least Squares Cubic Spline Approximation II—Variable Knots," Purdue Univ., CSD TR 21, Lafayette, Ind., 28 pp.

DRAPER, N. R. AND SMITH, H. (1966) *Applied Regression Analysis*. John Wiley & Sons, Inc., New York, 407 pp.

DYER, P., see Hanson, R. J.

DYER, P. AND MCREYNOLDS, S. R. (1969) "The Extension of Square-Root Filtering to Include Process Noise," *J. Opt.: Theory and Applications*, **3**, 92–105.

EBERLIN, P. J. (1970) "Solution to the Complex Eigenproblem by a Norm Reducing Jacobi Type Method," *Numer. Math.*, **14**, 232–245.

ECKART, C. AND YOUNG, G. (1936) "The Approximation of One Matrix by Another of Lower Rank," *Psychometrika*, **1**, 211–218.

FADDEEV, D. K., KUBLANOVSKAYA, V. N., AND FADDEEVA, V. N. (1968) "Sur les Systémes Linéaires Algébriques de Matrices Rectangularies et Mal-Conditionnées," Programmation en Mathématiques Numériques, *Editions Centre Nat. Recherche Sci.*, Paris, **VII**, 161–170.

FADDEEVA, V. N., see Faddeev, D. K.

FIACCO, A. V. AND MCCORMICK, G. P. (1968) *Nonlinear Programming: Sequential Unconstrained Minimization Techniques.* John Wiley & Sons, Inc., New York.

FLETCHER, R. (1968) "Generalized Inverse Methods for the Best Least Squares Solution of Systems of Non-Linear Equations," *Comput. J.*, **10**, 392–399.

FLETCHER, R. (1971) "A Modified Marquardt Subroutine for Nonlinear Least Squares," Report No. R-6799, *Atomic Energy Research Estab.*, Harwell, Berkshire, England, 24 pp.

FLETCHER, R. AND LILL, S. A. (1970) "A Class of Methods for Non-Linear Programming, II, Computational Experience," in *Nonlinear Programming*, ed. by J. B. Rosen, O. L. Mangasarian, and K. Ritter. Academic Press, New York, 67–92.

FORSYTHE, GEORGE E. (1957) "Generation and Use of Orthogonal Polynomial for Data-Fitting with a Digital Computer," *J. Soc. Indust. Appl. Math.* **5**, 74–88.

FORSYTHE, GEORGE E. (1967) "Today's Methods of Linear Algebra," *SIAM Rev.*, **9**, 489–515.

FORSYTHE, GEORGE E. (1970) "Pitfalls in Computation, or Why a Math Book Isn't Enough," *Amer. Math. Monthly*, **77**, 931–956.

FORSYTHE, GEORGE E. AND MOLER, CLEVE B. (1967) *Computer Solution of Linear Algebraic Systems.* Prentice-Hall, Inc., Englewood Cliffs, N.J., 148 pp.

FOX, L. (1965) *An Introduction to Numerical Linear Algebra.* Oxford Univ. Press, New York, 327 pp.

FRANCIS, J. G. F. (1961–1962) "The QR Transformation," Parts I and II, *Comput. J.*, **4**, 265–271, 332–345.

FRANKLIN, J. N. (1968) *Matrix Theory.* Prentice-Hall, Inc., Englewood Cliffs, N.J., 292 pp.

FRANKLIN, J. N. (1970) "Well-Posed Stochastic Extensions of Ill-Posed Linear Problems," *J. Math. Anal. Appl.*, **31**, 682–716.

GALE, DAVID (1960) *Theory of Linear Economic Models*, McGraw-Hill, 330 pp.

GALE, DAVID (1969) "How to Solve Least Inequalities," *Amer. Math. Monthly*, **76**, 589–599.

GARSIDE, M. J. (1971) "Some Computational Procedures for the Best Subset Problem," *Appl. Statist. (J. Roy. Statist. Soc., Ser. C)*, **20**, 8–15 and 111–115.

GASTINEL, NOEL (1971) *Linear Numerical Analysis.* Academic Press, New York, 350 pp.

GENTLEMAN, W. M. (1972a) "Least Squares Computations by Givens Transformations without Square Roots," *Univ. of Waterloo Report CSRR-2062*, Waterloo, Ontario, Canada, 17 pp.

GENTLEMAN, W. M. (1972b) "Basic Procedures for Large, Sparse or Weighted Linear Least Squares Problems," *Univ. of Waterloo Report CSRR-2068*, Waterloo, Ontario, Canada, 14 pp.

GILL, P. E. AND MURRAY, W. (1970) "A Numerically Stable Form of the Simplex Algorithm," *National Physical Lab. Math. Tech. Report No. 87*, Teddington, Middlesex, England, 43 pp.

GILL, P. E., GOLUB, G. H., MURRAY, W., AND SAUNDERS, M. A. (1972) "Methods for Modifying Matrix Factorizations," *Stanford Univ. Report No. CS-322*, Stanford, Calif., 60 pp.

GIVENS, W. (1954) "Numerical Computation of the Characteristic Values of a Real Symmetric Matrix," *Oak Ridge National Laboratory Report ORNL-1574*, Oak Ridge, Tenn.

GLASKO, V. B., see Tihonov, A. N.

GOLDSTEIN, A. A., see Cheney, E. W.

GOLUB, G. H. (1968) "Least Squares, Singular Values and Matrix Approximations," *Aplikace Mathematicky*, 13, 44–51.

GOLUB, G. H. (1969) "Matrix Decompositions and Statistical Calculations," in *Statistical Calculations*, ed. by R. C. Milton and J. A. Nelder. Academic Press, New York, 365–397.

GOLUB, G. H., see Bartels, R. H.

GOLUB, G. H., see Björck, Åke.

GOLUB, G. H., see Businger, P. A.

GOLUB, G. H., see Gill, P. E.

GOLUB, G. H., see Kahan, W.

GOLUB, G. H. AND BUSINGER, P. A. (1965) "Linear Least Squares Solutions by Householder Transformations," *Numer. Math.*, 7, H. B. Series Linear Algebra, 269–276.

GOLUB, G. H. AND KAHAN, W. (1965) "Calculating the Singular Values and Pseudoinverse of a Matrix," *SIAM J. Numer. Anal.*, 2, No. 3, 205–224.

GOLUB, G. H. AND PEREYRA, V. (1973) "The Differentiation of Pseudoinverses and Nonlinear Least Squares Problems Whose Variables Separate," *SIAM J. Numer. Anal.*, 10, 413–432.

GOLUB, G. H. AND REINSCH, C. (1971) "Singular Value Decomposition and Least Squares Solutions" in *Handbook for Automatic Computation*, II, *Linear Algebra* [see Wilkinson and Reinsch (1971)].

GOLUB, G. H. AND SAUNDERS, MICHAEL A. (1970) "Linear Least Squares and Quadratic Programming" in *Integer and Nonlinear Programming*, II, J. Abadie (ed.), North Holland Pub. Co., Amsterdam, 229–256.

GOLUB, G. H. AND SMITH, L. B. (1971) "Chebyshev Approximation of Continuous Functions by a Chebyshev System of Functions," *Comm. ACM*, 14, No. 11, 737–746.

GOLUB, G. H. AND STYAN, G. P. H. (to appear) "Numerical Computations for Univariate Linear Models," *J. Statist. Computation and Simulation.*

GOLUB, G. H. AND UNDERWOOD, RICHARD (1970) "Stationary Values of the Ratio of Quadratic Forms Subject to Linear Constraints," *Z. Angew. Math. Phys.,* **21,** 318–326.

GOLUB, G. H. AND WILKINSON, J. H. (1966) "Note on the Iterative Refinement of Least Squares Solution," *Numer. Math.,* **9,** 139–148.

GORDON, WILLIAM R., see Marcus, Marvin.

GORDONOVA, V. I., Translated by Linda Kaufman (1970) "Estimates of the Roundoff Error in the Solution of a System of Conditional Equations," *Stanford Univ. Report No. CS-164,* Stanford University, Stanford, Calif.

GRAYBILL, F. A., MEYER, C. D., AND PAINTER, R. J. (1966) "Note on the Computation of the Generalized Inverse of a Matrix," *SIAM Rev.,* **8,** 522–524.

GREVILLE, T. N. E. (1959) "The Pseudo-Inverse of a Rectangular or Singular Matrix and its Application to the Solution of Systems of Linear Equations," *SIAM Rev.,* **1,** 38–43.

GREVILLE, T. N. E. (1960) "Some Applications of the Pseudoinverse of a Matrix," *SIAM Rev.,* **2,** No. 1, 15–22.

GREVILLE, T. N. E. (1966) "Note on the Generalized Inverse of a Matrix Product," *SIAM Rev.,* **8,** 518–521.

HALMOS, PAUL R. (1957) *Introduction to Hilbert Space,* 2nd ed. Chelsea Pub. Co., New York, 114 pp.

HALMOS, PAUL R. (1958) *Finite Dimensional Vector Spaces,* 2nd ed. D. Van Nostrand Co., Inc., Princeton, N.J., 199 pp.

HANSON, R. J. (1970) "Computing Quadratic Programming Problems: Linear Inequality and Equality Constraints," *JPL Sec. 314 Tech. Mem. No. 240,* Jet Propulsion Laboratory, California Institute of Technology, Pasadena, Calif.

HANSON, R. J. (1971) "A Numerical Method for Solving Fredholm Integral Equations of the First Kind Using Singular Values," *SIAM J. Numer. Anal.,* **8,** No. 3, 616–622.

HANSON, R. J. (1972) "Integral Equations of Immunology," *Comm. ACM,* **15,** No. 10, 883–890.

HANSON, R. J. AND DYER, P. (1971) "A Computational Algorithm for Sequential Estimation," *Comput. J.,* **14,** No. 3, 285–290.

HANSON, R. J. AND LAWSON, C. L. (1969) "Extensions and Applications of the Householder Algorithm for Solving Linear Least Squares Problems," *Math. Comp.,* **23,** No. 108, 787–812.

HEALY, M. J. R. (1968) "Triangular Decomposition of a Symmetric Matrix," *Appl. Statist. (J. Roy. Statist. Soc., Ser. C),* **17,** 195–197.

HEMMERLE, W. J. (1967) *Statistical Computations on a Digital Computer*. Blaisdell Pub. Co., New York, 230 pp.

HERBOLD, ROBERT J., see Hilsenrath, Joseph.

HESTENES, M. R. (1958) "Inversion of Matrices by Biorthogonalization and Related Results," *J. Soc. Indust. Appl. Math.*, **6**, 51–90.

HESTENES, M. R. AND STIEFEL, E. (1952) "Methods of Conjugate Gradients for Solving Linear Systems," *Nat. Bur. of Standards J. Res.*, **49**, 409–436.

HILSENRATH, JOSEPH, ZIEGLER, GUY G., MESSINA, CARLA G., WALSH, PHILLIP J., AND HERBOLD, ROBERT J. (1966, Revised 1968) "OMNITAB, a Computer Program for Statistical and Numerical Analysis," *Nat. Bur. of Standards, Handbook 101*, 275 pp.

HOCKING, R. R., see La Motte, L. R.

HOERL, A. E. (1959) "Optimum Solution of Many Variable Equations," *Chemical Engineering Progress*, **55**, No. 11, 69–78.

HOERL, A. E. (1962) "Application of Ridge Analysis to Regression Problems," *Chemical Engineering Progress*, **58**, No. 3, 54–59.

HOERL, A. E. (1964) "Ridge Analysis," *Chemical Engineering Progress*, **60**, No. 50, 67–78.

HOERL, A. E. AND KENNARD, R. W. (1970a) "Ridge Regression: Biased Estimation for Nonorthogonal Problems," *Technometrics*, **12**, 55–67.

HOERL, A. E. AND KENNARD, R. W. (1970b) "Ridge Regression: Applications to Nonorthogonal Problems," *Technometrics*, **12**, 69–82.

HOFFMAN, A. J. AND WIELANDT, H. W. (1953) "The Variation of the Spectrum of a Normal Matrix," *Duke Math. J.*, **20**, 37–39.

HOUSEHOLDER, A. S. (1958) "Unitary Triangularization of a Nonsymmetric Matrix," *J. ACM*, **5**, 339–342.

HOUSEHOLDER, A. S. (1964) *The Theory of Matrices in Numerical Analysis*. Blaisdell Pub. Co., New York, 257 pp.

HOUSEHOLDER, A. S. (1972) "KWIC Index for Numerical Algebra," Report No. ORNL-4778, Oak Ridge National Laboratory, 538 pp.

JACOBI, C. G. J. (1846) "Über ein Leichtes Verfahren die in der Theorie der Säculärstörungen Vorkommenden Gleichungen Numerisch Aufzulösen," *Crelle's J.*, **30**, 297–306.

JENNINGS, L. S. AND OSBORNE, M. R. (1970) "Applications of Orthogonal Matrix Transformations to the Solution of Systems of Linear and Nonlinear Equations," *Australian National Univ. Computer Centre Tech. Rept.*, No. 37, Canberra, Australia, 45 pp.

JENNRICH, R. I. AND SAMPSON, P. I. (1971) "Remark AS-R3. A Remark on Algorithm AS-10," *Appl. Statist. (J. Roy. Statist. Soc., Ser. C)*, **20**, 117–118.

KAHAN, W. (1966a) "When to Neglect Off-Diagonal Elements of Symmetric Tri-Diagonal Matrices," *Stanford Univ. Report No. CS-42*, Stanford, Calif.

KAHAN, W. (1966b) "Numerical Linear Algebra," *Canadian Math. Bull.*, **9**, No. 6, 757–801.

KAHAN, W. M., see Davis, Chandler.

KAHAN, W. M., see Golub, G. H.

KALMAN, R. E. (1960) "A New Approach to Linear Filtering and Prediction Problems," ASME Trans., *J. Basic Eng.*, **82D**, 35–45.

KAMMERER, W. J. AND NASHED, M. Z. (1972) "On the Convergence of the Conjugate Gradient Method for Singular Linear Operation Equations," *SIAM J. Numer. Anal.*, **9**, 165–181.

KENNARD, R. W., see Hoerl, A. E.

KORGANOFF, ANDRÉ, see Pavel–Parvu, Monica.

KORGANOFF, ANDRÉ AND PAVEL–PARVU, MONICA (1967) *Eléments de Théorie des Matrices Carrées et Rectangles en Analyse Numérique.* Dunod, Paris, 441 pp.

KROGH, F. T. (1974) "Efficient Implementation of a Variable Projection Algorithm for Nonlinear Least Squares Problems," *Comm. ACM*, **17** (to appear).

KUBLANOVSKAYA, V. N., see Faddeev, D. K.

LA MOTTE, L. R. AND HOCKING, R. R. (1970) "Computational Efficiency in the Selection of Regression Variables," *Technometrics*, **12**, 83–93.

LAURENT, P. J., see Carasso, C.

LAWSON, C. L. (1961) "Contributions to the Theory of Linear Least Maximum Approximation," Thesis, UCLA, Los Angeles, Calif., 99 pp.

LAWSON, C. L. (1971) "Applications of Singular Value Analysis," *Mathematical Software*, ed. by John Rice. Academic Press, New York, 347–356.

LAWSON, C. L., see Hanson, R. J.

LERINGE, ÖRJAN AND WEDIN, PER–ÅKE (1970) "A Comparison between Different Methods to Compute a Vector x which Minimizes $\| Ax - B \|_2$ when $Gx = h$," Lund University, Lund, Sweden, 21 pp.

LEVENBERG, K. (1944) "A Method for the Solution of Certain Non-Linear Problems in Least Squares," *Quart. Appl. Math.*, **2**, 164–168.

LILL, S. A., see Fletcher, R.

LONGLEY, JAMES W. (1967) "An Appraisal of Least Squares Programs for the Electronic Computer from the Point of View of the User," *J. Amer. Statist. Assn.*, **62**, 819–841.

LYNN M. S. AND TIMLAKE, W. P. (1967) "The Use of Multiple Deflations in the Numerical Solution of Singular Systems of Equations with Applications to Potential Theory," IBM Houston Scientific Center, 37.017, Houston, Texas.

MARCUS, MARVIN AND GORDON, WILLIAM R. (1972) "An Analysis of Equality in Certain Matrix Inequalities, II," *SIAM J. Numer. Anal.*, **9**, 130–136.

MARQUARDT, DONALD W. (1963) "An Algorithm for Least-Squares Estimation of Nonlinear Parameters," *J. Soc. Indust. Appl. Math.*, **11**, No. 2, 431–441.

MARQUARDT, DONALD W. (1970) "Generalized Inverses, Ridge Regression, Biased Linear Estimation, and Nonlinear Estimation," *Technometrics*, **12**, 591–612.

MASON, J. C. (1969) "Orthogonal Polynomial Approximation Methods in Numerical Analysis," *Univ. of Toronto Tech. Report*, No. 11, Toronto, Ontario, Canada, 50 pp.

MCCORMICK, G. P., see Fiacco, A. V.

MCCRAITH, D. L., see Mitchell, W. C.

MCREYNOLDS, S. R., see Dyer, P.

MESSINA, CARLA G., see Hilsenrath, Joseph.

MEYER, CARL D., JR. (1972) "The Moore-Penrose Inverse of a Bordered Matrix," *Lin. Alg. and Its Appl.*, **5**, 375–382.

MEYER, C. D., see Graybill, F. A.

MITCHELL, W. C. AND MCCRAITH, D. L. (1969) "Heuristic Analysis of Numerical Variants of the Gram–Schmidt Orthonormalization Process," *Stanford Univ. Report No. CS-122*, Stanford, Calif.

MITRA, S. K., see Rao, C. R.

MOLER, CLEVE B. (1967) "Iterative Refinement in Floating Point," *J. ACM*, **14**, No. 2, 316–321.

MOLER, CLEVE B. AND STEWART, G. W. (1973) "An Algorithm for Generalized Matrix Eigenvalue Problem," *SIAM J. Numer. Anal.*, **10**, No. 2, 241–256.

MOLER, CLEVE B., see Forsythe, George E.

MOORE, E. H. (1920) "On the Reciprocal of the General Algebraic Matrix," *Bulletin, AMS*, **26**, 394–395.

MOORE, E. H. (1935) "General Analysis, Part 1," *Mem. Amer. Philos. Soc.*, **1**, 1–231.

MORRISON, DAVID D. (1960) "Methods for Nonlinear Least Squares Problems and Convergence Proofs," *Proc. of Seminar on Tracking Programs and Orbit Determination*, Jet Propulsion Lab., Pasadena, Calif., Cochairmen: Jack Lorell and F. Yagi, 1–9.

MORRISON, W. J., see Sherman, Jack.

MURRAY, W. (1971) "An Algorithm to Find a Local Minimum of an Indefinite Quadratic Program" (Presented at the VII Symposium on Mathematical Programming, The Hague, 1970) Report No. NPL-DNAC-1, National Physical Laboratory, Teddington, Middlesex, England, 31 pp.

MURRAY, W., see Gill, P. E.

NEWMAN, MORRIS AND TODD, JOHN (1958) "The Evaluation of Matrix Inversion Programs," *J. Soc. Indust. Appl. Math.*, **6**, No. 4, 466–476.

ODELL, P., see Boullion, T.

ORTEGA, JAMES M. AND RHEINBOLDT, WERNER C. (1970) *Iterative Solution of Nonlinear Equations in Several Variables.* Academic Press, New York, 572 pp.

OSBORNE, E. E. (1961) "On Least Squares Solutions of Linear Equations," *J. ACM*, **8**, 628–636.

OSBORNE, E. E. (1965) "Smallest Least Squares Solutions of Linear Equations," *SIAM J. Numer. Anal.*, **2**, No. 2, 300–307.

OSBORNE, M. R., see Jennings, L. S.

PAIGE, C. C. (1972) "An Error Analysis of a Method for Solving Matrix Equations," *Stanford Univ. Report No. CS-297*, Stanford, Calif.

PAINTER, R. J., see Graybill, F. A.

PARLETT, B. N. (1967) "The *LU* and *QR* Transformations," Chap. 5 of *Mathematical Methods for Digital Computers*, **II**. John Wiley & Sons, Inc., New York.

PAVEL–PARVU, MONICA, see Korganoff, André.

PAVEL–PARVU, MONICA AND KORGANOFF, ANDRÉ (1969) "Iteration Functions for Solving Polynomial Matrix Equations," *Constructive Aspects of the Fundamental Theorem of Algebra*, ed. B. DeJon and P. Henrici. Wiley–Interscience, New York, 225–280.

PENROSE, R. (1955) "A Generalized Inverse for Matrices," *Proc. Cambridge Phil. Soc.*, **51**, 406–413.

PEREYRA, V. (1968) "Stabilizing Linear Least Squares Problems," **68**, *Proc. IFIP Congress 1968*.

PEREYRA, V. (1969b) "Stability of General Systems of Linear Equations," *Aequationes Mathematicae*, **2**, Fasc. 2/3.

PEREYRA, V., see Golub, G. H.

PEREZ, A. AND SCOLNIK, H. D. (to appear) "Derivatives of Pseudoinverses and Constrained Non-Linear Regression Problems," *Numer. Math.*

PETERS, G., see Martin, R. S.

PETERS, G. AND WILKINSON, J. H. (1970) "The Least Squares Problem and Pseudoinverses," *Comput. J.*, **13**, 309–316.

PHILLIPS, D. L. (1962) "A Technique for the Numerical Solution of Certain Integral Equations of the First Kind," *J. ACM*, **9**, 84–97.

PLACKETT, R. L. (1960) *Principles of Regression Analysis.* Oxford Univ. Press, New York, 173 pp.

POWELL, M. J. D. (1968) "A Fortran Subroutine for Solving Systems of Non-Linear Algebraic Equations," Report No. R-5947, Atomic Energy Research Estab., Harwell, Berkshire, England.

POWELL, M. J. D. (1970) "A Survey of Numerical Methods for Unconstrained Optimization," *SIAM Rev.*, **12**, 79–97.

POWELL, M. J. D. AND REID, J. K. (1968a) "On Applying Householder's Method to Linear Least Squares Problems," Report No. T-P-332, Atomic Energy Research Estab., Harwell, Berkshire, England, 20 pp.

POWELL, M. J. D. AND REID, J. K. (1968b) "On Applying Householder's Method to Linear Least Squares Problems," *Proc. IFIP Congress, 1968*.

PRINGLE, R. M. AND RAYNER, A. A. (1970) "Expressions for Generalized Inverses of a Bordered Matrix with Application to the Theory of Constrained Linear Models," *SIAM Rev.*, **12**, 107–115.

PYLE, L. DUANE (1967) "A Generalized Inverse (Epsilon)-Algorithm for Constructing Intersection Projection Matrices, with Applications," *Numer. Math.*, **10**, 86–102.

RALSTON, A. AND WILF, H. S. (1960) *Mathematical Methods for Digital Computers.* John Wiley & Sons, Inc., New York, 293 pp.

RALSTON, A. AND WILF, H. S. (1967) *Mathematical Methods for Digital Computers,* **II**. John Wiley & Sons, Inc., New York, 287 pp.

RAO, C. R. AND MITRA, S. K. (1971) *Generalized Inverse of Matrices and its Application.* John Wiley & Sons, Inc., New York, 240 pp.

RAYNER, A. A., see Pringle, R. M.

REID, J. K. (1970) "A Fortran Subroutine for the Solution of Large Sets of Linear Equations by Conjugate Gradients," Report No. 6545, Atomic Energy Research Estab., Harwell, Berkshire, England, 5 pp.

REID, J. K. (1971a) "The Use of Conjugate Gradients for Systems of Linear Equations Possessing Property A," Report No. T-P-445, Atomic Energy Research Estab., Harwell, Berkshire, England, 9 pp. Also appeared in *SIAM J. Numer. Anal.,* **9**, 325–332.

REID, J. K., see Powell, M. J. D.

REID, J. K. (ed.) (1971b) *Large Sparse Sets of Linear Equations.* Academic Press, New York, 284 pp.

REINSCH, C., see Wilkinson, J. H.

RHEINBOLDT, WERNER C., see Ortega, James M.

RICE, JOHN R. (1966) "Experiments on Gram–Schmidt Orthogonalization," *Math. Comp.,* **20**, 325–328.

RICE, JOHN R. (1969) *The Approximation of Functions,* **2**—*Advanced Topics.* Addison Wesley Pub. Co., Reading, Mass., 334 pp.

RICE, JOHN R. (1971a) "Running Orthogonalization," *J. Approximation Theory,* **4**, 332–338.

RICE, JOHN R., see De Boor, Carl.

RICE, JOHN R. (ed.) (1971b) *Mathematical Software*. Academic Press, New York, 515 pp.

RICE, JOHN R. AND WHITE, JOHN S. (1964) "Norms for Smoothing and Estimation," *SIAM Rev.*, **6**, No. 3, 243–256.

ROSEN, EDWARD M. (1970) "The Instrument Spreading Correction in GPC III. The General Shape Function Using Singular Value Decomposition with a Nonlinear Calibration Curve," Monsanto Co., St. Louis, Mo.

ROSEN, J. B. (1960) "The Gradient Projection Method for Nonlinear Programming, Part I, Linear Constraints," *J. Soc. Indust. Appl. Math.*, **8**, No. 1, 181–217.

RUTISHAUSER, H. (1968) "Once Again the Least Squares Problem," *J. Lin. Alg. and Appl.*, **1**, 479–488.

SAMPSON, P. I., see Jennrich, R. I.

SAUNDERS, M. A. (1972a) "Large-Scale Linear Programming Using the Cholesky Factorization," *Stanford Univ. Report CS-252*, Stanford, Calif., 64 pp.

SAUNDERS, M. A. (1972b) "Product Form of the Cholesky Factorization for Large-Scale Linear Programming," *Stanford Univ. Report No. CS-301*, Stanford, Calif., 38 pp.

SAUNDERS, M. A., see Bartels, R. H.

SAUNDERS, M. A., see Gill, P. E.

SAUNDERS, M. A., see Golub, G. H.

SCHUR, I. (1909) "Uber die Charakteristischen Wurzeln einer Linearen Substitution mit einer Anwedung auf die Theorie der Integralgleichungen," *Math. Ann.*, **66**, 488–510.

SCOLNIK, H. D., see Perez, A.

SHERMAN, JACK AND MORRISON, W. J. (1949) "Adjustment of an Inverse Matrix Corresponding to the Changes in the Elements of a Given Column or a Given Row of the Original Matrix," *Ann. Math. Statist.*, **20**, 621.

SHERMAN, JACK AND MORRISON, W. J. (1950) "Adjustment of an Inverse Matrix Corresponding to a Change in one Element of a Given Matrix," *Ann. Math. Statist.*, **21**, 124.

SMITH, GERALD L. (1967) "On the Theory and Methods of Statistical Inference," *NASA Tech. Rept. TR R-251*, Washington, D.C., 32 pp.

SMITH, H., see Draper, N. R.

SMITH, L. B., see Golub, G. H.

STEWART, G. W. (1969) "On the Continuity of the Generalized Inverse," *SIAM J. Appl. Math.*, **17**, No. 1, 33–45.

STEWART, G. W. (1970) "Incorporating Origin Shifts into the QR Algorithm for Symmetric Tridiagonal Matrices," *Comm. ACM*, **13**, No. 6, 365–367, 369–371.

STEWART, G. W. (1973) *Introduction to Matrix Computations*. Academic Press, New York, 442 pp.

STEWART, G. W., see Moler, C. B.

STIEFEL, E., see Hestenes, M. R.

STOER, JOSEPH (1971) "On the Numerical Solution of Constrained Least Squares Problems," *SIAM J. Numer. Anal.*, **8**, No. 2, 382–411.

STRAND, OTTO NEAL, see Westwater, Ed R.

STRAND, OTTO NEAL AND WESTWATER, ED R. (1968a) "Statistical Estimation of the Numerical Solution of a Fredholm Integral Equation of the First Kind," *J. ACM*, **15**, No. 1, 100–114.

STRAND, OTTO NEAL AND WESTWATER, ED R. (1968b) "Minimum-RMS Estimation of the Numerical Solution of a Fredholm Integral Equation of the First Kind," *SIAM J. Numer. Anal.*, **5**, No. 2, 287–295.

STYAN, G. P. H., see Golub, G. H.

SWERLING, P. (1958) "A Proposed Stagewise Differential Correction Procedure for Satellite Tracking and Prediction," *Rand Corp. Report P-1292*, Santa Monica, Calif. [also in *J. Astronaut. Sci.*, **6** (1959)].

THOMAS, D. H., see Buchanan, J. E.

TIHONOV, A. N. AND GLASKO, V. B. (1964) "An Approximate Solution of Fredholm Integral Equations of the First Kind," *Zhurnal Vychislitel'noi Matematiki i Matematicheskoi Fiziki*, **4**, 564–571.

TIMLAKE, W. P., see Lynn, M. S.

TODD, JOHN, see Newman, Morris.

TORNHEIM, LEONARD (1961) "Stepwise Procedures Using Both Directions," *Proc. 16th Nat. Meeting of ACM*, 12A4.1–12A4.4.

TUCKER, A. W. (1970) "Least Distance Programming," *Proc. of the Princeton Sympos. on Math. Prog.*, Princeton Univ. Press, Princeton, N.J., 583–588.

TURING, A. M. (1948) "Rounding off Errors in Matrix Processes," *Quart. J. Mech.*, **1**, 287–308.

TWOMEY, S. (1963) "On the Numerical Solution of Fredholm Integral Equations of the First Kind by Inversion of the Linear System Produced by Quadrature," *J. ACM*, **10**, 97–101.

UNDERWOOD, RICHARD, see Golub, G. H.

VAN DER SLUIS, A. (1969) "Condition Numbers and Equilibration of Matrices," *Numer. Math.*, **14**, 14–23.

VAN DER SLUIS, A. (1970) "Stability of Solutions of Linear Algebraic Systems," *Numer. Math.*, **14**, 246–251.

VARAH, J. M. (1969) "Computing Invariant Subspaces of a General Matrix When the Eigensystem is Poorly Conditioned," *Math. Res. Center Report 962*, University of Wisconsin, Madison, Wis., 22 pp.

WALSH, PHILLIP J., see Hilsenrath, Joseph.

WAMPLER, ROY H. (1969) "An Evaluation of Linear Least Squares Computer Programs," *Nat. Bur. of Standards J. Res.*, Ser. B Math. Sci., **73B**, 59–90.

WEDIN, PER–ÅKE (1969) "On Pseudoinverses of Perturbed Matrices," Lund University, Lund, Sweden, 56 pp. Also appeared with revisions and extensions as "Perturbation Theory for Pseudo-Inverses," *BIT*, **13**, 217–232 (1973).

WEDIN, PER-ÅKE (1972) "Perturbation Bounds in Connection with Singular Value Decomposition," *BIT*, **12**, 99–111.

WEDIN, PER-ÅKE (1973) "On the Almost Rank Deficient Case of the Least Squares Problem," *BIT*, **13**, 344–354.

WEDIN, PER–ÅKE, see Leringe, Örjan.

WESTWATER, ED R., see Strand, Otto Neal.

WESTWATER, ED R. AND STRAND, OTTO NEAL (1968) "Statistical Information Content of Radiation Measurements Used in Indirect Sensing," *J. Atmospheric Sciences*, **25**, No. 5, 750–758.

WHITE, JOHN S., see Rice, John R.

WIELANDT, H. W., see Hoffman, A. J.

WILF, H. S., see Ralston, A.

WILKINSON, J. H. (1960) "Error Analysis of Floating-Point Computation," *Numer. Math.*, **2**, 319–340.

WILKINSON, J. H. (1962) "Householder's Method for Symmetric Matrices," *Numer. Math.*, **4**, 354–361.

WILKINSON, J. H. (1963) *Rounding Errors in Algebraic Processes*. Prentice-Hall, Inc., Englewood Cliffs, N.J., 161 pp.

WILKINSON, J. H. (1965a) *The Algebraic Eigenvalue Problem*. Clarendon Press, Oxford, 662 pp.

WILKINSON, J. H. (1965b) "Convergence of the *LR*, *QR* and Related Algorithms," *Comput. J.*, **8**, No. 1, 77–84.

WILKINSON, J. H. (1968a) "Global Convergence of *QR* Algorithm," *Proc. IFIP Congress 1968*. North-Holland Pub. Co., Amsterdam.

WILKINSON, J. H. (1968b) "Global Convergence of Tridiagonal *QR* Algorithm with Origin Shifts," *J. Lin. Alg. and Appl.*, **1**, 409–420.

WILKINSON, J. H. (1970) "Elementary Proof of the Wielandt–Hoffman Theorem and its Generalization," *Stanford Univ. Report No. CS-150*, Stanford, Calif.

WILKINSON, J. H. AND REINSCH, C. (ed. by F. L. Bauer, et al.) (1971) *Handbook for Automatic Computation*, **II**, *Linear Algebra*. Springer–Verlag, New York, 439 pp.

WILKINSON, J. H., see Golub, G. H.

WILKINSON, J. H., see Peters, G.

WILLOUGHBY, RALPH A. (ed.) (1968) "Sparse Matrix Proceedings," *Thomas J. Watson Research Center Report RA1-(11707)*, White Plains, N.Y., 184 pp.

WOLFE, P. (1965) "The Composite Simplex Algorithm," *SIAM Rev.*, **7**, 42–54.

YOUNG, G., see Eckart, C.

ZIEGLER, GUY G., see Hilsenrath, Joseph.

INDEX

Page numbers in bold type refer to primary or definitional occurrences of the item referenced.

M

N